Y0-AAX-517

Lecture Notes in Physics

269

PDMS and Clusters

Proceedings of the
1st International Workshop on the Physics of Small Systems
Held on the Island of Wangerooge, Germany
September 8–12, 1986

Edited by E. R. Hilf, F. Kammer and K. Wien

Springer-Verlag
Berlin Heidelberg New York London Paris Tokyo

Editors

Eberhard R. Hilf
Friedrich Kammer
Fachbereich Physik, Universität Oldenburg
Postfach 2503, D-2900 Oldenburg

Karl Wien
Institut für Kernphysik, TH Darmstadt
Schloßgartenstraße 9, D-6100 Darmstadt

ISBN 3-540-17209-2 Springer-Verlag Berlin Heidelberg New York
ISBN 0-387-17209-2 Springer-Verlag New York Berlin Heidelberg

Printed in Germany

Printing: Druckhaus Beltz, Hemsbach/Bergstr.;
Bookbinding: J. Schäffer GmbH & Co. KG., Grünstadt
2153/3140-543210

Beach cluster, desorbed from the mudflat

Preface

This volume contains the proceedings of the First International Workshop on the Physics of Small Systems held on the Isle of Wangerooge, F.R.G., 8 - 12, September 1986. Cluster physics and fast-ion-induced desorption of particles are independent fields of research that have shown rapid progress in recent years. The two topics are descended from quite different branches of physics — solid-state physics on the one hand, radiation and nuclear physics on the other — so why deal with them together in one joint workshop ?

After MacFarlane's basic discovery of PDMS (plasma desorption mass spectrometry) this particle desorption induced by high-energy ions from accelerators or from fission events in a ^{252}Cf-source was at first studied for its own sake. Recently it has emerged that it has the potential to become an extremely precise and practicable tool for mass spectrometry with unique, new and surprising practical applications primarily in cluster physics, biology and chemistry, for example:

* clusters with high internal excitation
* intact organic and even biological molecules

Each of these applications will open up a new field. A further motivation for the Workshop was the growing interest in the physics of the transient region between the world of single atoms and the one of extended solids, together with related interest in the properties of and processes at surfaces.

The Workshop was suggested by the cooperation of the organizing groups in the field of cluster desorption and by the fact that there is now a tradition of annual meetings in PDMS/SIMS research. We are thankful to the contributors who proved that the idea was worth putting into practice.

Our thanks go to Mrs. Petra Schellnhuber, who succesfully undertook the local organization of the meeting at short notice, to Mrs. Theresia Meyer, who set many of the manuscripts with a textprocessor system, and to Michael Wendel for succeeding in implementing it on a portable PC with an ordinary matrix dot printer.

The Workshop was financed generously by the Stiftung Volkswagenwerk; additional support from the company BAWI GmbH in Wilhelmshaven is also acknowledged.

Oldenburg in December 1986, E.R.Hilf
H.F.Kammer
K.Wien

TABLE OF CONTENTS

P D M S INTRODUCTION

Karl Wien

Institut für Kernphysik
Technische Hochschule Darmstadt

1. Introduction

PDMS is an abreviation for Plasma Desorption Mass Spectrometry. The items, which are processed by means of mass spectrometry, are secondary ions ejected from a solid by fast heavy ion impact. The term "plasma desorption" has been selected by R.D. Macfarlane and co-workers[1)] in order to indicate that the emission of the secondary ions occurs from a microplasma generated in the close vicinity of the heavy ions path through the solid. The question, however, whether a rapid volatilisation from a plasma is the appropriate desorption-ionisation mechanism or an other process, is still open.

In 1973 Macfarlane and co-workers[2)] irradiated an organic layer deposited onto an Al foil by fission fragments of a ^{252}Cf source and investigated the secondary ions ejected from the target surface by means of a simple time-of-flight (=TOF) technique. They stated two important findings:

1. This method of secondary ion mass spectrometry can be applied to solid organic material - even if the compound is involatile.
2. The mass spectra yield information about the molecular mass and the molecular structure. The undestroyed molecular ion forms usually a distinct mass line, fragment ions show a molecule specific pattern.

Since 1973 PDMS has gained increasing interest because of its capability for chemical analysis. Today a dozen groups are working in the field. Their attention is mainly focussed on mass spectrometry of large thermally labil biomolecules beyond mass number 1000 or even 10000. It has been found that the quality of the mass spectra can be considerably improved by means of the chemical environment and the substrate of the considered substance.

2. Related Phenomena

In this section a few phenomena concerning the passage of heavy ions through solids and related probably to secondary ion emission from the surface are called in mind.

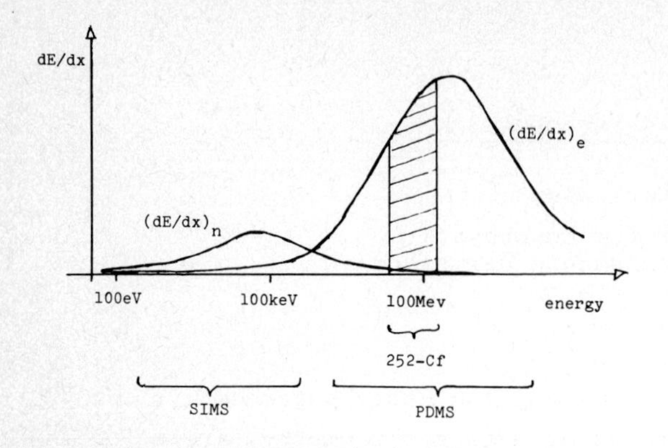

Fig. 1 Specific energy loss of a heavy ion traversing a solid as function of energy

Depending on its energy a heavy ion penetrating a solid looses energy either by atomic collissions or by electronic excitation. As illustrated in Fig.1, up to several 100 KeV the specific energy loss dE/dx is dominated by nuclear stopping, above an energy corresponding to a velocity of 0.2 cm/ns (Bohr velocity) by electronic stopping. In both regimes secondary ion emission from solids occurs ; the corresponding mass spectrometry in the atomic collision regime is called SIMS (= Secondary Ion Mass Spectrometry) and in the electronic excitation regime PDMS (or HIID = Heavy Ion Induced Desorption). Mass spectra of involatile organic material have been obtained by both methods, but with increasing mass - in particular above mass number 5000 - PDMS turned out to have a much higher desorption-ionisation effectiveness than SIMS (or FAB). Whereas at low bombarding energies the desorption process is well described by the theory of linear atomic collision cascades or in the non-linear case by collision spikes[3], at high energies the problem, how energy is transferred from the electronic system into atomic motion, is not yet solved. Desorption related to electronic excitation is often designated as electronic sputtering.

As an example of a ^{252}Cf-fission fragment, a Xe ion is considered with an energy of 100 MeV (see Fig.2a). Such an ion has in Aluminium a range of 13 µm. That means, it traveses a typical PDMS target - a 2000 Å thick organic layer deposited onto a 5 µm Al foil - and leaves it at the back side. The Xe ion produces inside the target close to its path primarily electronic excitation and δ-electrons, which carry further excitation far into the surrounding volume. Due to the maximum range of the δ-electrons perpendicular to the ions path, this volume has a diameter of about 840 Å (see Fig.2c). It has been experimentally observed[4] that the atomic or molecular excitation decreases like $1/r^2$, when r is the distance from the ions path. If the total energy is assumed to appear as electronic excitation, one would evaluate about 80 eV per molecule in the center of the track, when the aminoacid phenylalanine is used as a target. At the surface this value is certainly lower because several 100 δ-electrons carrying cinetic energy escape into vacuum.

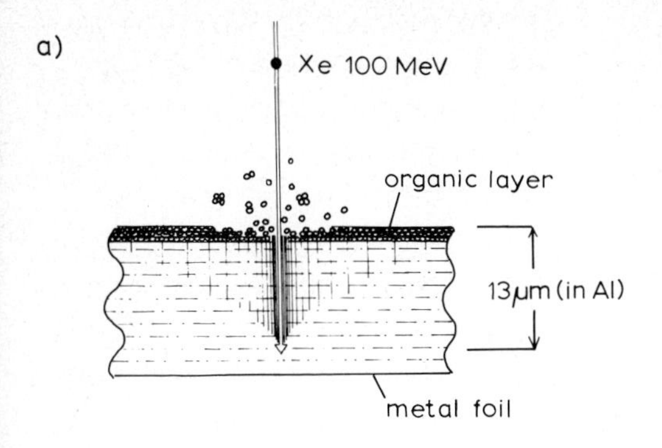

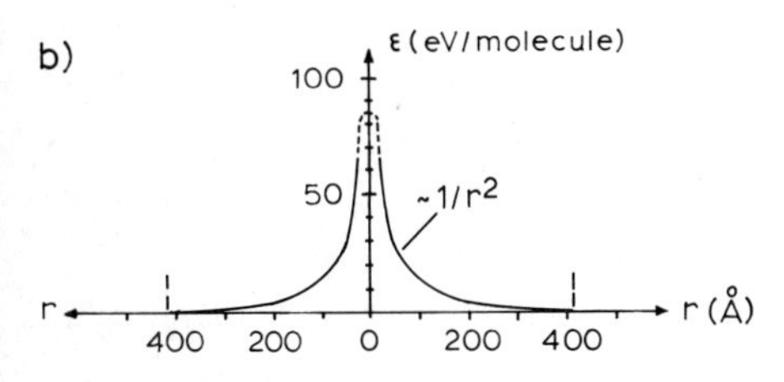

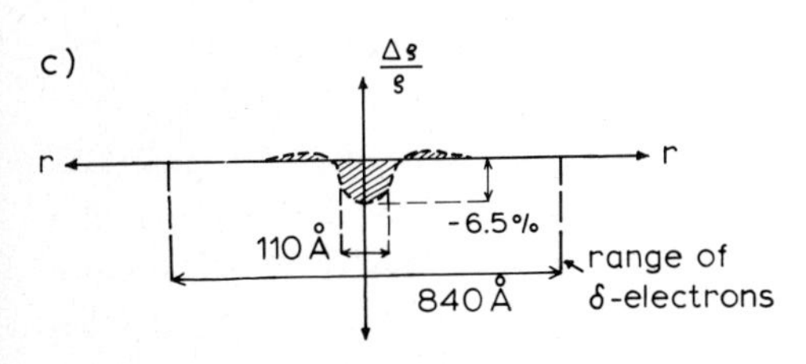

Fig. 2 a) A 100 MeV-Xe ion penetrating a PDMS target. b) Energy distribution as function of the radial distance r to the ions path. c) Relative number density in the vicinity of the nuclear track.

In principle desorption of atoms or molecules can be caused by the decay of excited states or excitons at the surface via Frank-Condon transitions. It is, however, hard to understand, how the decay of individual states can detach large intact molecules beyond 1000 mass units.

After the heavy ion impact, metals show only little defects of the crystalline structure. In insulating solids, however, a relatively wide zone of permanent destruction has been observed by neutron[5] and X-ray[6] scattering. According to Ref. 5, in case of 100 MeV-Xe the zone would have a diameter of about 110 Å. Inside the zone the number density of atoms is reduced up to 6.5 %. At the intersection of the damage zone with the surface, obviously a certain amount of material should be expelled into vacuum (see Fig.2a). Again, the observed vigorous damage is contrary to desorption of labil molecules.

3. Techniques

Since the very first PDMS experiments at Texas A&M University[2] the TOF technique turned out to be the most comprehensive and flexible method to study fast heavy ion induced desorption. In fact, all PDMS groups use this method. Several variations of the TOF technique have been developed in order to investigate, for instance, the energy and angular distributions of secondary ions or the metastable decay of molecules.

The Fig.3 shows a ^{252}Cf-PDMS apparatus. ^{252}Cf dissociates by spontaneous nuclear fission in two high energetic fragments ejected into opposite directions. Fission fragment A is used to generate via δ-electrons a start signal for the TOF measurement, fission fragment B desorbs secondary ions from the target. These ions are accelerated by an electric potential of 10 KV which is much higher than the original secondary ion

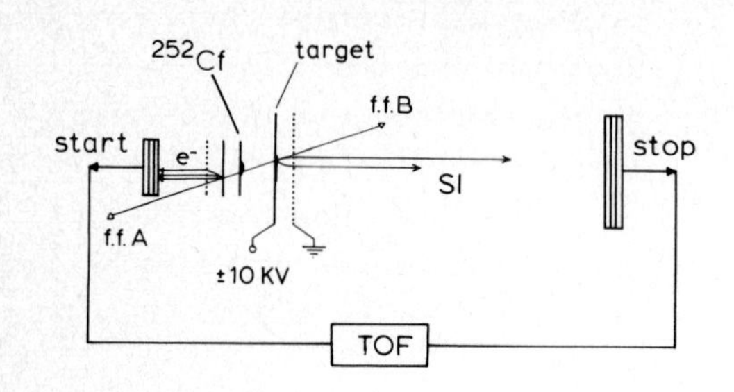

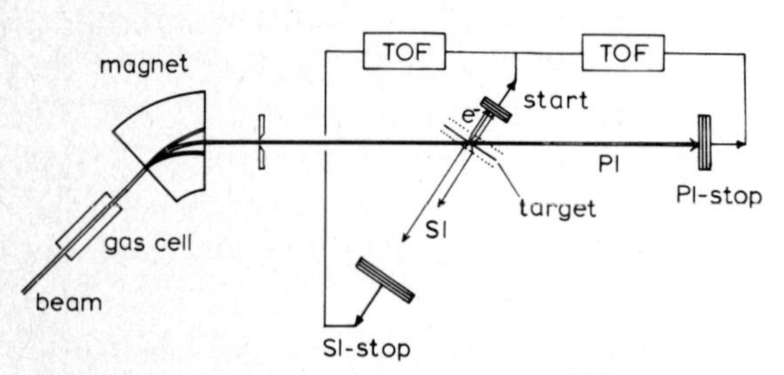

Fig. 3 a) Schematic drawing of a ^{252}Cf-PDMS TOF spectrometer (f.f.=fission fragment).
b) Set-up of a HIID spectrometer at a heavy ion accelerator (SI=secondary ion, PI=primary ion).

energy (1-3 eV). The ions drift over a distance of typical 0.5 m to the stop detector of the spectrometer. The time difference between the start and stop signals is proportional to the square-root of the ion mass. The TOF distributions measured and collected by electronic devices deliver directly the mass spectrum.

A similar arrangement of target, accelerating grids and start/stop counters is used at heavy ion accelerators as illustrated in Fig.3. A narrow beam of heavy ions traverses again the target generating δ-electrons as well as secondary ions. This time energy, mass and angle of incidence have definite values contrary to ^{252}Cf-fission fragments which have a wide spectrum of masses and energies. Also the charge state of the beam can be chosen by means of a gas cell which generates a charge distribution and a bending magnet selecting one charge state. A second TOF branch is used to measure the beam energy and to gate the SI-TOF measurement for normalisation.

The mass resolution of a PDMS apparatus depends particularly on the ratio of acceleration- and drift-distance[7]. Values of 3000 have been reached, usually the resolution is in the order of 500 to 1000. An advantage of the method is its high detection efficiency for the ejected secondary ions. The detection efficiency can be 80% for low mass ions, but it decreases at higher masses, because the momentum transfer needed to produce secondary electrons in the stop detector decreases with mass.

4. Investigations of Mass Spectra

The aim of this section is to deduce some fundamental aspects of heavy ion induced desorption from a few mass spectra taken under certain surface conditions and with three types of target material. It is not possible to present here results of the numerous sophisticated experiments concerning the functual dependence of secondary ion emission on beam parameters, the energy distributions or the influence of the chemical environment and the substrate.

4.1 Secondary ion emission from metals

As mentioned above the heavy ion impact generates no permanent damage zone in metals, but secondary ion mass spectra can be obtained as demonstrated in Fig.4. Here, a pure Al foil cleaned by low energy ion etching in UHV has been irradiated by a 65 MeV-Ni beam. Three dominant mass lines of Al ions are observed, all showing a tail of events at shorter flight-times. This line shape broadening is due to the original energy distribution of the Al ions. They are ejected from the surface with energies up to several keV. Contrary to the metal ions the mass lines of surface impurities like H^+, H_2^+ and K^+ have a much narrower line width. The mean energy of these ions is about one order of magnitude smaller than that of the metal ions.

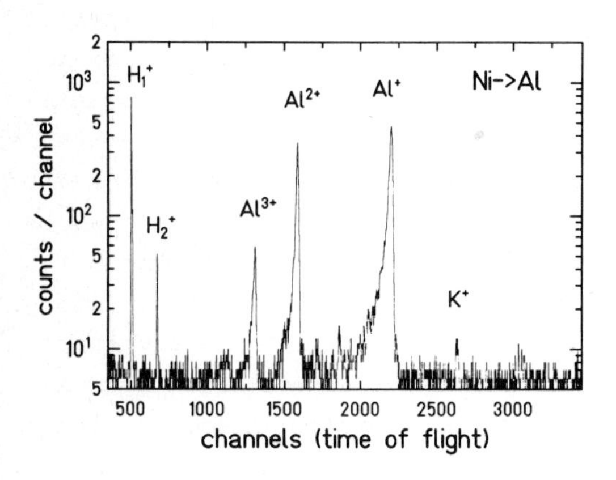

Fig. 4 Time-of-flight mass spectrum of secondary ions emitted from an Al surface by 65 MeV-Ni ion impact

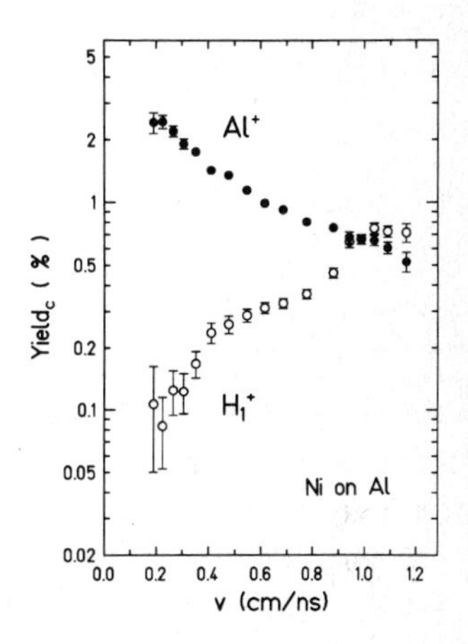

Fig. 5 Secondary ion yields as function of projectile velocity (Al bombarded by Ni ions)

An analysis of the energy distributions indicates that the metal ions are generated by atomic collision cascades. This conclusion is strongly supported by the dependence of the ion yields on primary ion velocity. In Fig. 5 the Al^+ yield decreases with velocity as expected from sputter theory[3)]. On the other hand the H_1^+ yield increases in accordance to the increase of the electronic energy loss. The conclusions from these experiments with weakly contaminated metals are:

1. Electronic sputtering effects small amounts of impurities located on a metal surface, but not the metal matrix itself.
2. Secondary ion emission related to the electronic energy loss is not restricted to dielectric material.

4.2 Secondary ion emission from a CaF_2 crystal

In order to produce a surface free of impurities and with epitaxial structure a CaF_2 crystal was cleaved in vacuum (10^{-8} mbar) and then irradiated by 70 MeV-Ni and 290 MeV-U beams. The resultant sprectrum is given in the middle of Fig.6. All mass lines can be easily assigned to combinations of F and Ca atoms, whereas in a spectrum taken from the crystal surface before cleaving, only two ions could be indentified as CaF_2 specific ions (see Fig. 6a). In the lower mass range the spectrum taken with the only chemicaly clean surface is dominated by ions of organic contaminants.

A third spectrum of the same crystal was measured after destroying the epitaxial surface structure by means of 4 KeV Ar-ion etching. The pattern of the CaF_2 specific mass lines was not changed substantially, a few background lines appeared. A comparison of ion yields obtained under the three different conditions is presented in Fig.7.

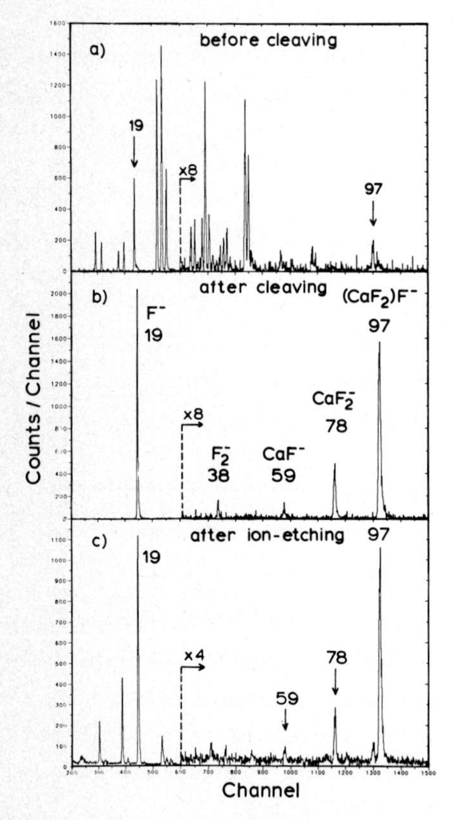

Fig. 6 Mass spectra of secondary ions ejected from CaF_2 by a 290 MeV-U beam. a) chemically clean surface b) surface of a cleavec crystal c) surface structure destroyed by ion etching

The conclusions from these experiments are: In case of CaF_2 a clean and epitaxial surface does not hinder or reduce secondary ion emission induced by fast heavy ions. The high energy impact breaks up the compound of the solid and it is the matrix material blown into vacuum, from which secondary ions are formed. The desorption process is probably not supported by preaggregation of atoms at the surface.

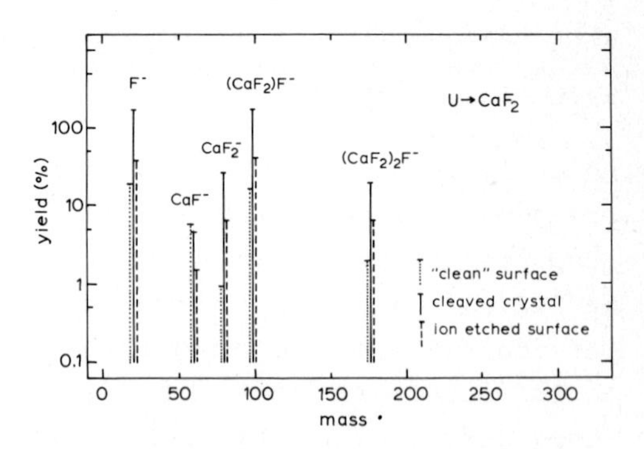

Fig. 7 Comparison of secondary ion yields measured with CaF_2 which was irradiated by a 290 MeV-U beam

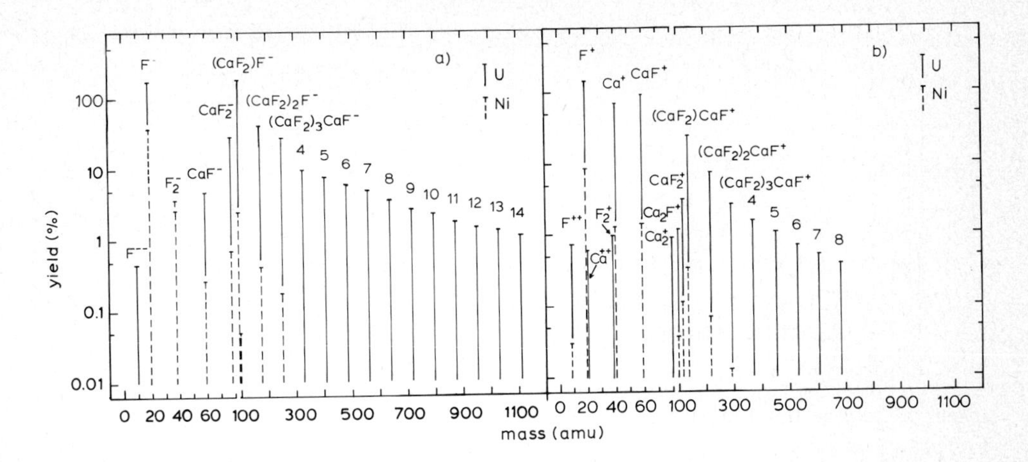

Fig. 8 Schematic mass spectra of a) negative and b) positive secondary ions ejected from CaF_2 (cleaved crystal) by irradiation with 65 MeV-Ni and 290 MeV-U ions

Also the emission of cluster ions is not supported by preformation of aggregates at the surface. Evidence for this statement is obtained from the high mass frange in the CaF_2 spectra. Here, a series of cluster ions appears emitted from the cleaved as well as from the ion etched crystal. In the later case the yields are lower. Schematic mass spectra of all negative and positive ions ejected from the cleaved crystal by U- and Ni-beams are presented in Fig.8. The cluster series are formed by $(CaF_2)_nF^-$ ions and $(CaF_2)_mCaF^+$ ions, respectively. From these two series one could conclude that the production of cluster ions occurs by breaking the CaF-F bond in a $(CaF_2)_mCaF\text{-}F(CaF_2)_n$ aggregate. That means, originally excited $(CaF_2)_{n+m+1}$ cluster are generated, which dissociate into two fragments. In case of U-irradiation the yields of the negative and positive cluster ions scale like $n^{-1.9}$ and $m^{-2.0}$, respectively. In case of Ni-irradiation the scaling as function of n or m is steeper. Similar series of cluster ions having, however, lower yields are also observed with evaporated CaF_2 targets.

4.3 Secondary ion emission from organic films

Targets of organic metarial are usually produced by electrospraying or drying of solutions of the substance. Contaminations are hardly avoidable. In Fig.9 a TOF spectrum of the involatile compound bovine insulin is presented, which has a molecular mass of 5734.3 amu. The low mass region contains separated but mostly unspecific mass lines. A wide continuum of events follows, which even extends beyond the molecular peak. This peak is a convolution of many unresolved mass lines due to the ^{13}C content of the organic compound. Depending on the substance the pattern of molecular fragments can be more pronounced as it is shown in Fig.9.

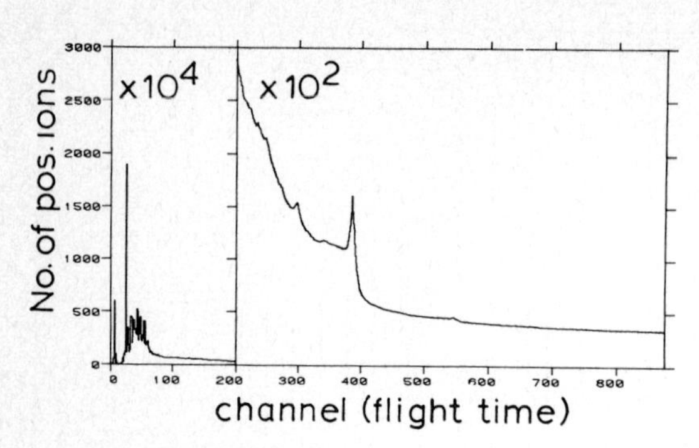

Fig. 9 Time-of-flight spectrum of positive ions from bovine insulin (Ref. 8)

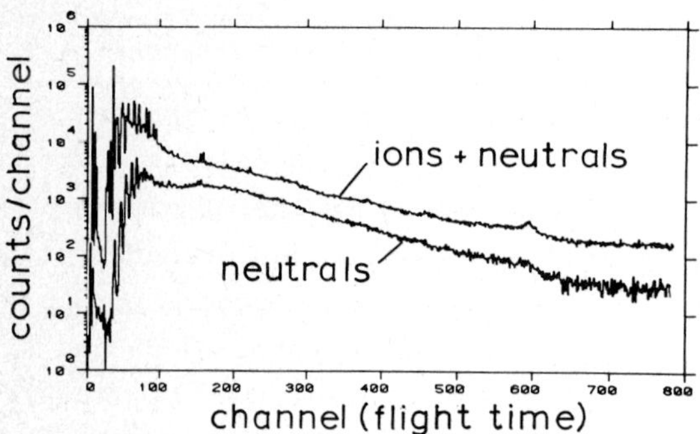

Fig. 10 Positive ion spectrum of bovine insulin. Upper curve is a complete spectrum while the lower curve is only due to neutrals (Ref. 8)

The enormous continuum is caused by metastable decay of molecular ions in flight. To generate a TOF continuum, the decay has to occur in the acceleration region between the target and the acceleration grid (see Fig.3). Behind the grid decay products change their velocity only a little, as the dissociation energy is generally small compared to the acceleration potential. The decay of a molecular ion leads usually to a neutral and a charged particle. If the later is deflected out off the flight path by an electric field, only the neutral species reach the stop detector and generate a TOF spectrum as illustrated in Fig.10. Comparing this spectrum with the complete spectrum in the same figure obviously a considerable part (30-40%) of the complete spectrum consists of neutral particles.

Metastable decay behind the acceleration grid is regognized by a line structure in the neutral spectrum. By means of electric reflection and a second stop detector for charged particles, it is possible to identify pairs of decay products and to study metastable molecular ions[9].

Certain large molecular ions are able to carry two or more charges without decomposition. As an example, a spectrum of multicharged ions ejected from porcine proinsulin by ^{225}Cf fission fragments is shown in Fig.11. The spectrum also contains the molecular dimer $2M^+$ at mass number 18180. Because of their higher energy multicharged ions have a higher detection probability.

Many fundamental studies have been performed with low mass aminoacids. As illustrated in Fig.12, the fragmentation pattern of a Phenylalanine mass spectrum is very characteristical for the molecular structure.

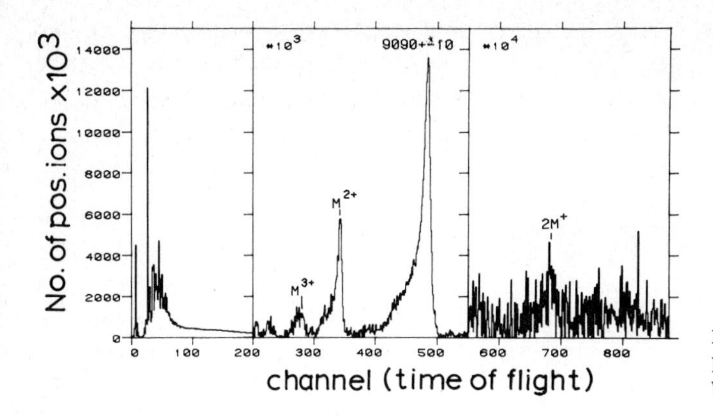

Fig. 11 A mass spectrum of porcine proinsulin measured with 252-Cf PDMS (Ref. 8)

Following certain cleavage rules most of the more intensive mass lines could be attributed to molecule specific groups. Quite informative is a comparison of Phenylalanine mass spectra taken with a U-beam and a N-beam at same projectile velocity. The specific energy losses differ by a factor of 17. In case of N-ions the secondary ion yields are extremely low, but the spectrum consists preferentially of molecule-specific mass lines. For instance, the peak at mass number 18 represents the amonium ion NH_4^+. The spectrum changes drastically, when the same target is irradiated by U-ions. The yields of molecule specific ions are increased by two or three orders of magnitude, but many of them can hardly be detected in the enormous amount of unspecific hydro-carbon ions. These groups of hydro-carbon ions are comparable with pyrolyse products of organic material. That indicates that close to the heavy ions path a hot -eventually plasma like- zone is formed, from where destruction residues are released into the gas phase by rapid volatilization.

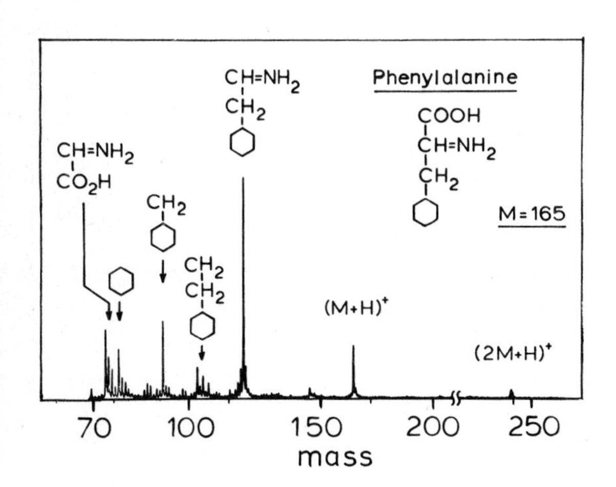

Fig. 12 Upper part of a mass spectrum measured with Phenylalanine by 252-Cf PDMS

5. Total Desorption Yields

In order to determine the relative production probability of charged species, also total desorption yields were measured using evaporated Eu_2O_3 and CsI targets[10]. The material sputtered by the incident heavy ion beam (Kr,Xe,U) was collected onto thin car-

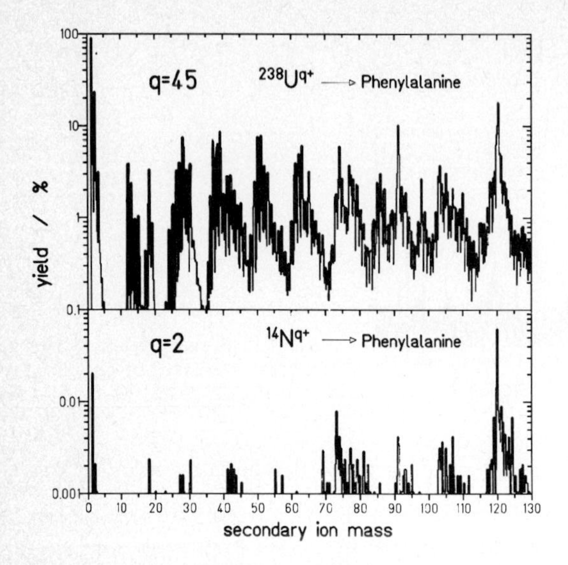

Fig. 13 Mass spectra of Phenylalanine irradiated by U- and N-beams having constant velocity and the charge state q=45 and q=2, respectively (low mass region)

bon foils. The number of Eu-and Cs-atoms was then determined by means of neutron activation analysis. The total yields were measured as function of projectile velocity (0.2 to 1.3 cm/ns). Some results are plotted in Fig.14 as function of the specific energy loss dE/dx together with data of experiments performed by Meins et al.[11]. They used UF_4 as target and irradiated it with relatively light ions (O,F,Cl) having a low specific energy loss. As seen in Fig.14 the total yields can be as high as 8000 per incident heavy ion. Similar results concerning electronic sputtering from organic material will be reported by P. Håkansson at this workshop. They found yields of intact molecules ($m \simeq 1000$amu) of about 1000 per impact.

From Fig.14 a simple relation between a certain power of dE/dx and the measured yields can not be deduced. The yield curves measured with individual beams tend to saturate before dE/dx reaches its maximum value. This phenomenon is most clearly displayed in the O- and F-curves. In these cases, since the maximum of dE/dx fell into the available energy range, a certain value of dE/dx had been reached at two energies (see Fig.1). The measured yield corresponding to the low energy value is generally higher than that of the high energy value. This effect is probably caused by the energy distribution in the vicinity of the heavy ion track (see Fig.2). At the low energy value the excitation energy is concentrated in a smaller volume than that at the high energy value, because the range of δ-electrons increases continously with projectile velocity.

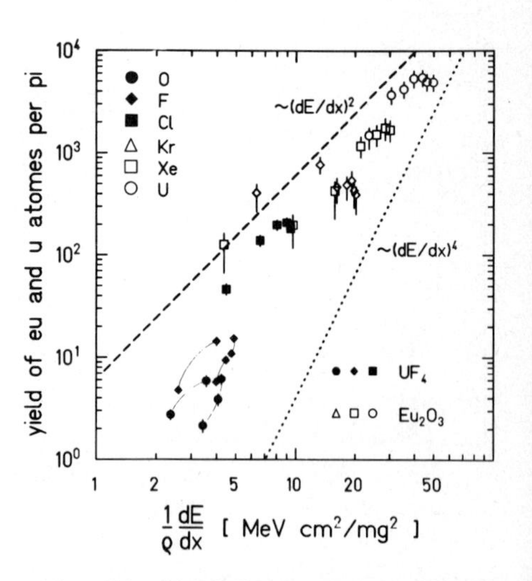

Fig. 14 Total desorption yields of Eu_2O_3 and UF_4 (Ref.11)

When desorption occurs only above a certain energy density, the yield should be enhanced in the low energy case.

6. Ionisation

The yields of ions ejected from Eu_2O_3 were measured using the same target as for the total yield experiments. A comparison of ion and total yields is presented in Fig.15. Here, the data points of the ions have been evaluated by summing up all yields of mass lines which contain Eu atoms. The production rate of the positive ions is 3 orders of magnitude smaller than that of the total number of desorbed Eu atoms.

Nevertheless, quite often data obtained with secondary ions are taken to draw conclusions about the desorption mechanism assuming that the ion yields are proportional to the total desorption rate. Ionisation, however, can be influenced by surface-composition,-structure and -contamination. A drastical example is shown in Fig.16. Here, two mass spectra both measured with CaF_2 are presented: In the first case, CaF_2 was evaporated onto an Al foil, in the second case, a cleaved CaF_2 crystal was used. CaF_2 is a rather inert compound, but the evaporated target had been obviously contaminated, when it was transported through air. The spectrum in Fig.16 consists of several mass lines due to organic impurities. The more striking result is that the F^+ ion in Fig. 16a is suppressed

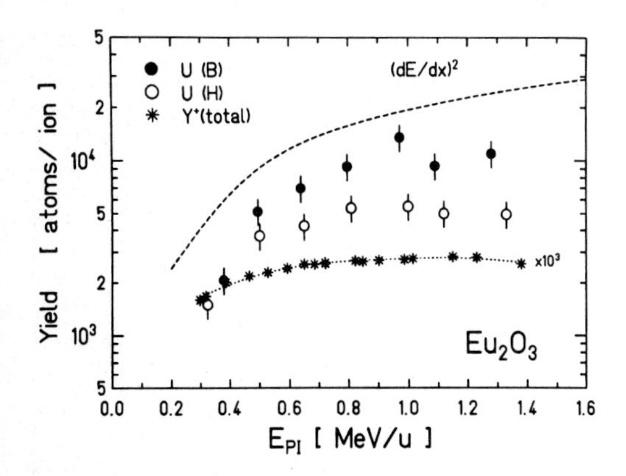

Fig. 15 Comparison of total desorption yields and secondary ion yields of Eu_2O_3 irradiated with an U-beam. U(B): desorption from the target backside. U(H): desorption form the target frontside. Y^+(total): yields of all ions containing Eu-atoms have been summed up.

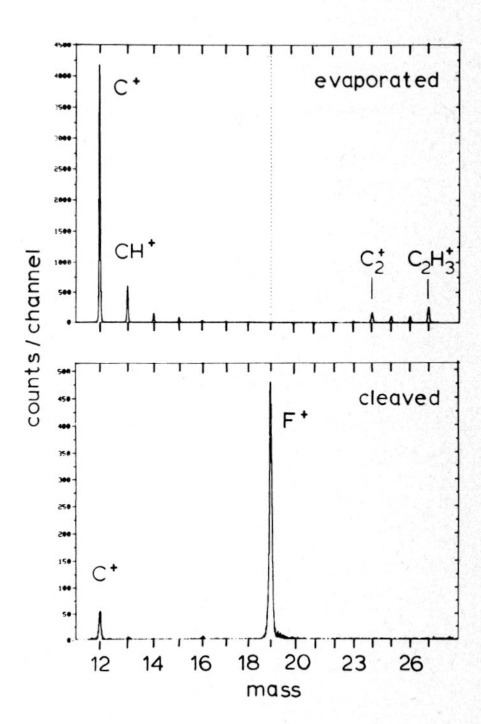

Fig. 16 Low mass part of TOF spectra measured with CaF_2 a) evaporated onto Al b) cleaved crystal

by 4 orders of magnitude compared to the spectrum of the cleaved crystal in Fig. 16b. Probably a small amount of contamination in the evaporated material has reduced the ionisation probability.

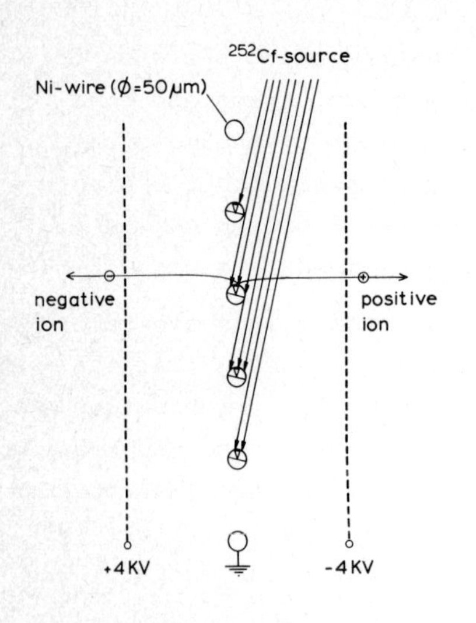

Fig. 17 Arrangement of target carrier (Ni-wire) and acceleration grids in order to detect ion pairs (Ref.12)

As will be discussed by Y. Hoppilliard at this workshop, ionisation can occur as a result of certain chemical reactions. B. Reimann[12] has studied ion pairs produced by decomposition of valine dimers. He evaporated Valine onto thin Ni-wires (Φ = 50 μm) and extracted positive and negative ions simultaneously out off the vicinity of the wire by means of an electric field as illustrated in Fig.17. In fact, depending on the field direction and the ion energy, many ions move back to the wire and are not detectable. By means of trajectory calculations using known energy and angular distributions of the ejected ions, the detection efficiency for ion pairs was evaluated. Under the given geometrical conditions of the experiment (see Fig.17), the effective surface area on each wire, from where ion pairs can be detected, is about 27% of the irradiated area (marked in Fig.17 by the small triangle in the cross section of the wire). Reimann obtained the following fractions of correlated events:

$$(M+1)^+ - (M-1)^- \quad : \quad 2.5\ \%$$
$$(M-COOH)^+ - (M-1)^- \quad : (12 \pm 4)\ \%$$
$$(\text{all ions})^+ - (\text{all ions})^- \quad : \ < 35\ \%$$

That means, ionisation by decomposition of molecules or aggregates of molecules into pairs of positive and negative ions is of minor importance for ion emission from valine and probably other aminoacids.

7. Summary

PDMS has become an important tool for chemical analysis. So far, it is mainly used for mass spectrometry of large thermally labil molecules. An easy to handle TOF-technique enables the users to generate mass spectra of solid targets over the complete mass range with relatively high effectiveness . Information on mass and structure of the compounds can be obtained up to molecular weights of 20000. Desorption and ionisation mechanisms are still under investigation. Experimental observation leads to the following conclusions:

Fast heavy ions induced desorption is due to electronic excitation.This so called electronic sputtering does not affect the matrix of clean metals ; small amounts of impurities on the metal surface, however, are desorbed. Contrary to metals the high energy impact on an insulator leads to an erosion like damage. Depending on the electronic energy loss several 1000 atoms or molecules can be ejected from the surface; about 0.1% are charged and can be utilized for mass spectrometry. The matrix material itself is transfered into the gas phase forming fragment ions, molecular ions and large aggregates of atoms or molecules. In accordance to three types of secondary ions observed in the mass spectra, three different types of electronic sputter mechanisms seem to exist:

1. The decay of individual excited states, excitons or electron-hole pairs ejects single atoms or molecules, as it has been found for contaminated metals.
2. The nuclear track zone in an insulator expelles residues of a vigorous destruction process into vacuum. Organic material releases numerous unspecific hydrocarbon ions, when the traversing heavy ion has a high specific energy loss. The pattern in the low mass region of the spectrum implies a fast evaporation process from a hot zone.
3. Over the whole energy loss region fragment ions, intact molecular ions and cluster ions are desorbed, which are specific for the molecular structure. The total number of the large undestroyed species (neutrals and ions) can be in the order of 1000 per impact. The corresponding desorption mechanism can be neither a decay of localized states nor a highly destructive evaporation process. A kind of ablation seems to occur, where certain parts of the surface are lifted up having a relatively low internal energy, which leads to further gentle decomposition.

References

1) MacFarlane, R. D., and D.F. Torgerson, Science 191 (1976) 920

2) Torgerson, D.F., R.P. Skowronski and R.D. MacFarlane, Biochem. Biophys. Res. Commun. 60 (1974) 616

3) Behrisch, R. (ed.), Sputtering by Particle Bombardment I, Topics in Appl. Phys., Vol. 47 Springer Verlag (1981)

4) Katz, R., B. Ackerson, M. Homayoonfar and S.C. Sharma, Radiat. Res. 47 (1971) 402

5) Albrecht, D., P. Armbruster and R. Spohr, Appl. Phys. A37 (1985) 37

6) E. Dartyge, M. Lambert, Radiat. Eff. 21 (1974) 71

7) Becker, O., and K. Wien, Nucl. Instr. and Meth. B16 (1986) 456

8) Sundqvist, B., I. Kamensky, P. Håkansson, J. Kjellberg, M. Salehpour, S. Widdiyasekera, J. Fohlman, P.A. Peterson and R. Roepstorff, Biomed. Mass Spectr., Vol. 11, No.5 (1984) 242

9) Della Negra, S., and Y. LeBeyec, Analyt. Chem. accepted for publication
B. Chait, Int. J. of Mass Spectr. and Ion Phys. 53 (1983) 227

10) Guthier, W., Thesis at Institut für Kernphysik TH Darmstadt 1986 and Springer Proceedings in Physics 9 (1986) 17

11) Meins, C. K., J.E. Griffith, Y. Qiu, M. H. Mendenhall, L.E. Seiberling and T.A. Tombrello, Rad. Eff. 71 (1983) 13

12) Reimann, B., Diploma Thesis at Institut für Kernphysik TH Darmstadt 1986

SPECTROSCOPY OF METAL CLUSTERS*

W. D. Knight, Walt A. de Heer, and Winston A. Saunders**

Physics Department, University of California, Berkeley CA 94720

INTRODUCTION

The optical spectra of dimers [1-3] and trimers [4-6] of alkali and other [7] metal clusters have been explored in molecular beams with considerable success. It was expected that the spectra of larger clusters would be complex and congested with thermal broadening by vibrational and rotational effects. Few data are yet available for larger clusters except for measurements of ionization potential. The IP represents transitions to an ion state from the highest occupied state of a neutral cluster. The variation of IP with cluster size is related to the behavior of the highest occupied levels in a series of clusters. These transitions form an important first chapter in the development of cluster spectroscopy [7-11]. Experiments on series of clusters demonstrate the existence of energy levels which may be investigated within individual clusters, and represent the uniqueness of the metal cluster state.

IONIZATION POTENTIALS AND ABUNDANCE SPECTRA

Ionization potentials (IP) have recently been measured, Fig. 1, for K clusters containing from N = 3 to 101 atoms [9-10]. While local maxima in the curve appear at the spherical clusters n_s, local minima occur for all the clusters n_s+1 immediately succeeding the spherical

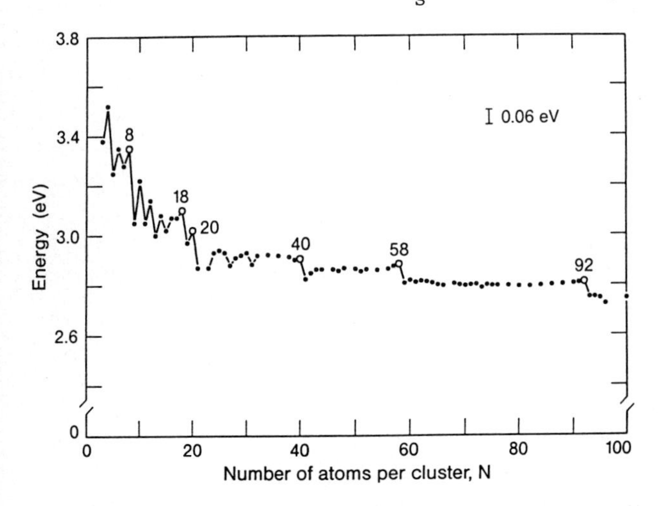

Fig. 1 Ionization potentials for potassium clusters N = 3 - 101, showing major shell closings and fine structure.

ones. These adjacent maxima and minima result in the observed sharp drops at the shell closings. These features in the IP curve correlate well with the patterns in the abundance spectra, Fig. 2, and are related to the relative binding energies of neighboring clusters [12].

The abundance spectra were taken by sweeping the mass selective detector at fixed detector ionizing photon energy. A cluster ionization potential was derived from the threshold of a PIE [13] curve of detected intensity vs photon energy for constant mass. The location of the IP discontinuity at a spherical cluster in the series may also be found [10] by taking a succession of abundance spectra over a given mass range, with a slightly different photon energy for each spectrum. In Fig. 3 we can see the general shift in IPs between the adjoining lg

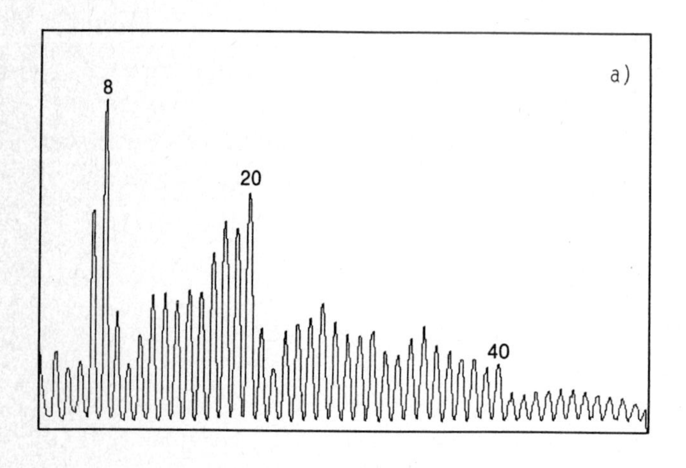

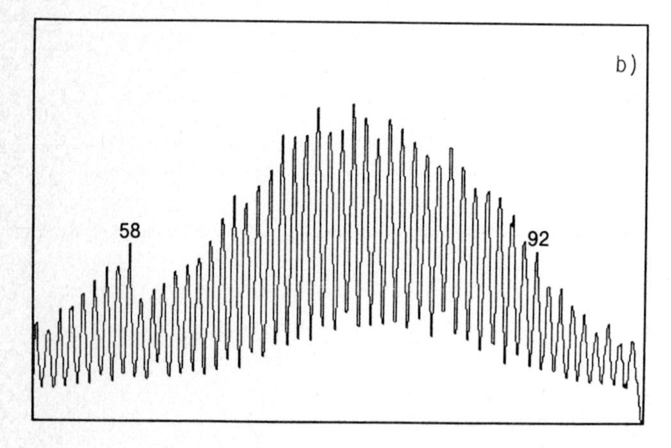

Fig. 2 Abundances for potassium clusters (a) N = 3 - 51, ionization by filtered xenon arc lamp; (b) N = 50 - 98, ionization by CW dye laser.

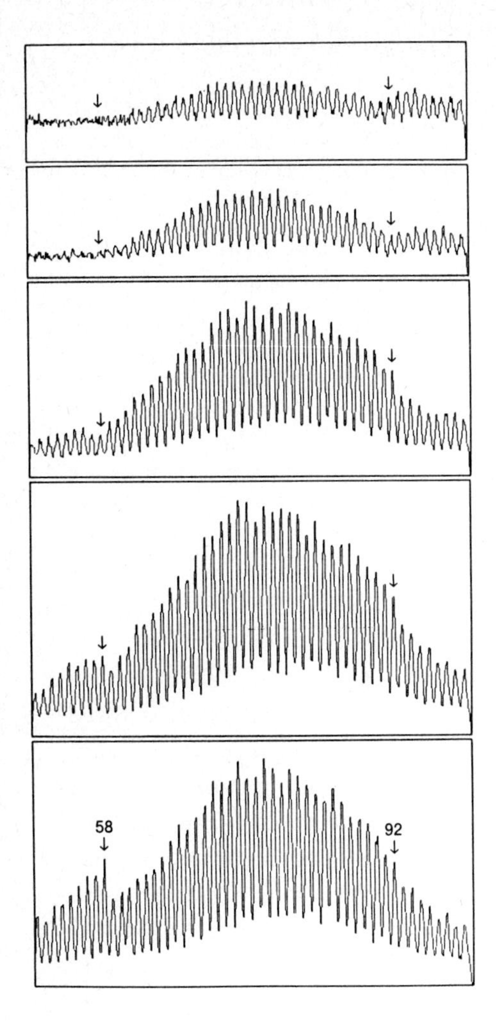

Fig. 3 Abundances for potassium clusters N = 50 - 98 for several ionization energies. Laser photon energy is above the ionization potential for N = 58 in the bottom spectrum and decreases to approximately the ionization potential for N = 92 in the top spectrum.

and 2d shells around N = 58. As ionizing wavelength increases in successive mass scans the drop in IP between shells is signaled by the decrease in intensities for the smaller clusters (higher IP) below the shell closing. The IP for the spherical cluster 58 is close to the wavelength for which its intensity is a minimum in the above series of mass scans. The same effect can be seen around both 58 and 92 in a sequence of mass scans [10].

FINE STRUCTURE AND THE ELLIPSOIDAL MODEL

In addition to the IP discontinuities at the spherical closings, new patterns emerge for the non-spherical clusters. Odd-even alternations in both the IP thresholds [9] and PIE shapes [9,11] occur for $N < 20$ and variations are seen for the following sequences of four: 15-18, 23-26, and 27-30. The available IP data and the abundance patterns indicate further sequences at 31-34 and 41-44. The magnitudes of the fine structure features are smaller than the IP jumps at spherical clusters, but they are seen consistently and reflect fine structure of the energy level system of the clusters.

The fine structure features are also consistent with odd-even and fourfold patterns in the level structure derived from the Nilsson theory [14,12,15], which predicts spheroidally distorted clusters [16]. Solution of the free electron angular momentum problem in a spherical potential well gives a well known set of energy levels with $2(2l+1)$ degeneracy. The spherical clusters possess relatively high stabilities which correspond to the filling of shells. The high stabilities are reflected in the relatively large abundances and IPs of the spherical clusters. If the spherical potential well is spheroidally distorted the spherical levels split into a series of sublevels with two-fold ($m_l=0$) and fourfold ($m_l \neq 0$) degeneracies. Physically an axial distortion results in a spheroid for which the states with reversed poles are indistinguishable. Thus a system including both spin and shape degeneracies can exhibit fourfold structures. These are directly reflected as distinct features in the abundance and IP patterns of alkali clusters. Current experiments and the spheroidal model are in qualitative agreement on points which depend on basic symmetries. The qualitative comparisons are made with no adjustable parameters, and the odds of finding such agreements accidentally are very small.

RESOLVED LEVEL SPECTROSCOPY

While ionization thresholds represent an important part of cluster spectroscopy, resolved single particle levels permit the observation of further fine structure in a more complete spectroscopy of individual clusters. In fact resolvable peaks [10,15,17] are observed near the thresholds of clusters which initially populate shells which follow the spherical clusters N = 40, 58 and 92. For example the intensities of the peaks at N = 59, 60, 61, 62, Fig. 4, are in the ratios 1:2:3:4, which suggests that every electron at the beginning of the new shell contributes identically to the PIE curve. Similar behavior

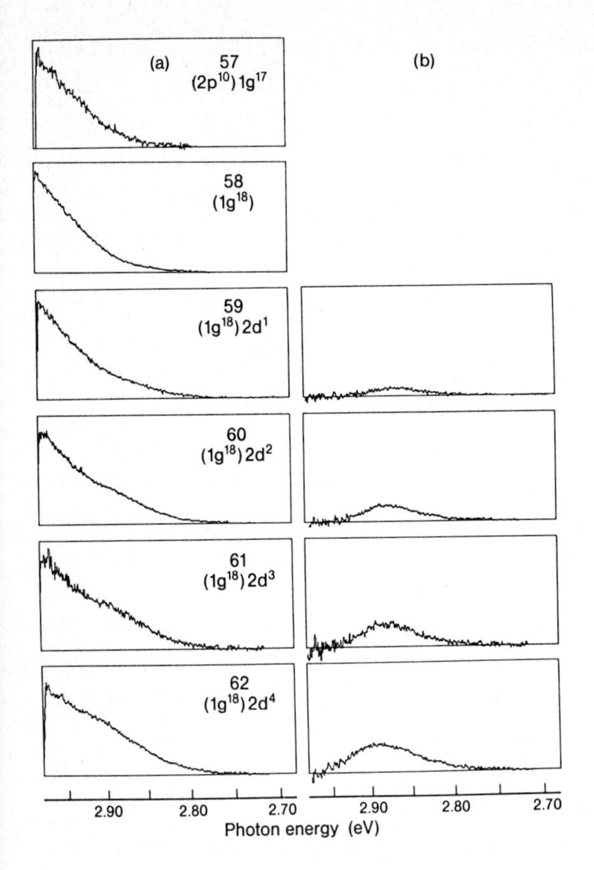

Fig. 4 (a) Level structure at the thresholds of clusters 59 - 62. (b) Intensity remainder after subtraction of N = 58 profile from profiles for 59 - 62.

is observed at 93-96 and indicated at 41-44. Single particle levels begin to emerge in the PIE curves for the smaller clusters $N < 30$, as was noticed earlier [9]. The small shifts of the above peaks with cluster electron number gives further evidence for fine structure in the electronic level system.

DISCUSSION

The observation of the abundance patterns, corresponding IP drops at the spherical clusters, and the additional twofold and fourfold fine structure patterns suggests that the ellipsoidal model is a good first approximation. The additional observation of resolved spectral peaks assures the identification of single particle energy levels. With the employment of methods to cool [6] the clusters and improve resolution the prospects for comprehensive spectroscopic analysis are good.

A further spectroscopic exploration of clusters has been carried out with longitudinal beam depletion spectroscopy (LBDS) [18]. A highly collimated beam of sodium clusters was irradiated longitudinally with laser photons of ~2.0 eV energy. Detected intensities of clusters in the range N = 2 - 21 were examined as a function of light intensity and wavelength. Reduction of detected intensity of a particular cluster indicates a process in which transverse recoil momentum from a photon induced cluster reaction causes an appreciable depletion of that cluster population in the beam. Relative depletion of neighboring clusters in the distribution indicates the relative stabilities against photo-dissociation. The preliminary result of these experiments is that clusters 8 and 20 are especially stable, in agreement with the abundance and ionization potential results. The cross sections for these processes are large and fine structures are present in the depletion spectra. Further experiments are in progress to determine the dependences on photon energy and intensity [19].

Individual cluster spectroscopy may include fluorescence, ionization, dissociation, and plasma resonance. The energy gaps indicated by the IP sequence measurements represent corresponding energy levels of individual clusters. Direct excitation of individual clusters to excited states corresponding to these energy gaps is feasible, depending mainly on the availability of suitable light sources. The energy gaps associated with the IP discontinuities are ~0.1 eV for potassium clusters and will be larger for sodium and lithium since the gap scales [12] approximately as $E_F/N^{1/3}$, where E_F is the Fermi energy.

Detection methods in cluster spectroscopy rely heavily on processes which identify the presence or absence of a cluster as a result of a specific treatment. The most familiar example is the presence of ions identified with parent neutrals which were ionized in a mass selective detector. Although the probability for direct detection of absorption or scattering of radiation by clusters is small because of small cross sections and small cluster concentrations, detection can rely instead on the absorption of energy by individual clusters followed by evaporation or dissociation. In the calculations of energies and relative stabilities against evaporation it was assumed [12] that internal energy was equally distributed among the dynamical modes as in the Dulong Petit model for heat capacity. The above experiments on recoil photo-fragmentation are an example of this behavior. Other similar processes may be exploited, for example it is expected that vibrationally absorbed IR energy will equilibrate [18,20] among the many modes and raise the temperature, causing evaporation which is observable in depletion spectroscopy because of recoil effects.

EXTENSION TO OTHER METALS AND CLUSTER IONS

The possibilities for cluster spectroscopy are not limited to the alkali metals. It is known [21] that cluster ions as large as $N \sim 200$ can be made of the noble metals, and significant results have been published for divalent [22] and trivalent [23] metals as well. The abundance spectra [21-25] of cluster ions formed in secondary ion mass spectroscopy (SIMS) experiments show relative stabilities reflecting a shell structure like that of the alkali metals. Both positive and negative ions can be manipulated and trapped increasing the concentrations of clusters to be investigated. Cluster ions can be produced by several methods including ion bombardment [26] and laser vaporization [27].

While the abundance spectra of the noble metals are similar to those of the alkali metals, significant differences are observed. For example, the odd-even features in the cluster ion abundance spectra persist to large sizes, and no fourfold fine structure has yet been observed. The odd-even behavior is typical of cluster ions [26,24]. Experiments have also been performed on doubly ionized clusters [25]. The lack of other fine structure could be accounted for by the fact that noble metal cluster ions tend to be spherical, in contrast to the ellipsoidal shapes observed in the open shell alkali metal clusters. We may anticipate that further information gained by studying spectra of alkali and other simple metals will provide the clues required to interpret the more complex spectra of transition metals.

CONCLUSIONS

At the present time there is some controversy concerning the relative importance of shell structure and geometrical structure of clusters. The experimental results for the alkalies indicate that for $N > 7$ knowledge of detailed geometrical structure is not required in first order. The symmetry constraints of the ellipsoidal model are sufficient to give major and minor observed abundance and IP features. Other models fail to account for the main experimental results for clusters containing up to 100 atoms. Comparison has mainly been made for clusters containing fewer than 8 atoms, a range in which the shell model is hardly expected to apply. Actually it may be inferred that any crystal field splittings associated with ion geometry must be smaller than implied by the observed fine structure in abundances and IPs, i.e. < 0.1 eV for K.

The ellipsoidal shell model is a first approximation and only qualitative agreement with experiment is expected. The self-consistent jellium model is an improvement but predictions of IP jumps are too large. However the jellium model employed thus far assumes spherical clusters. In fact the IPs predicted [28] for spherical closed shell potassium clusters 8 and 20 are within 15% of the observed values. The IP jump from a spherical cluster to the next open shell cluster will be reduced by the ellipsoidal distortion [14]. Consequently we may expect a self-consistent ellipsoidal jellium calculation to agree with experiment rather well, and we hope that such a calculation will soon be available.

The evidence indicates that individual clusters are amenable to detailed spectroscopic analysis. In addition to studies of direct absorption, scattering, and fluorescence, recoil beam depletion involving evaporation, dissociation, and chemical interactions between particles of intersecting beams, the surface plasma resonance will be observable in clusters and will provide a direct measure of the dynamical polarizability and photoemission properties [29,30] of clusters. It will be possible to study the interactions of the single particle states with the collective plasma resonance. The total spectral range will be wide, including for example UV inner core excitations [31], optical excitations of plasma resonances and of ions, and IR excitations of cluster electronic and vibrational states. Further experiments on magnetic spectroscopy will extend from the microwave region of the spectrum to DC [32,33]. Although magnetic resonance experiments may be limited [33] to the smaller clusters, hyperfine structure effects in magnetic resonance (so far investigated in matrix isolated clusters) [34,35] will yield important structural information for free clusters. Similarly extensions of electric deflection [36] spectroscopy are promising. Finally it is expected that collective vibrational modes analogous to those in nuclei will occupy an important segment of the observable spectrum.

The results of the possible spectroscopies will assist in answering many questions such as: are metal clusters like liquid drops, what is the relative importance of ion crystal structure, how far can the ellipsoidal model be used with selfconsistent treatments including spin, at what size does the cluster state merge into the crystalline condition of the bulk material, when do metal clusters act like bulk metals? Other yet unasked questions will arise. The prospects for spectroscopic study of clusters appear at this time to be very rich.

REFERENCES

* This work has been supported by the Materials Research Division of the U. S. National Science Foundation under Grant #DMR 84-17823.

** Present address, Laboratoire de Physique Expérimentale, Ecole Polytechnique Fédérale de Lausanne, Switzerland.

1. D. L. Feldman, R. K. Lengel, and R. N. Zare, Chem. Phys. Lett. 52, 413 (1977).
2. A. Hermann, S. Leutwyler, E. Schumacher, and L. Wöste, Chem. Phys. Lett. 52, 418 (1977).
3. M. Broyer, J. Chevalyre, G. Delacrétaz, P. Fayet, and L. Wöste, Chem. Phys. Lett. 114, 477 (1985).
4. A. Hermann, M. Hofmann, S. Leutwyler, L. Wöste and E. Shumacher, Chem. Phys. Lett. 62, 216 (1979).
5. J. L. Gole, G. J. Green, and D. R. Preuss, J. Chem. Phys. Lett. 26, 2247 (1982).
6. G. Delacretaz and L. Wöste, Surf. Sci. 156, 770 (1985).
7. M. D. Morse, J. B. Hopkins, P.R.R. Langridge-Smith, and R. E. Smalley, J. Chem. Phys. 79, 5316 (1983).
8. A. Hermann, E. Schumacher, and L. Wöste, J. Chem. Phys. 68, 2327 (1978).
9. Winston A. Saunders, Keith Clemenger, Walt A. de Heer and W. D. Knight, Phys. Rev. B32, 1366 (1985).
10. Winston A. Saunders, Ph.D. Thesis, University of California, Berkeley, 1986.
11. C. Bréchingnac, Ph. Cahuzac and J. Roux, Chem. Phys. Lett. 127, 445 (1986).
12. Walt A. de Heer, Solid State Physics (Academic Press 1986), in press.
13. (PIE) Photoionization efficiency.
14. Keith Clemenger, Phys. Rev. B32, 1359 (1985).
15. W. D. Knight, Walt A. de Heer and Winston A. Saunders, International Symposium on Metal Clusters, Heidelberg, (April 1986), Springer (in press).
16. Triaxial distortions have been included in the calculations for ellipsoidal clusters [10]. The conclusions drawn from the Nilsson diagram [14] remain qualitatively the same, save for cluster 12, which becomes more stable, in agreement with experiment.
17. Similar partly resolved peaks were reported earlier [9].
18. Walt A. de Heer, to be published.
19. It is worth mentioning that size effects are also observed in TOF spectroscopy in which the velocities of sodium clusters 8 and 20 are distinguished from the rest.
20. N. Bloembergen and E. Yablonovitch, Physics Today 31, 23 (1978).
21. I. Katakuse, I. Ichihara, Y. Fujita, T. Matsuo, T. Sakurai and H. Matsuda, Int. J. Mass Spect. Ion Proc. 67, 229 (1985).
22. I. Katakuse, I. Ichihara, Y. Fujita, T. Matsuo, T. Sakurai, and H. Matsuda, Int. J. Mass. Spect. Ion Proc. 69, 109 (1986).
23. W. Begemann, K. H. Meiwes-Broer, and H. G. Lutz, International Symposium on Metal Clusters, Heidelberg, 1986. Springer Verlag, in press.
24. G. Hortig and M. Müller, Z. Physik 221, 119 (1969).
25. P. Joyes and P. Sudraud, Surf. Sci. 156, 451 (1985).
26. P. Joyes, J. Phys. Chem. Solids 32, 1269 (1971).
27. T. G. Dietz, M. A. Duncan, D. E. Powers, and R. E. Smalley, J. Chem. Phys. 74, 6511 (1981).
28. M. Y. Chou, Andrew Cleland, and Marvin L. Cohen, Solid State Commun. 52, 643 (1984).

29. W. Ekardt, Phys. Rev. Lett. 52, 1925 (1984).
30. W. Ekardt, Phys. Rev. B31, 6360 (1985).
31. C. Bréchignac, M. Broyer, Ph. Cahuzac, G. Delacrétaz, P. Labastie, and L. Wöste, Chem. Phys. Lett. 120, 559 (1985).
32. W. D. Knight, Helv. Physica Acta 56, 521 (1983).
33. Walt A. de Heer, Ph.D. Thesis, University of California, Berkeley, 1985.
34. D. M. Lindsay, D. R. Herschbach, and A. L. Kwiram, Mol. Phys. 32, 1199 (1976).
35. D. A. Garland and D. M. Lindsay, J. Chem. Phys. 80, 4761 (1984).
36. W. D. Knight, K. Clemenger, W. A. de Heer, and W. A. Saunders, Phys. Rev. B31, 2539 (1985).

CLUSTER EVAPORATION AND DESORPTION: EXPERIMENTAL DETERMINATION OF BINDING ENERGIES

Michael Vollmer and Frank Träger
Physikalisches Institut der Universität Heidelberg
D-6900 Heidelberg, Philosophenweg 12, Federal Republic of Germany

Introduction

The thermodynamic properties of clusters exhibit interesting changes as a function of size. However, only few experiments have been undertaken to investigate these characteristics in detail. For example, the evaporation of lead clusters was observed directly by electron microscopy to test the validity of the Kelvin equation [1]. Another example for thermodynamic properties is the work by Buffat and Borel who determined the melting temperature of metal clusters as a function of size [2]. In the present paper we report on thermal desorption experiments with sodium clusters, i.e. on evaporation of atoms from the surface of clusters. Furthermore, scattering experiments have been carried out for cluster size determination. The results do not only give insight into evaporation processes on a microscopic scale, but also permit to extract cluster binding energies as a function of size, and follow their convergence to the properties of the bulk. In addition, fractional order thermal desorption was analysed in detail. Our experiments also yield information on cluster nucleation associated with atom diffusion on surfaces.

Thermal desorption

In thermal desorption, also called temperature programmed desorption or flashdesorption [3] a sample consisting of a substrate with one or more adsorbates on the surface is heated uniformely with constant rate and the rate of desorbing particles is monitored with a mass spectrometer as a function of the surface temperature T. Depending on the combination of substrate and adsorbate at least one pronounced maximum is observed. The position and shape of the spectrum contain information on the kinetics of the desorption and on the binding energies of the adsorbate. Theoretically, the description of thermal desorption is commonly made with an Arrhenius-law:

$$-dn(T)_{des}/dt = n(T)^{x} \nu \exp(-E/kT) \qquad (1)$$

where $dn(T)_{des}/dt$ stands for the rate of desorbing particles and n(T) for the coverage at the temperature T. ν is the frequency factor for vibration in the surface potential of typically 10^{12}/s, E the activation energy for desorption and x the formal order. x is usually interpreted in such a way, that $n(T)^{x}$ gives the number of particles which participate in the critical step of the desorption process. Most commonly, first and second order processes are observed [3,4]. Zero order thermal desorption has also been investigated (see e.g. [5]).

The order is called fractional if $x<1$. Such a case, however, has only been observed in few experiments [6,7]. One would expect fractional order desorption for metal deposits on insulator surfaces. Here, surface diffusion occurs, and the individual particles may collide and form clusters. In epitaxy this type of nucleation is called Volmer-Weber mode. If the support temperature is rised desorption does not take place directly from the substrate. Rather atoms located at the surface of the clusters will desorb. This results in a fractional order process.

Equation (1) is based on several assumptions [3] the validity of which has to be checked carefully for the adsorbate/substrate system under study. In the present study, the crucial point is that the binding energy E in Equ.(1) is independent of coverage. We expect, however, that the desorption energy increases with cluster size, i.e. with coverage. Moreover, during a desorption experiment the evaporating clusters become smaller so that the binding of surface atoms is weakened. Still, the Arrhenius law can be a good description for the low temperature part up to the maximum of each individual spectrum [12].

Cluster formation and cluster size determination

When atoms impinge on a cold insulator surface they are trapped in the surface potential. Their residence time depends on the surface temperature and is given by $\tau_{res} = \tau_o \exp(E/kT)$. Since the energy barrier for surface diffusion is typically only about one half of the binding energy in the surface potential, the atoms can easily diffuse. If diffusing sodium atoms occasionally collide with each other and form clusters or if they meet surface defects, the binding energy is increased. Therefore, the nucleation of atoms to clusters is energetically favourable. These clusters are likely bound to surface defects which act as nucleation centers. The growth of metal clusters on surfaces and their size distributions have been studied extensively, in particular by electron microscopy (see e.g. [8]).

In the present experiment the growth of clusters during the exposure of the surface to a thermal atomic beam has been monitored by scattering measurements. The sodium beam is directed onto the surface and the rate of scattered atoms is measured as a function of time with a quadrupole mass spectrometer. The time dependence of the scattering signal reflects the increase of the area on the surface which is covered with clusters. Since atoms bind more strongly to these clusters than to the alkali halide crystal, the scattering signal decreases during the deposition. Within the framework of an atomistic model [9] the time dependence of the scattering signal can be calculated. A fit to the experimental data then gives the mean cluster size for a given deposition time and a given constant flux of the atomic beam. The model assumes that Na atoms which are initially adsorbed with 100% probability on the cooled LiF(100) surface may either desorb after a mean residence time or form clusters via surface diffusion. Furthermore, each nucleus or growing cluster is considered to be surrounded by a capture zone of radius R. The surface diffusion of adatoms can be described as a random walk. During the diffusion the atoms cover an area R^2. It is related to the time of diffusion τ_{diff} to the nearest capture site. Under our experimental conditions the probability that an adatom is captured by the nearest cluster is of the order of 1. From nucleation experiments it is known that for large deposition times, but well before coalescence occurs, the number density of clusters reaches a constant value, the saturation density of nuclei. Any further increase of the integral coverage then merely results in further growth of the clusters. Nucleation of new clusters does not take place. Under these preconditions the scattering of atoms from the surface, the diffusion of adatoms and the nucleation of clusters can be described quantitatively. We obtain for the scattering signal:

$$S(t)=C\cdot R_{1/2}^{2}(t)\cdot h(y(t)) \qquad \text{with} \qquad h(y)=\frac{\left(\frac{3}{4}-y+\frac{\pi}{12}y^{3}\right)^{2}}{\left(1-\frac{\pi}{4}y^{2}\right)}.$$

The factor C can be determined with a least square fit of the scattering rates to the above formula. S(t) and the coverage N(t) permit to extract the mean cluster radius $\langle R(t)\rangle$ and half the mean cluster distance $R_{1/2}(t)$. y denotes $\langle R(t)\rangle/R_{1/2}(t)$.

Experiment

The experimental arrangement consists of an ultrahigh vacuum system with a base pressure of typically 2×10^{-10} mbar. A quadrupole mass spectrometer serves to detect desorbing sodium atoms. A glass cell with high purity metallic sodium is attached to the system. Through heating sodium atoms diffuse out of a small orifice, and pass a number of liquid nitrogen cooled diaphragms to form a well collimated beam. It impinges either on the substrate or on a quartz crystal microbalance for flux or coverage measurements. Both are attached to a manipulator and can be cooled to liquid nitrogen temperature. The substrate consists of a LiF(100) single crystal which is pressed against a small molybdenum oven to permit heating. A NiCrNi thermocouple is integrated in the crystal holder and measures the surface temperature of the crystal. The frequency of the quartz crystal microbalance, the surface temperature of the LiF crystal, and the sodium ion counting rate of the mass spectrometer are stored in a microcomputer.

Before each desorption experiment the LiF crystal was heated to 700 K for about 2 hours. This procedure serves to remove contaminations on the surface and to anneal active sites for subsequent adsorption of residual gases [10]. Between two runs, the crystal was cleaned by heating to 700 K for more than 30 minutes.

Measurements and Results

a) Scattering experiments

The result of a measurement of the scattering rate as function of time is shown in Fig. 1. As expected, the signal gradually decreases. The solid line refers to the analytical dependence for S(t) which is based on the atomistic model sketched above. As can be seen from the figure, the calculated scattering signal S(t) fits the experimental data very well. Even the amplitude of the signal for the first few seconds of Na deposition is in excellent agreement with the theoretical fit. This indicates that the number density of clusters saturates very rapidly. We obtain the value of 5×10^{8} clusters/cm^{2}, which is comparable to the number of defects on well annealed alkali halide surfaces. Therefore, it is very likely that the defects act as nucleation centers. Unfortunately, the number of clusters on the surface is relatively small so that the cluster diameters quickly reach values of the order of 100 Å. We find that clusters with radii between 100 Å and 1500 Å were investigated.

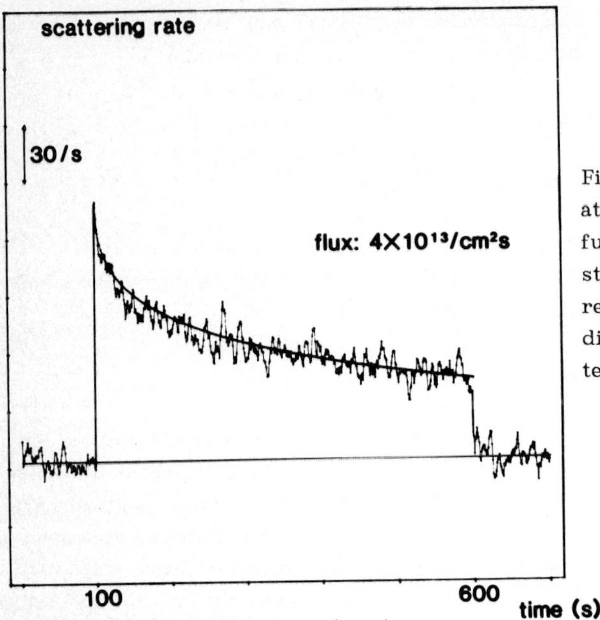

Fig. 1: Scattering rate of sodium atoms from a LiF(100) surface as a function of deposition time at a constant Na flux. The solid line represents a least square fit according to the theory explained in the text.

b) Thermal desorption experiments

Thermal desorption spectra for different integral coverages are displayed in Fig. 2. The most characteristic feature is the shift to larger desorption temperatures with increasing coverage. According to the description of the desorption process with an Arrhenius equation, this implies fractional order thermal desorption.

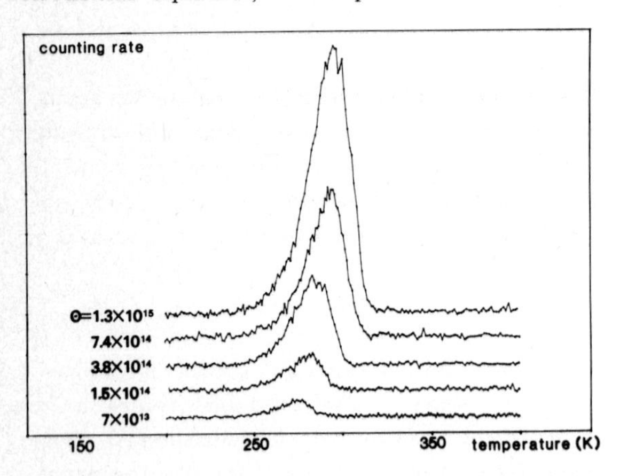

Fig. 2: Thermal desorption spectra of sodium atoms evaporating from Na_n clusters for different coverages on the surface, i.e. different cluster sizes.

An Arrhenius plot with log(dn/dt) versus 1/T gives the energy E , if the change in $n(T)^x$ is negligible. This condition holds if the analysis is restricted to that part of the signal where the desorption rate gradually starts to increase, i.e. where the coverage decreases by at most 4% of its initial value [11]. Such an analysis is easily possible for the good signal-to-noise ratios associated with large coverages, i.e. large clusters. For low coverages, however, E can only be determined with reasonable accuracy if a decrease of up to 40% enters the analysis (see Fig. 3). This introduces an error in E. We account for this error in the following way. The order x is calculated from the previously derived energy value E with a relation that can be calculated from the analytical expression Equ. (1) and the condition that the derivative of the rate of desorbing particles is zero at the temperature T_p where the signal is maximum, i.e. $d/dT[dn/dt]=0$. The value for x is then used for a corrected Arrhenius diagram with $\log[(dn/dt)/n^x]$ versus 1/T, from which a new E value is computed and so forth until selfconsistency is obtained.

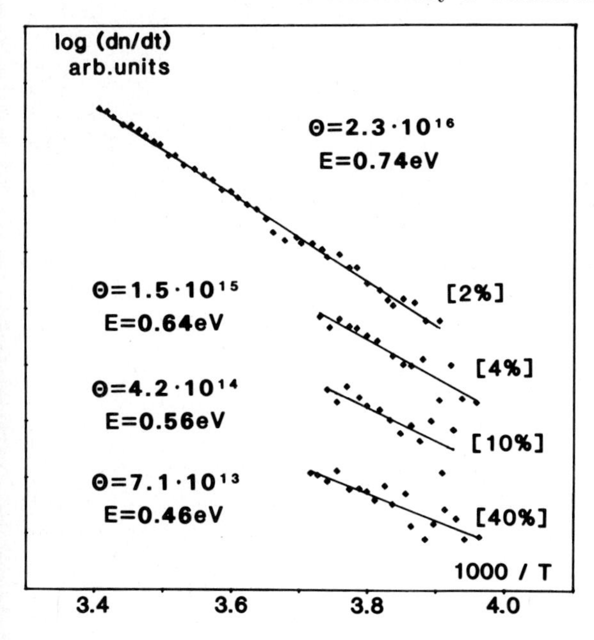

Fig. 3: Arrhenius diagrams for different sodium coverages. The values in brackets give the decrease of the initial coverage on which the plots are based.

As a result of the thermal desorption spectra, we finally obtain the binding energy E as a function of the average cluster size and the order of desorption. The dependence of the binding energies on the cluster radius is displayed in Fig. 4. The order of desorption is x=0.79(8) and is almost independent of coverage. For more details on the analysis of fractional order thermal desorption the reader is referred to ref. [12].

Discussion

a) Order of desorption

The order of x=0.79(8) can be explained with desorption of atoms from clusters. In a simple picture of f hemispherical clusters of only one size R where atoms can only desorb from the cluster surface, the order of desorption is given by

$$x = \log(n_{sur}) \;/\; \log(n_{tot}).$$

Here, $n_{tot}=(2\pi/3)f\times(R/r_o)^3$, $n_{sur}=2\pi f\times(R/r_o)^2$ and $n_{per}=2\pi f\times(R/r_o)$ denote the total number of atoms in the cluster, at the cluster surface and at the cluster perimeter, respectively. r_o denotes the lattice constant of 3.7 Å for solid sodium. With the experimental value of $f = 5\times10^8$ (see above) the order x has been calculated for cluster radii of R=100Å and R=1000Å. We find x_{sur}(100Å)=0.93 whereas x_{per}(100Å)=0.82. Similarly, x_{sur}(1000Å)=0.82 and x_{per}(1000Å)=0.73. With a Gaussian cluster size distribution similar values for x are obtained. Therefore, it seems that desorption primarly takes place from the cluster perimeters. In a more realistic picture of a cluster the atoms would then desorb from edges of similar binding energy.

b) Binding energies

Fig. 4 displays the cluster binding energies as a function os size. They slowly approach a saturation value of approximately 0.8 eV. A similar increase of the binding energies with coverage was also observed for Cu and Au on graphite substrates [6]. However, no attempt was made to relate the energies to certain cluster sizes. The saturation value of 0.8 eV observed here is smaller than the heat of sublimation of 1.113 eV and the heat of vaporisation of 0.93 eV. At a first glance one would expect that the measured binding energies should approach one of these two values. We will therefore have to discuss the physical meaning of the extracted energies.

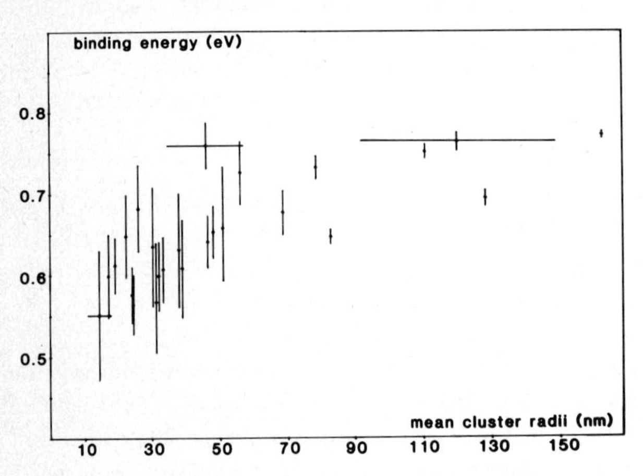

Fig. 4: Binding (desorption) energies of sodium clusters as a function of mean cluster size.

In the present experiment the clusters are relatively large and cold so that the desorption of atoms should be governed by the heat of sublimation. Even effects of finite particle size which lead to a decrease of the melting temperature [2] do not play a significant role here since the particles have sizes of at least 100 Å. We therefore have to explain the apparent difference of the sublimation energy of 1.113 eV and the binding energies of maximum 0.8 eV. For this purpose the kinetics of the desorption process is helpful. Since the order is in agreement with the assumption of evaporation from the cluster perimeters, the binding energies of such atoms can help to understand the observed saturation value. Clearly, the number of nearest neighbours influences the binding energy an effect which is particularly important at the cluster perimeter. In a simple geometrical model one expects the number of nearest neighbours for perimeter atoms of a large cluster to vary between about 0.65 and 0.7 of the corresponding number for atoms on a flat surface. Thus, in agreement with our results, the bulk sublimation energy of 1.113 eV should decrease to ≈0.8 eV for perimeter atoms. These more or less qualitative arguments illustrate the important influence of the coordination number on the binding energies. Comparison to theoretical work is not possible since most calculations are limited to smaller clusters and, in addition, the theoretical binding energies are usually given as average values for the whole cluster (see e.g. [13]).

In conclusion, we have analysed fractional order desorption kinetics and determined binding energies of sodium clusters. The average size and the number density of the clusters could be derived from scattering experiments. The model developed here for surface diffusion and cluster formation is supported, for example, by the result of 5×10^8 clusters/cm^2 which likely coincides with the number of defects on the surface. Measurements with clusters of smaller size and comparable signal-to-noise ratio should be possible by artificially increasing the number of defects to which the clusters stick. This can be accomplished e.g. by electron bombardment of the surface after the heat treatment. For a given coverage the average cluster size will then be smaller than in the present work. Actually, it should be possible to study the evaporation of particles as small as 10 Å. For future measurements electron microscopy for direct cluster size determination would also be desirable.

References:

[1] J.R. Sambles, L.M. Skinner, N.D. Lisgarten, Proc. Roy. Soc. London **A318**, 507 (1970)

[2] Ph. Buffat, J.-P. Borel, Phys. Res. **A13**, 2287 (1976)

[3] D. Menzel, "Thermal Desorption", in Springer Ser. Chem. Phys. **20**, Chem. & Phys. of Solid Surf. **IV** (1982)

[4] C.M. Chan, R. Aris, W.H. Weinberg; Appl. Surf. Sci. **1**, 360 (1978)

[5] K. Nagai, T. Shibanuma, M. Hashimoto; Surf. Sci. **145**, L459 (1984)

[6] J.R. Arthur, A.Y. Cho, Surf. Sci. **36**, 641 (1973)

[7] L. Chan, G.L. Griffin; Surf. Sci. **145**, 185 (1984)

[8] H. Schmeisser, Thin Solid Films **22**, 83 (1974)

[9] M. Vollmer, F. Träger, Z. Phys. D - Atoms, Molecules and Clusters **3**, 291 (1986)

[10] J. Estel, H. Hoinkes, H. Kaarmann, H. Nahr, H. Wilsch, Surf. Sci. **54**, 393 (1976)

[11] E. Habenschaden, J. Küppers, Surf. Sci. **138**, L147 (1984)

[12] M. Vollmer, F. Träger, to be published

[13] J. Koutecky, P. Fantucci, Chem. Rev. **86**, 539 (1986)

A PENNING TRAP FOR STUDYING CLUSTER IONS

H.-J. Kluge
CERN, Geneva, Switzerland
and
für Physik, Universität Mainz, D-65 Mainz, Fed. Rep. Germany

H. Schnatz, L. Schweikhard
Institut für Physik, Universität Mainz, D-65 Mainz, Fed. Rep. Germany

Abstract: We propose to use a Penning trap for spectroscopy of stored cluster ions. A similar device has been built by us for the purpose of mass measurements of short-lived nuclei produced at the on-line isotope separator ISOLDE/CERN. A resolving power of 500,000 in a mass measurement of ^{39}K and an accuracy of 2×10^{-7} for the $^{85}Rb/^{39}K$ mass ratio were obtained. An efficiency for in-flight capture as high as 70% was achieved. A method of very high sensitivity is realized since typically only 10 to 100 ions are stored in the trap. We intend to perfom laser spectroscopy on trapped Na cluster as a first application of the trap technique.

1. INTRODUCTION

Although enormous progress has been made in cluster spectroscopy during the past years, only very little is known about, for example, electronic structures of clusters, life times, and reaction, catalysis, or electron detachment processes. The reasons can be found in essentially two general drawbacks of all techniques applied so far to the study of clusters:

(i), the limited time of observation because atomic or ionic beams are used,

(ii), the difficulty of preparing a sample of one specific cluster size with sufficient purity and abundance.

Both problems can be solved by confining and accumulating the cluster ions of the wanted size in an ion trap. The device best suited for this purpose is the Penning trap where a combination of static magnetic and electric fields is used to establish a three-dimensional trapping potential: In such a trap [1,2] the cluster ions would be almost at rest in space, confined in a small volume, essentially free of any undesired perturbation, and at hand for almost infinite times limited only by the life time of the cluster itself. The high resolving power of mass measurements in a Penning trap would allow a precise identification of the cluster size and even enable studies of isotope effects if they exist at all.

In the past many experiments made use of the ion trap technique for the investigation of charged particles and led to a variety of high-precision experiments [3]. These experiments include laser spectroscopy on single ions [4,5], accurate mass measurements [6,7], ultra-high resolution microwave spectroscopy [8], and determination of the g factors for the electron and positron [9], which represents the most accurate fundamental constant known today.

All charged particles investigated so far in an ion trap were created inside the potential well produced by the quadrupole field. However in case of cluster ions it is desirable to produce these species by well established techniques [10] outside the trap, eventually to separate them in mass and to guide them with high efficiency into a trap.

Recently we reported on the first in-flight capture of ions in a Penning trap starting from a continuous ion beam [11]. Simultaneously and independently Alford et al. [12] were successful in capturing bunches of cluster ions in a similar device which differs from our set up mainly with respect to the geometry and the detection scheme of the cyclotron resonance.

In the following, our method and our apparatus will be described which was developed for precise mass measurements of short-lived isotopes. Emphasis will be put on the discussion of the geometry of the Penning trap, the harmonic motion of the charged particles in the trap, and the detection scheme which differ from those described in [12]. The performance of the existing device and finally its potential for a future application for cluster spectroscopy will be discussed.

The very same article is published in the proceedings of the International Symposium on Metal Clusters, Heidelberg, and will appear in Z. Physik D.

2. PRINCIPLE OF A PENNING TRAP

Fig. 1 shows the essential layout of a Penning trap. In the homogeneous magnetic field B directed along the z axis particles with charge e and mass m perform a cyclotron motion with a frequency given by

$$\omega_c = (e/m)\, B \,. \tag{1}$$

In order to obtain a restoring force along the z axis in addition to the confinement by the magnetic field in x-y direction, a positive potential for positively charged ions is applied to the upper and lower electrodes (so-called endcaps, see Fig. 1) relative to the ring electrode. Due to the special (hyperbolical) shape of the electrodes the motion of a stored ion in the z direction is decoupled from those in the x-y directions and furthermore all motions are harmonic. They can be separated into an oscillation ω_z along the z direction and the modified cyclotron frequencies ω_+ and ω_-. These frequencies deviate from the pure cyclotron frequency ω_c due to the presence of the electrostatic field. The harmonic oscillations are described by the frequencies

$$\omega_z = \sqrt{(2eU/mr_0^2)} \tag{2}$$

$$\omega_+ = \omega_c/2 + \sqrt{(\omega_c^2/4 - \omega_z^2/2)} \tag{3}$$

$$\omega_- = \omega_c/2 - \sqrt{(\omega_c^2/4 - \omega_z^2/2)} \quad . \tag{4}$$

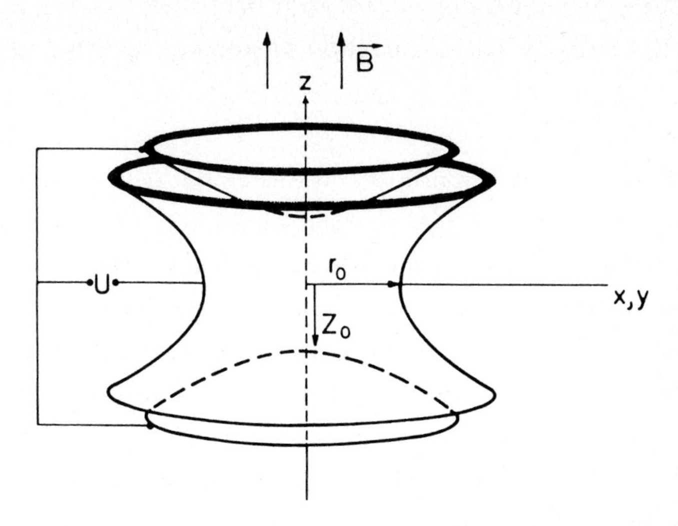

Fig. 1 Scheme of a Penning trap. The shapes of the electrodes are hyperbolical with rotational symmetry around the z axis which is also the direction of the magnetic field. The distance of the endcaps $2z_0$ is related to the minimum radius of ring electrode by $r_0^2 = 2z_0^2$.

The voltage U and the radius r_0 are explained in Fig. 1. In case of an ion with mass m = 40 amu, B = 5.87 T, U = 8 V and r_0 = 0.8 cm the frequencies are ν_c = 2253 kHz, ν_z = 124 kHz, ν_+ = 2250 kHz, and the magnetron frequency ν_- = 3.4 kHz almost independent of the mass of the stored particle.

Note that ω_z as well as the modified cyclotron frequencies depend on the voltage U applied to the electrodes. These frequencies correspond to one-quantum transitions in the harmonic oscillators. However, the sum frequency $\omega_+ + \omega_-$ equals to ω_c. This frequency depends only on B (see Eq.(1)) and represents a two-quantum transition. Although the ω_c resonance has a small strength because it is a two-quantum transition, it has the advantage of being independent of the trapping voltage. Even more important, this resonance is in contrast to ω_+ insensitive to space charge effects caused by a larger number of stored ions and is not very much influenced by imperfections of the hyperbolical shape of the electrodes caused, for example, by the necessary holes for capture and ejection of the ions.

3. DETECTION OF THE CYCLOTRON RESONANCE BY TIME OF FLIGHT

In resonance the trapped ions gain energy out of the applied radio frequency (RF) field. This is detected by a time-of-flight method [6]. The charged particles are ejected out of the trap by applying an electrical pulse to the ring electrode. For a short time the potential between the endcaps and the ring electrode is lowered so that the ions with highest energy can just escape the trap. These ions drift to a channel plate detector and their time of flight is determined. This procedure is repeated by continuously lowering the trapping potential as shown in Fig. 2 until the trap is empty. The mean time of flight as a function of the RF frequency shows a resonance which gives the mass of the stored ions (Fig. 3).

The basic principle of this detection technique is easily understood. In resonance the cyclotron orbit becomes larger. As a consequence the orbital magnetic moment μ_L increases, leading to a higher energy of the charged particle in the magnetic field. If the ions are ejected out of the trap into a region of lower magnetic field, the ions experience a force $\mu_L \delta B/\delta z$ in the direction of magnetic field gradient. Conservation of energy demands that the radial energy $\mu_L B$ is transformed into longitudinal kinetic energy. This is just the reverse of the magnetic-bottle effect. Hence in resonance the ions reach the detector faster than out of resonance.

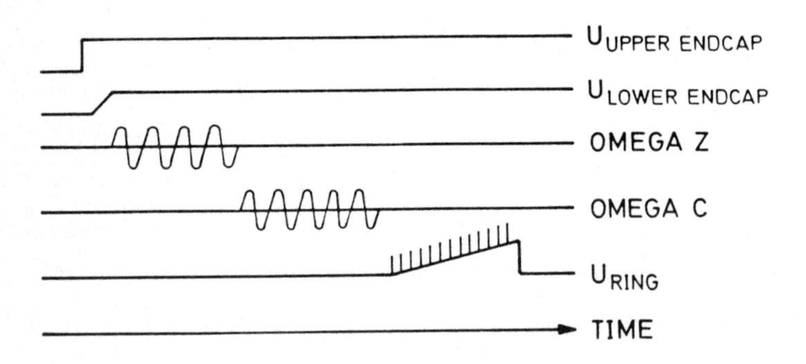

Fig. 2 Timing sequence for trapping charged particles, inducing the radio frequency, and for ejection of the ions. This sequence was used for mass measurements in a Penning trap using the time-of-flight technique.

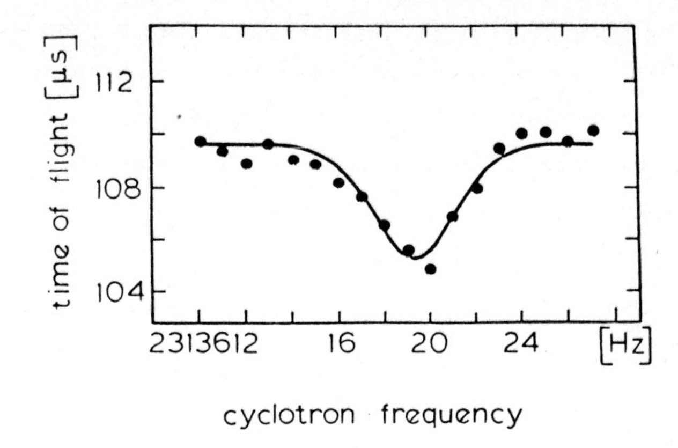

Fig. 3 Cyclotron resonance of potassium ions as obtained by the mean time of flight as a function of the frequency of the applied RF field. The resolution observed is 5×10^5.

Fig. 4 shows the time-of-flight effect for the case of a mixture of N and N_2 ions. If the frequency ν_c (N_2) is applied, the mean time of flight is shifted to smaller values. This is seen in the bottom part of Fig. 4 which shows the difference between the time-of-flight pattern in and off resonance.

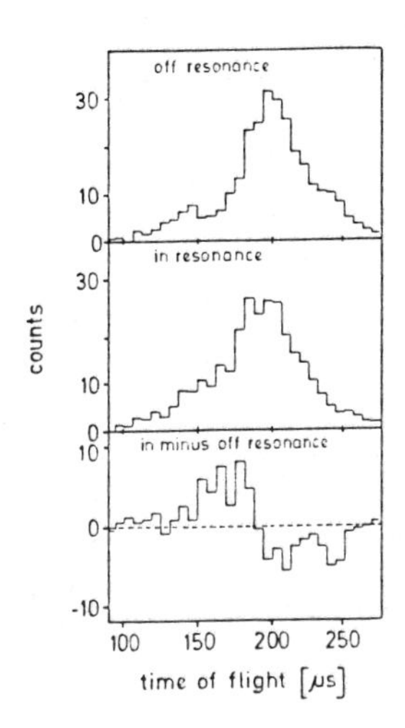

Fig. 4 Time-of-flight spectrum of N and N_2 ions. At t = 0 the ions are ejected out of the trap and are then detected by the channel-plate detector. Top: Spectrum obtained with the radio frequency off resonance. Middle: Spectrum obtained in resonance at $\nu = \nu_c$. Bottom: Difference spectrum obtained by substracting the spectra in resonance and off resonance.

The change in the mean time of flight has to be kept small if high resolution of the mass determination is required. Hence the change in time of flight seen in Fig. 4 amounts only to a few percent. If the strength of the RF field is increased, the ν_c resonance is power broadened. Finally at very high RF power, the ions gain such a high orbital energy that they can no longer escape through the hole of the ion trap. In this case, they strike the electrodes and are lost (Fig. 5). This effect can be used for purification of the sample, i.e. to get rid of unwanted contaminating species in the trap.

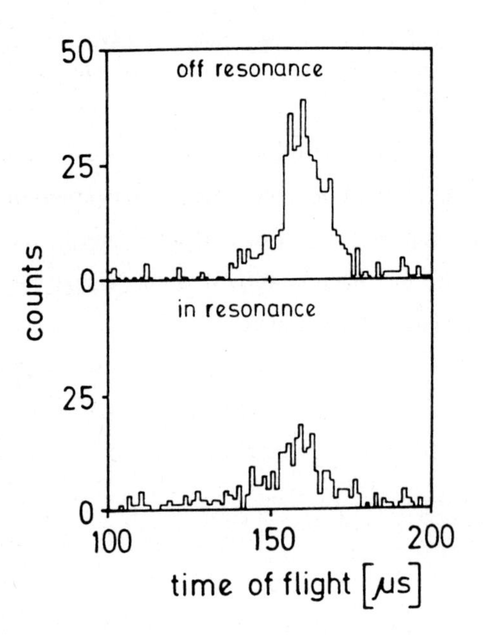

Fig. 5 Time-of-flight spectrum as in Fig. 4 but for He ions. In this case, the radio frequency power is much higher. Hence a large fraction of the ions do not reach the detector when the resonance frequency ν_c is induced.

4. EXPERIMENTAL SETUP

The experimental setup for the mass measurements is shown in Fig. 6. The apparatus consists of an alkali ion source, a Penning trap (trap 1) placed in the pole gap of an electro magnet, a transfer line with a number of electrostatic lenses and deflectors, a Penning trap (trap 2) placed in the stable and homogeneous field of a superconducting magnet, a drift tube and finally a channel plate detector.

The first trap is essentially a bunching device to collect the continuous beam of the ion source. The second trap is used for high-precision measurements of the cyclotron frequency. Hence ultra-high vacuum, excellent homogeneity of the magnetic field and perfect geometry of the second trap are essential. Since the operating conditions of a bunching trap are incompatible with those of a precision trap, the layout of Fig. 6 was chosen with two completely separated traps connected only by a transfer tube. This configuration allows efficient differential pumping between traps 1 and 2 and the installation of ion optical devices for steering the ion beam.

BUNCHING TRAP: Alkali ions delivered by the ion source are implanted into a tungsten foil mounted in one endcap of the Penning trap, then surface ionized by heating the foil, trapped [13], [14], and finally ejected by a sudden change of the trapping potential. In case of the study of cluster ions this trap can be abandoned because it is possible to create cluster ions directly in a bunched mode.

TRANSFER TUBE: After leaving the trap the ion bunch is accelerated to 1 keV by some electrodes. A system of electrostatic lenses guides the bunch into the fringe field of the superconducting magnet in such a way that the radial energy is not increased. Extensive ion optical calculations have been performed to find an optimum design [15]. This is most important for mass measurements,which require a very small energy spread of the trapped ions. The transmission of the transfer tube has been measured to be 80 % (including the transmission through the holes in the endcaps of trap 2). At the entrance of trap 2 the length of the bunch is shorter than 30 μs.

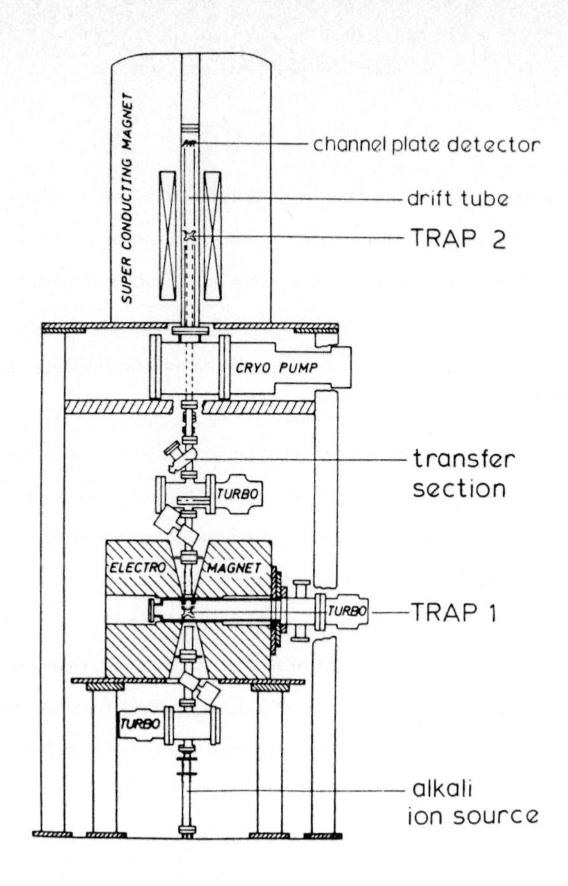

Fig. 6 Experimental set up for direct mass determination of externally created ions.

PRECISION TRAP: Here the bunched ion beam is captured in flight by retarding it to a few eV just before entering the trap. The potential of the lower endcap of trap 2 is switched to the potential of the ring electrode just at the moment the ion bunch arrives. When the ions pass the center of trap 2 the potential of the endcap is raised again and the ions can no longer escape. More details of the apparatus built for mass measurements are given in Ref. [11].

5. PERFORMANCE

TRAPPING EFFICIENCY: It has been determined that up to 70% of an ion bunch ejected out of trap 1 can be retarded and captured in flight in trap 2. More details can be found in Ref. [11].

RESOLUTION: A resolving power of 500,000 has been observed in a mass measurement of ^{39}K. Here a line width of 4.4 Hz was obtained at a resonance frequency of 2.3 MHz (Fig. 3). Studies to increase further the resolving power are under way.

ACCURACY: The ratio of the cyclotron frequencies of, for example, the Rb/K pair coincide with the tabulated mass ratio within 2×10^{-7}. This deviation corresponds approximately to the statistical uncertainty of the centroid of the resonance (Fig. 3) which is about 10% of the line width. Therefore the ratios of unknown to previously-known masses can be determined with an accuracy of the order of 2×10^{-7}.

STORAGE TIME: The storage time of our Penning trap has not yet been investigated systematically. We can give only a lower limit: Up to a delay of 1.5 s in respect to the capture of the ions in trap 2 no decrease in the number of stored ion was observed. It can be expected that the life time of the charged particles is much longer because the ions already survived a large number of oscillations (see Eqs. (3) - (5)) during this time of storage. The storage time depends strongly on the vacuum in the apparatus which was $<10^{-9}$ mbar in the experiments reported here.

SENSITIVITY: Typically 10 to 100 ions were stored in trap 2 at a time. Since the detection scheme used by us is destructive, the trap has to be refilled after each cycle. This involves filling the trap, inducing the RF, ejecting the ions, and measuring the time of flight. The cycle time is about 0.5 s. For the mass measurement of ^{39}K (Fig. 3) 100 cycles per frequency value were performed, corresponding to a total time of 20 min needed for the measurement of the ^{39}K mass.

6. DISCUSSION

It is very attractive to apply the Penning trap technique to the study of clusters. In comparison to investigations on atomic or ionic beams one gains several orders of magnitude in observation time by using stored cluster ions. The ultra-high mass resolution achievable will even permit investigations of clusters of very large size. These advantages, together with the possibility of preparing pure samples of one cluster size only, lead us to expect that new fields in cluster spectroscopy will open up. All those techniques developed for the study of atomic ions can, in principle, be applied also to a study of cluster ions. These methods include cooling of the ions in the trap, detection of their motion, investigations using single ions, ultra-high resolution spectroscopy, and spectroscopy in the microwave as well as in the optical region. A quite comprehensive overview on the status of trapping atomic ions can be found in the contributions to the 1984 International Conference on Atomic Physics [16-18].

We intend to combine a Penning trap with a cold metal cluster beam produced in a supersonic expansion. Bunched ions can be obtained by pulsed photoionization. Appropriate timing of the opening of the trap can be used to pre-purify the cluster bunch in respect to mass before the clusters are captured in the Penning trap. It should also be possible to accumulate a larger number in the trap than obtained by a single injection. For this purpose the potential of the ring electrode can be continuously decreased while the lower endcap is pulsed with constant amplitude in order to allow the pass of additional bunches. In this way, stacks of cluster bunches can be accumulated in the trap.

As we discussed briefly in the Introduction, a large number of experiments can be performed with stored clusters. The most challenging might be to investigate Na clusters by laser spectroscopy and to search for optical resonances. Recently, Knight et al. [19] found peaks in the mass spectra of Na clusters at cluster sizes of n = 8, 20, 40, 58, and 92. These "magic" numbers can be explained in a one-electron shell model in which independent delocalized atomic 3s electrons are bound in a spherically symmetric potential well [19]. The calculation yields discrete electronic energy levels and a shell structure which reproduces the peaks observed in the mass spectra. The authors conclude that the good correspondence between the experimental results and the model calculations suggests that there are no perturbations large enough to distort the main features of the level structure. Hence, discrete resonances should also be observable by laser spectroscopy stored Na ions. Such an experiment which presents a stringent test of the model seems only feasible using the trapping technique.

This work was supported by the Bundesministerium für Forschung und Technologie. We would like to thank G. Bollen, P. Dabkiewicz, P. Egelhof, F. Kern, H. Kalinowsky, H. Stolzenberg and H. Stürmer for their collaboration in realizing the Penning trap designed to measure the masses of short-lived nuclei.

References

[1] Dehmelt, H.G.: Adv. At. and Mol. Phys. 3, 53 (1967)

[2] Brown, L.S., Gabrielse, G.: Rev. Mod. Phys. 58, 233 (1986)

[3] Wineland, D.J., Itano, W.M.: Adv. At. and Mol. Phys. 19, 135 (1983)

[4] Neuhauser, W., et al.: Phys. Rev. A22, 1137 (1980)

[5] Nagourney, W., Janik, G., Dehmelt, H.: Proc. Natl. Acad. Sci. USA 80,643 (1983)

[6] Gräff, G., Kalinowski, H., Traut, J.: Z. Physik A297, 35 (1980)

[7] Van Dyck, R.S.Jr., et al.: Int. J. Mass Spectr. and Ion Proc. 66, 327 (1985)

[8] Blatt, R., Schnatz, H., Werth, G.: Phys. Rev. Lett. 48, 1601 (1982)

[9] Schwinberg,P.B., Van Dyck, R.S.Jr., Dehmelt, H.G.: Phys. Rev. Lett. 47, 1679 (1981)

[10] Recknagel, E.: Proc. 9th Intern. Conf. Atomic Physics, Seattle 1984 (ed. by R.S. van Dyck, Jr. and E.N. Fortson) "Atomic Physics 9", World Scientific: Singapore (1984)

[11] Schnatz, H. et al.: Submitted to Nucl. Instr. and Meth.

[12] Alford, J. M., et al.: submitted to Int. J. Mass Spectr. and Ion Proc.

[13] Coutandin, J., Werth, G.: Appl. Phys. B29, 89 (1983)

[14] Bonn, J., et al.: Appl. Phys. B30, 83 (1983)

[15] Stürmer, H.: Diploma work, Mainz, unpublished (1984)

[16] van Dyck, R.S.Jr., Church, D.A.:Proc. 9th Intern. Conf. Atomic Physics, Seattle 1984 (ed. by R.S. van Dyck, Jr. and E.N. Fortson) "Atomic Physics 9", World Scientific: Singapore (1984)

[17] Werth, G.: Proc. 9th Intern. Conf. Atomic Physics, Seattle 1984 (ed. by R.S. van Dyck, Jr. and E.N. Fortson) "Atomic Physics 9", World Scientific: Singapore (1984)

[18] Wineland, D.J., et al.: Proc. 9th Intern. Conf. Atomic Physics, Seattle 1984 (ed. by R.S. van Dyck, Jr. and E.N. Fortson) "Atomic Physics 9", World Scientific: Singapore (1984)

[19] Knight, W.D., et al.: Phys. Rev. Lett. 52, 2141 (1984)

NEUTRAL YIELDS IN ELECTRONIC SPUTTERING OF BIOMOLECULES

P Håkansson, A Hedin, M Salehpour* and B U R Sundqvist
Tandem Accelerator Laboratory
University of Uppsala
Box 533, S-751 21 Uppsala, Sweden

**Present address: Argonne National Laboratory*
Argonne IL 60439, USA

When primary ions hit a target surface, neutral and charged molecules, fragments and atoms can be sputtered away. For fast ions i.e ions with a velocity larger than the Bohr velocity the sputtering phenomenon is active only in insulators and has been shown to be connected to the electronic part of the stopping power of the primary ion [1, 2]. This process is therefore called electronic sputtering. In general much more neutrals than ions are ejected in sputtering processes and one goal of the present work is to find out if that also holds for electronic sputtering.

To measure the absolute neutral yield of an amino acid, a collector method have been used. The primary ion beam hits a target and the ejected molecules are collected on a silicon slice. The number of intact molecules on the collector is then determined by an amio acid analysis technique [3]. The result for leucine [MW131] is a yield of 2150 +- 480 intact molecules if an isotropic angular distribution is assumed and 1180 +- 280 for a $cos\theta$ distribution [4]. The yield of positive molecular ions is 0.13 which gives a ratio of 10 000 for neutral/charged intact leucine molecules.

The relative yield of molecules can be obtained by analyzing the collector in a ^{252}Cf ToF spectrometer with the so called PDMS technique [5]. This can be done directly after the irradiation without letting air into the experimental chamber. Because of the low coverage on the collector (submonolayer) the measured yield of positive ions is assumed to be proportional to the number of neutral molecules on the collector.

To investigate how the yield of neutral molecules scale with the stopping power of the primary ion, a leucine target was irradiated with beams of ^{32}S, ^{58}Ni, ^{79}Br and ^{127}I ions. For each ion several irradiations were made and the collector was analyzed with the PDMS technique inbetween each irradiation. The doses were choosen to avoid damage in the target [6] and the primary ion energies were choosen so that the velocity was keept constant. This implies that the energy density in the track around the primary ion path varies linearly with the electronic stopping power, dE/dX. The charge states of the primary ions were choosen to be the equilibrium ones.

For comparison we have also measured how the yield of positive and negative leucine molecular ions scale with the energy density in a parallel experiment. The results are that the yield of $[M+H]^+$ is proportional to dE/dX, the yield of $[M-H]^-$ is proportional to $[dE/dX]^2$ wheras the neutal yield seems to scale faster than $[dE/dX]^2$. [7]

Not only amino acids can be sputter deposited in vaccum but also a relativly large peptide, LHRH, with a molecular weight of 1182 amu has been studied. By comparing the spectra from a collector with that from a target with known target thickness it was estimated that the sputtering yield for LHRH with 90 MeV ^{127}I ions is of the order of 900 if an isotropic angular distribution is assumed and 500 for a $cos\theta$ distribution.

References:

1 P Håkansson and B Sundqvist
Rad Eff **61** (1982) 179-193

2 A Albers, K Wien, P Dück, W Treu and H Voit
Nucl Instr Meth **198** (1982) 69

3 D H Spackman, W H Stein and S Moore
Anal Chem **30** (1958) 1190

4 M Salehpour, P Håkansson, B Sundqvist and S Widdiyasekera
Nucl Instr Meth **B13** (1986) 278

5 D F Torgerson, R P Skowronski and R D Macfarlane
Biochem Biophys Res Commun **60** (1974) 616

6 M Salehpour, P Håkansson and B Sundqvist
Nucl Instr Meth **B2** (1984) 752

7 M Salehpour, A Hedin, P Håkansson and B Sundqvist
TLU 138/86, Tandem Laboratory Report, Uppsala, Sweden 1986
To be published

THE ELECTRONIC STRUCTURE OF As_4S_4, S_4N_4 AND RELATED CLUSTERS: A GAS PHASE UPS AND SCC-Xα MO STUDY/1/

S. Elbel and M.Grodzicki
Institut für Anorganische und Angewandte Chemie der Universität Hamburg
Martin-Luther-King-Platz 6, D-2000 Hamburg 13, West Germany
and
I.Institut für Theoretische Physik der Universität Hamburg
Jungiusstr.9, D-2000 Hamburg 36, West Germany

1. Introduction

Tetraarsenic tetrasulfide (As_4S_4) and tetrasulfur tetranitride (S_4N_4) both belong to the most fascinating cluster species as offered by inorganic chemistry/2,3/ because of their unusual structural properties and the outstanding chemistry and physics of at least the thermolabile N_4S_4 and its known subunits $(NS)_x$, x=1,2,.., /4-6/. As_4S_4 has recently been identified by mass spectrometry as the essential building block in the formation of larger van der Waals clusters/7/. Although both molecules belong to the same molecular point group D_{2d} and are composed of an equal number of group 15 and group 16 atoms, they nevertheless escape a direct comparison: whereas As_4S_4 possesses a square array of S atoms with the As atoms above and below located at the corners of a distorted tetrahedron (bisphenoid), and resembling thus the tetrahedral geometry found in As_4, the atomic arrangement in S_4N_4 is just reversed with the group 15 atoms spanning now the plane. This structural phenomenon had been subject to theoretical analyses/8-10/ and has recently been interpreted as the consequence of 'topological charge stabilization'/9/.

Both D_{2d} clusters have been investigated theoretically several times, cf. ref. 11-14 for S_4N_4 and ref.15-17 for As_4S_4. The same holds for the UV-photoelectron(p.e.) spectrum of S_4N_4/13,14,18/, however, its interpretation remained vague due to the poor agreement between experimental ionization potentials and calculated orbital energies, and presumably due to unknown amounts of thermal decomposition products being present under measuring conditions. The valence bands of solid As_4S_4 have been investigated by X-ray photoemission spectroscopy/15/.

The present study is devoted to the common features of and the differences between S_4N_4 and As_4S_4 on the basis of their valence electron ionization potentials, the p.e. spectrometric assignment of S_4N_4 being revised and revisited at the same time.

2. Charge Distribution Analysis

The assignment of the respective p.e. spectra and the charge density analysis has been accomplished on the basis of ab initio and CI calculations available from literature /14,16/ as well as by semiempirical SCC-Xα calculations /19/. The latter method is described in detail elsewhere in this volume /20/. Discussing at first the various theoretical results of P_4S_4 not investigated experimentally so far, it is seen from table 1 that the SCC-Xα and CI results agree remarkably well while ab initio results do not give even the correct ordering in all cases. This confirms our

Table 1. Calculated and experimental ionization potentials of investgated compounds

S_4N_4			As_4S_4			P_4S_4			Se_4N_4	As_4Se_4
exp.	SCC-Xα[a]	CI[b]	exp.	SCC-Xα[c]	ab initio[d]	SCC-Xα[e]	ab initio[f]	CI[g]	SCC-Xα[c]	SCC-Xα[c]
9.36	9.24(4b$_2$)	10.09(4A$_2$)	8.58	8.49(2a$_2$)	8.61(4b$_2$)	8.83(a$_2$)	9.67(a$_2$)	8.76(A$_2$)	9.15(b$_2$)	7.77(a$_2$)
	9.48(2a$_2$)	10.15(4B$_2$)		9.11(4b$_2$)	8.92(2a$_2$)	9.27(b$_2$)	9.75(b$_2$)	9.00(B$_2$)	9.16(a$_2$)	8.71(e)
10.60	10.66(4a$_1$)	10.60(2B$_1$)	9.45	9.40(5e)	9.65(5e)	9.87(e)	10.57(e)	9.82(E)	10.50(e)	8.87(b$_2$)
10.92	11.36(2b$_1$)	10.62(4A$_1$)	9.85	10.03(3b$_2$)	9.49(4a$_1$)	10.67(b$_2$)	10.96(a$_1$)	10.72(B$_2$)	10.74(a$_1$)	9.35(b$_2$)
11.42	11.14(5e)	12.33(5E)		10.71(4a$_1$)	9.93(3b$_2$)	10.75(a$_1$)	11.22(b$_2$)	10.74(A$_1$)	10.93(b$_1$)	10.08(e)
12.74	12.60(3b$_2$)	12.04(3B$_2$)	10.33	10.78(2b$_1$)	10.15(2b$_1$)	11.01(e)	11.60(b$_1$)	11.03(E)	11.85(e)	10.35(a$_1$)
13.66	12.59(4e)	13.64(4E)	10.67	10.74(4e)	10.47(4e)	11.17(b$_1$)	11.63(e)	11.36(B$_1$)	12.64(b$_2$)	10.46(b$_1$)
15.27	15.09(3e)		12.06	11.90(3a$_1$)	11.82(3e)	12.09(a$_1$)	14.07(e)		15.00(e)	11.35(a$_1$)
	15.66(3a$_1$)			12.05(3e)	12.26(3a$_1$)	12.39(e)	14.64(a$_1$)		15.62(a$_1$)	11.91(e)
16.93	16.64(2b$_2$)	17.25(3A$_1$)	12.62	13.70(1a$_2$)	13.34(1a$_2$)	13.91(a$_2$)	14.65(a$_2$)	13.91(A$_2$)	16.44(b$_2$)	13.44(a$_2$)
	18.03(1a$_2$)			14.01(2b$_2$)	13.75(2b$_2$)	14.08(b$_2$)	16.01(b$_2$)		16.94(a$_2$)	13.81(b$_2$)

[a] this work, all values reduced by 0.4eV; [b] ref.14, all values reduced by 0.4eV; [c] this work; [d] ref.16, all values reduced by 1.0eV; [e] this work, all values reduced by 0.2eV; [f] ref.16, with d-functions; [g] ref.16

experience from other SCC-Xα calculations e.g. on titanium compounds /21/ where the same is true, and we expect a similar behaviour when turning later to the assignment of the p.e. spectrum of S_4N_4 and As_4S_4 on the basis of various theoretical results, intensity arguments as well as qualitative considerations. The necessity of applying also qualitative arguments follows from the fact that for S_4N_4 even the results from CI calculations are not conclusive for a complete and unique assignment of the p.e. spectrum which in turn suffers also under the abovementioned possible presence of decay products.

The qualitative discussion starts from the charge distribution analysis as given in table 2. In all cases, the atoms forming the planar subunit of the molecular skeletons (S in As_4S_4 and P_4S_4, N in N_4S_4) are negatively charged supporting the qualitative topological charge rules /9/. This is equivalent with an energetic destabili-

Table 2. Charge distribution analysis of investigated molecules

E_4Ch_4		$\lvert Q_{eff}\rvert$	atomic orbital occupation numbers						overlap population		
			Ch(n_s)	Ch(n_p)	Ch(n_d)	E(n_s)	E(n_p)	E(n_d)	E-Ch	bisph.	squ.
N_4S_4	SCC-Xα	0.441	1.614	3.945	-	1.267	4.174	-	0.801	0.231	-0.082
	ab init.	0.455	1.697	3.529	0.319	1.607	3.848	-			
As_4S_4	SCC-Xα	0.415	1.648	4.767	-	1.492	3.093	-	0.703	0.650	-0.102
	ab init.	0.307	1.913	4.394	-	2.199	2.895	9.60	0.452	0.464	-0.090
P_4S_4	SCC-Xα	0.231	1.640	4.591	-	1.556	3.213	-	0.709	0.612	-0.134
	ab init.	0.197	1.818	4.303	0.075	1.668	2.953	0.182			
Se_4N_4	SCC-Xα	0.605	1.658	3.737	-	1.315	4.290	-	0.773	0.285	-0.055
As_4Se_4	SCC-Xα	0.177	1.686	4.491	-	1.519	3.304	-	0.726	0.668	-0.097

zation of those molecular orbitals that are sulfur based in As_4S_4 and P_4S_4 or are N based in S_4N_4, respectively, that are therefore expected in the low energy part of the spectrum. The reverse situation should be valid for molecular orbitals with large coefficients at the positively charged atoms of the bisphenoid subunits.

Continuing with discussing the charge distribution in table 2, it can be seen that the gross features of SCC-Xα and ab initio calculations coincide with each other for S_4N_4 and P_4S_4 but there are some pecularities in the ab initio results for As_4S_4. First of all, there is an As(4s) valence shell occupancy of 2.199 which is unresonable and might be related to the apparent omission of 4d functions on As or other defects in the chosen basis set. Furthermore, there are considerable differences in the effective charges and the overlap populations between both calculations. Accordingly, we consider the ab initio results for As_4S_4 as less conclusive in assigning the p.e. spectrum.

As far as the bonding properties are concerned of the E_4Ch_4 series, the E-Ch bond turns out to be similar in all compounds, since only slight variations in the overlap population for the E-Ch bonds are obtained from the SCC-Xα calculations. Secondly, it can be concluded definitely that no bonding interactions exist between the atoms of the square subunit. The bond strength of adjacent atoms of the bisphenoid subunit, however, depends strongly on the type of atom occupying this position that is almost three times stronger for the group 15 elements As and P than for the chalcogens S,Se in S_4N_4 and Se_4N_4, respectively. This is also displayed in fig.1 where the total valence electron density is presented for S_4N_4 and As_4S_4 as derived from the SCC-Xα calculations.

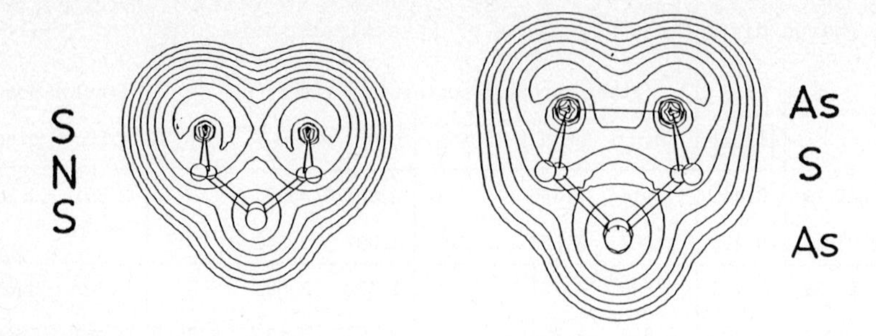

Figure 1: Total valence electron density for S_4N_4 and As_4S_4. Contour values are (in a_0^{-3}): 0.512, 0.256, 0.128, 0.064, 0.032, 0.016, 0.008, 0.004, 0.002.

3. Assignment of spectra

The He(I) spectra of As_4S_4 and N_4S_4 are presented in fig.2 and are assigned on the assumption of D_{2d} molecular symmetry, and according to respective SCC-Xα orbital energies as well as to the results from ab initio calculations available from literature /14,16/. Our final suggestion for the interpretation of these spectra is displayed in table 1 together with results for parent P_4S_4, Se_4N_4 and As_4Se_4 clusters, their p.e. data being subject to future investigations. It should also be mentioned that to make the assignment more obvious, in some cases the calculated eigenvalues are shifted by a constant amount.

Starting with S_4N_4, the first peak in the spectrum at 9.36 eV is correlated with two ionization events of symmetry b_2 and a_2 being purely(a_2) or predominantly(b_2) nitrogen p-orbitals. The second band is separated from the first one by a gap of 1.24eV occurring also in the SCC-Xα eigenvalue spectrum but not in the CI results. (A small peak at 10.11eV is ignored here because there is some experimental evidence for this to originate from decay products.) Intensity arguments additionally favour an assignment of the second and third band to calculated $4a_1$ and $2b_1$ levels where a reversed ordering might also be possible. The fourth broad band at 11.42eV is supposed to correspond to 5e again for intensity reasons and as a compromise between CI and SCC-Xα results, and the same arguments hold for the next two bands assigned to $3b_2$ (12.74eV) and to 4e (13.66eV). Finally, the interpretation of the high energy end of the p.e. spectrum remains unclear as far, since the CI calculation is not completed in this part, the detected peaks are broad and low in intensity and, moreover, a breakdown of the one particle approximation cannot be excluded, as it is usually the case for ionization of electrons from higher energy regions /22/. Therefore, the proposed assignment of table 1 has to be regarded as merely tentative. To draw the final conclusion, it has to be stated that both theoretical results differ considerably from each other and none of them agrees with the experimental spectrum in such a way that a definite assignment of the spectrum can be given.

The situation does not seem that bad for As_4S_4 though some discrepancies exist

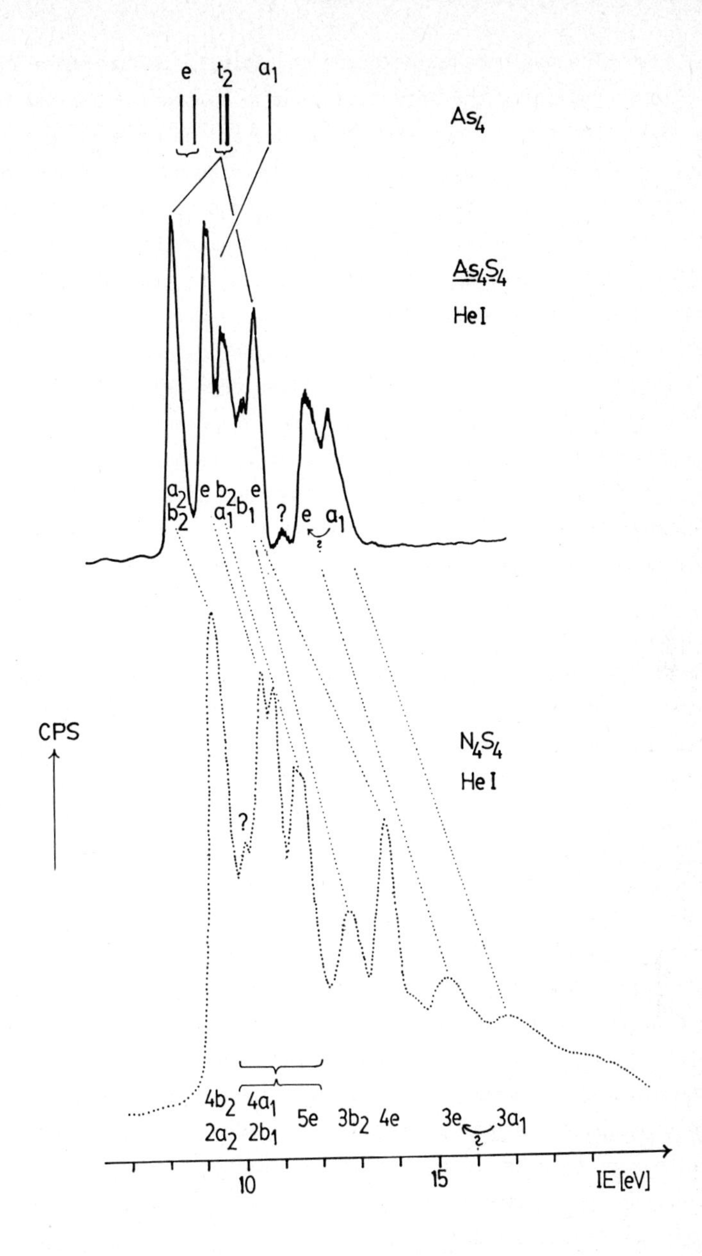

Figure 2: He(I) photoelectron spectra of As_4S_4 and N_4S_4 /13/ assigned according to SCC-Xα orbital energies and correlated with experimental ionization potentials of As_4.

in this case, too. The assignment, as given in table 1, is therefore again a result of considerations concerning the intensities and a compromise between both calculations. As in S_4N_4, the first band is correlated with the $4b_2$ and $2a_2$ eigenvalues, while the second one is assumed to be the 5e according to the SCC-Xα results as well as with respect to its high intensity. The third band is assigned to $3b_2 + 4a_1$ where only the position of the $4a_1$ level is questionable since both calculations differ by more than 1.2eV in this case. On one side, according to previous experience with the SCC-Xα method the totally symmetric orbitals tend to have too large ionisation potentials, on the other side the ab initio result gives an anomalously large S(3s) contribution to this orbital (cf. section 4) arising from As(4p)-S(3s) antibonding interactions making the low value for this ionization potential suspect, too. The compromise between both theoretical results seems thus to be the assignment to the third band. The next three bands can be correlated without ambiguities since both calculations yield similar results. In the high energy part of the spectrum, however, serious difficulties are met again since only one band is detected at 12.62eV and two other weak features at 15.8eV and 16.7eV are seen in the He(II) spectrum, whereas both calculations yield orbital energies between 13.3 and 14 eV. It is not yet clear, how to resolve this discrepancy apart from assuming again a breakdown of the single particle picture. Altogether it can be said that both calculations are in reasonable agreement, except the $4a_1$ level, making thus a reliable assignment for the first six bands of the p.e. spectrum possible.

4. Comparison between S_4N_4 and As_4S_4

Comparing both p.e. spectra of S_4N_4 and As_4S_4, respectively, at first sight the overall band patterns confirm the intuitive expectation: Namely, since both species have the same point group and the same number of valence electrons, the spectra should resemble each other with the modification that only due to the more effective interaction between adjacent orbitals in S_4N_4 the p.e. spectrum should cover a larger energy range with bands being better resolved. This superficial resemblance of both spectral patterns is apparent and indicated by the dotted lines in figure 2.

However, a more detailed quantitative analysis of the electronic structure proofs this simple picture of limited validity only, as must be concluded from a comparative discussion of the molecular orbital decomposition as displayed in table 3. A first understanding of the As_4S_4 electronic structure can be obtained from a correlation with tetrahedral As_4 since only a slight modification of those molecular orbitals is expected that are parallel in direction to the S_4 plane. These molecular orbitals are in fact identified as $4b_2$, $4a_1$ and 4e in the SCC-Xα result while the other five low energy levels are of S-character. This clear structure is approximately reflected in the ab initio result, too, with the difference of As(4p) orbitals of e-symmetry contributing with nearly equal amount to 4e and 5e, and the $4a_1$ containing an anomalously large S(3s) orbital participation as already mentioned above.

Table 3. Calculated percentage contribution of atomic orbitals to low lying molecular orbitals of As_4S_4 and S_4N_4

	As_4S_4								S_4N_4			
	SCC-Xα [a]				ab initio [b]				SCC-Xα [a]			
	As		S		As		S		S		N	
	s	p	s	p	s	p	s	p	s	p	s	p
$2a_2$	-	-	-	100	-	-	-	100	-	-	-	100
$4b_2$	4	88	-	8	5	93	-	2	2	22	-	76
5e	1	10	-	89	8	36	-	57	-	34	5	61
$3b_2$	4	12	-	84	2	18	-	80	4	84	-	12
$4a_1$	5	61	6	28	4	71	19	6	-	30	11	59
$2b_1$	-	7	8	85	-	11	7	82	-	2	6	92
4e	18	45	2	35	10	28	1	61	-	14	12	74
3e	5	21	14	60	-	20	12	68	17	69	1	13
$3a_1$	17	35	12	36	16	51	4	29	15	78	3	4
$1a_2$	-	38	-	62	-	40	-	60	-	56	-	44
$2b_2$	2	63	-	35	-	48	-	52	5	77	-	18
2e	40	14	23	23	49	10	19	21	37	27	16	20
$2a_1$	31	26	29	4	35	20	41	3	31	45	21	3

[a] this work; [b] M.H.Palmer, private communication, and ref.17

Turning next to similarities and analogies among As_4S_4 and S_4N_4, atomic orbital contributions of symmetry equivalent sites have to be compared, i.e. As in As_4S_4 with S in S_4N_4 and S in As_4S_4 with N in S_4N_4. There is one molecular orbital ($2a_2$) being unique in character since it possesses only orbital contributions from the atoms in square position. According to the strongly destabilizing through space interactions of its four p-type components that are tangentially oriented with respect to the square, this molecular orbital is expected to be the HOMO in qualitative agreement with all calculations. At the same time however, the calculations show that the $4b_2$ state, almost degenerate with $2a_2$, is considerably different in its composition in both compounds, namely of As-character in As_4S_4 but N-type in S_4N_4. Hence the $4b_2$ orbital in As_4S_4 has to be correlated with $3b_2$ in S_4N_4 and vice versa. In the same way $4a_1$ in As_4S_4 corresponds to $3a_1$ in S_4N_4 and vice versa and the same applies at least approximately for the 3e and 4e levels. The SCC-Xα correlations for the first few mo-

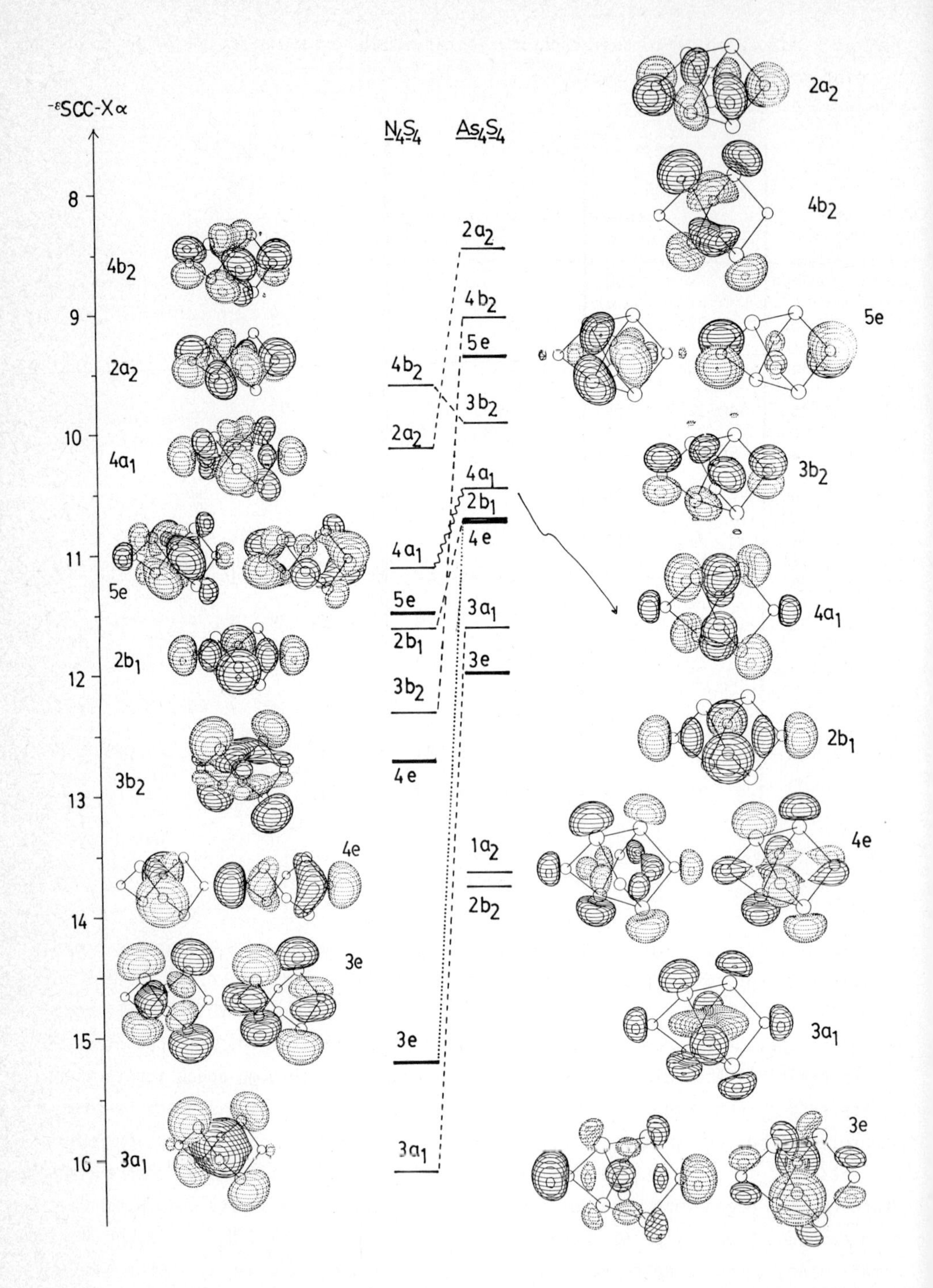

Figure 3: SCC-Xα eigenvalue and eigenfunction correlation diagram for the As_4S_4/N_4S_4 pair. The correlation lines connect molecular orbitals of equal symmetry and comparable character.

lecular orbitals of both clusters is depicted in figure 3 showing that a simple band correlation as performed in figure 2 is not at all as straightforward as suggested on the basis of qualitative considerations.

5. References

/1/ Part XVII of the series: "Photoelectron spectra of group V compounds"; for part XVI see S.Elbel, H.Egsgaard, L.Carlsen, JCS Dalton 1986 (submitted)
/2/ A.F.Holleman, E.Wiberg, N.Wiberg, Lehrbuch der Anorganischen Chemie, de Gruyter, Berlin 1985
/3/ F.Kober, Chemiker Zeitg. 105,199 (1981) and A.J.Banister, Nature 239,69 (1972)
/4/ H.J.Emelius, Endeavour 32,76 (1973)
/5/ T.Chivers, Chem.Rev. 85,341 (1985) and literature cited therein
/6/ T.Yamabe, K.Tanaka, A.Tachibana, K.Fukui, H.Kato, J.Phys.Chem. 83,767 (1979)
/7/ T.P.Martin, Angew.Chemie 98,197 (1986) and J.Chem.Phys. 80,170 (1984)
/8/ R.Gleiter, Angew.Chemie 93,442 (1981)
/9/ B.M.Gimarc, J.J.Ott, J.Am.Chem.Soc. 108,4298 (1986)
/10/ T.H.Tang, R.F.Bader, P.J.McDougall, Inorg.Chem. 24,2047 (1985)
/11/ R.Gleiter, J.Chem.Soc. A1970,3174
/12/ K.Tanaka, T.Yamabe, A.Tachibana, H.Kato, K.Fukui, J.Phys.Chem. 82,2121 (1978)
/13/ R.H.Findlay, M.H.Palmer, A.J.Downs, R.Evans, Inorg.Chem. 19,1307 (1980)
/14/ M.H.Palmer, Z.Naturforsch. 38a,378 (1983)
/15/ W.R.Salaneck, K.S.Liang, A.Paton, N.O.Lipari, Phys.Rev. B12,725 (1975)
/16/ M.H.Palmer, R.H.Findlay, J.Molec.Struct.(THEOCHEM) 104,327 (1983)
/17/ S.Elbel, M.Grodzicki, H.J.Lempka, M.H.Palmer, Chem.Ber. (submitted)
/18/ W.R.Salaneck, J.W.P.Liu, A.Paton, C.B.Duke, G.P.Cesar, Phys.Rev. B13,4517 (1976)
/19/ M.Grodzicki, J.Phys. B13,2683 (1980)
/20/ O.Kühnholz, M.Grodzicki, 'Nickel clusters as surface models for adsorption' (in this volume)
/21/ M.Hütsch, M.Grodzicki, 'SCC-Xα calculations on small titanium compounds and TiO_2 clusters' (in this volume)
/22/ L.S.Cederbaum, W.Domcke, J.Schirmer, W.v.Niessen, Phys.Scr. 21,481 (1980)

TREATMENT OF SMALL METALLIC CLUSTERS WITH QUANTUM CHEMICAL METHODS

J. Koutecký, I. Boustani, V. Bonačić-Koutecký
P. Fantucci* and W. Pewestorf
Institut für Physikalische und Theoretische Chemie
Freie Universität Berlin
D-1000 Berlin 33

Abstract:
The quantum chemical treatment of neutral and cationic lithium clusters reveals some very general basic features which exhibit in a simple way why quite different quantum mechanical approaches yield very similar results concerning the cluster structure. An insight in the reasons for the occurence of "magic numbers" is given.

1. Introduction

Various theoretical methods used both in quantum chemistry and solid state theory have been recently employed for the investigation of the electronic and geometrical structure of metal clusters.[1-18] The comparison and evaluation of the studies carried out with different approaches is very instructive because it yields better insight in the problem of cluster structure. Moreover, the results of the theoretical treatments of the interactions leading to the cluster stability can be also valuable as bench marks proving some advantages and disadvantages of the theoretical methods themselves. In this contribution we are paying our attention to the alkali metal clusters. The reason for this emphasis on such relatively very simple systems is that the very small number of electrons involved makes feasible an application of a variety of methods characterized by very different approaches and by a quite different degree of sophistication.

*Present address: Università di Milano, Dipartimento di Chimica Inorganica e Metallorganica, Via Venezian, 21; I-20133 Milano, Italy

Our own investigation[19] on Li_n (n=1-8) has been carried out with a LCAO Hartree-Fock treatment combined with the gradient search of geometry with the lowest Hartree-Fock energy. The MRD-CI calculations have been performed in the neighbourhood of the Hartree-Fock local minima in multiplying by a common factor all interatomic distances within the topology corresponding to the HF local minima and in determining the minimal MRD-CI energy due to this "scaling" process. Very often the MRD-CI energy and HF energy sequences of the isomeric cluster topologies corresponding to the various HF local minima are different.

2. All-electron HF and MRD-CI investigation of Li_n clusters

The dependences of the binding energy per atom (or atomization energy) for neutral Li clusters

$$BE/n = - E_n/n + E_1$$

as well as for cationic Li_n^+ clusters

$$BE/n^+ = - (E_n^+ -(n-1)\ E_1 - E_1^+)/n$$

are shown in Fig. 1.

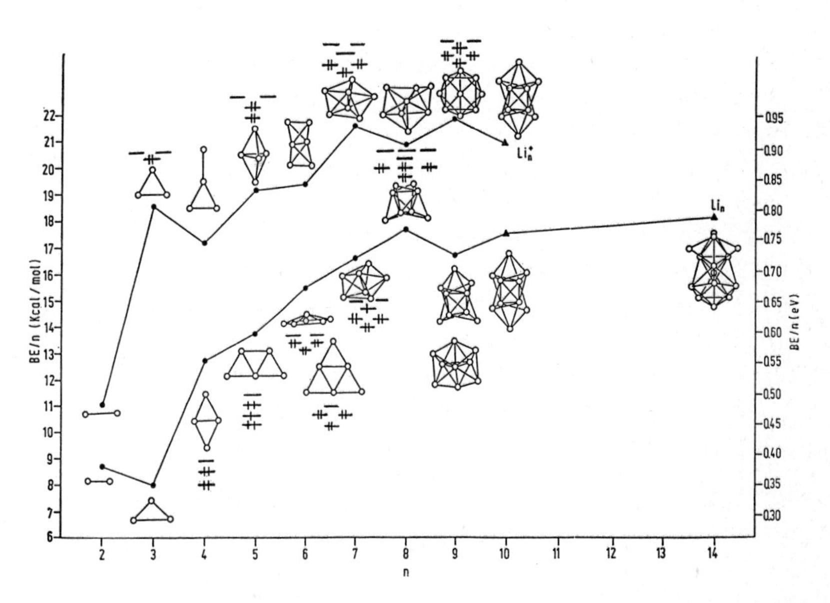

Fig. 1: The binding energy per atom (BE/n) of the neutral and (BE/n^+) (cationic Li clusters) as a function of the number of atoms n. The optimal geometries are shown as well as the schematic MO (or NO) spectrum.

E_k and E_k^+ means the energy of a neutral and cationic Li cluster with Li atoms, respectively. The following characteristic properties of the dependence of BE/n and BE/n^+ upon n can be pointed out:

1) BE/n is generally an increasing function of n but there exhibit characteristic minima for n=3 and n=9 as well as a maximum for n=8.

2) BE/n^+ shows maxima for n=3,7 and 9 and minima for n=4,8 and 10.

3) The smalles Li_n clusters with $n \leqslant 6$ have twodimensional forms which can be considered as deformed sections of (111) fcc crystal plane. Starting with Li_6 the compact three-dimensional structures with condensed tetrahedra as building elements are energetically favorable.

4) The optimal geometries of the cationic Li_n^+ clusters differ usually from the optimal structures of the neutral Li_n clusters.

5) In agrrement with the general rules upon the electronic structure of the alkali metal clusters (compare ref. 20) very symmetric compact geometries are favorable only if closed shell character of the electronic structure is guaranteed. This explains the change from the planar cluster geometries toward the three-dimensional ones for the nuclearity around n=6.

The schematic MO (or NO) schemes for some typical cluster topologies in the Fig. 1 demonstrate the tendency towards closed shell structures in a very considering way.

As an other measure of cluster stability can be considered the second difference of the cluster energy as a function of the nuclearity

$$\Delta_n = E_{n+1} + E_{n-1} - 2E_n$$

and the corresponding quantity for cationic clusters

$$\Delta_n^+ = E_{n+1}^+ + E_{n-1}^+ - 2E_{n-1}^+.$$

The first function exhibits pronounced maxima for n=4 and 8, the second on the contrary for n=3, 7 and 9.
The maxima of Δ_n for n=4 and 8 can be put in connection with the "magic numbers" in abundances of alkali metal clusters if the occurence of neutral clusters plays a predominant role in the complicated process of their preparation and detection. If the experimental conditions cause that the occurence frequency of ionic alkali metal clusters is the most important factor, then the odd magic numbers n=3,9 and probably n=7 can be expected. It is very encouraging that inspite of the complexity of the investigated phenomenon of the measured cluster abundances the experimental findings of Schumacher[20] and Brechignac[21] on Na and K are in complete agreement with these theoretical predictions. According Brechignac[21] the occurence of neutral clusters is dominant factor if the energy of photons used in the photoionization suffizes only to ionize the cluster. On the contrary, if the photons cause the cission of alkali metal clusters then the stability of the cationic form is essential. The

other factors are also very important, of course, but this explanation justifies the occurence of the even and odd "magic numbers" making this agreement with the maxima of $\Delta^2 E_n$ or $\Delta^2 E_n^+$, respectively, understandable.

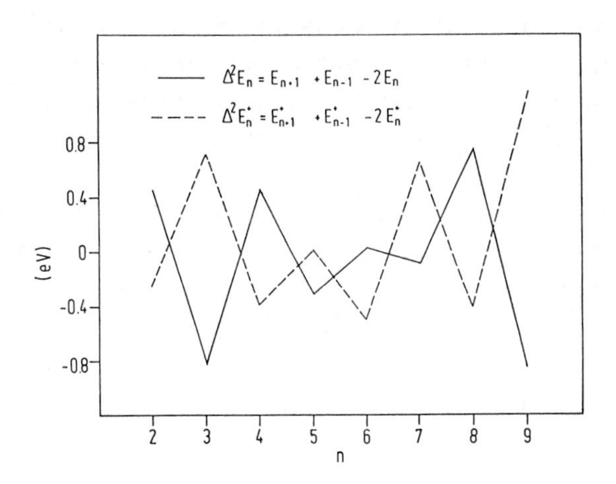

Fig. 2: Second differences of neutral (Δ_n) and of cationic clusters (Δ_n^+).

3. Comparison with other theoretical investigations.

In a systematic study on the electronic and structural properties of sodium clusters by Martins et al.[10] the authors have used pseudopotential method combined with the local-spin-density approximation to search the optimal cluster geometries considering the Hellmann-Feynman forces. Their optimal topologies of the neutral Na_n clusters for $n \leq 7$ agrees with the optimal Li_n cluster topologies found in our investigation. There are discrepancies between the results of Martins et al.[10] and our results for cationic clucters (n=4,6,8) and for the neutral and cationic octamers. The all electron unrestricted Hartree-Fock method sometimes combined with the Møller-Plesset procedure has been employed by Rao and Jena[12-15] on very small Li_n clusters (n=2-5) with the geometries determined by Hellmann-Feynman forces. Again the obtained topologies are the same as those in this contribution. On the other hand, they predict in contrast to our results that Li_6 is magnetic.[15] McAdon and Goddard III[16] have carried out the generalized-valence-bond calculation on small Li_n(n=4,6) as well as Li_{13}^+. They have obtained planar Li_4 (D_{2h}) and Li_6 (D_{3h}) forms as very stable.

Lindsay et al.[17] have considered the topology of Li_n clusters using the Hückel-type approach and also here many of the optimal geometrical forms agree with the results shown in the present contribution.

The jellium model[11] is characteristic for an other quite different group of investigations upon alkali metal clusters. In this model the spherical (or more generally elipsoid) form of the constant positive charge determines the main important features of the electronic structure of alkali metal cluster. The nodal properties of the one electron functions and their occupation numbers are here predominant and cause the similarities with the regularities known from the atomic physics. The "magic numbers" 2,8 and 20 follows necessarily from the concept of the closed shell.

Already Martins, Buttet and Car[10] have mentioned the evident analogy with the properties of molecular orbitals for three-dimensional (and two-dimensional) cluster geometries. The importance of the pseudo-Jahn-Teller effect has been already few times mentioned in the literature (e.g. ref. 5,7,18,19). It is not surprising that very different methods give in general (with some exception, of course) qualitatively very similar results for very small clusters because their basic electronic properties depend on very basic elementary features which any reasonable theoretical method should exhibit. For this reason also, the intuitive investigation based on chemical inituition has predicted all stable geometries for Li_n ($n \leqslant 7$) correctly.[2-7]

4. Conclusions

The proper consideration of the electronic structure and of the distribution of atomic nuclei of small alkali metal clusters with quantum chemical methods is feasible and yields interesting predictions on cluster properties. This approach gives some results which are quite similar to the results of the superatom model (e.g. a special stability of a neutral alkali metal octamer and of a cationic nonamer) but gives additional predictions (e.g. a stability of a neutral alkali metal tetramer, and of a cationic trimer and pentamer). Moreover, the quantum chemical methods do not use any ad hoc model simplifications and can be consequently considered as a justification of the use of the "super atom" model as well as of the limits of its applicability.

Acknowledgement:

This work has been supported by Deutsche Forschungsgemeinschaft (Sfb 6) and Consiglio Nationale delle Recerche (CNR).

Literature:

1) J. Flad, H. Stoll and H. Preuss, J. Chem. Phys. 71, 3042 (1979).

2) H.-O. Beckmann, J. Koutecký and V. Bonačić-Koutecký, J. Chem. Phys. 73, 5182 (1980).

3) G. Pacchioni, D. Plavšić and J. Koutecký, Ber. Bunsenges. Phys. Chem. 87, 503 (1983).

4) D. Plavšić, J. Koutecký, G. Pacchioni and V. Bonacić-Koutecký, J. Phys. Chem. 87, 1096 (1983).

5) J. Koutecký and G. Pacchioni, Ber. Bunsenges. Phys. Chem. 88, 233 (1984).

6) P. Fantucci, J. Koutecký and G. Pacchioni, J. Chem. Phys. 80, 325 (1984).

7) G. Pacchioni and J. Koutecký, J. Chem. Phys. 81, 3588 (1984).

8) A. K. Ray, J. L. Fry and C.W. Myles, J. Phys. B18, 381 (1985).

9) J. L. Martins, R. Car and J. Buttet, J. Chem. Phys. 78, 5646 (1985).

10) J. L. Martins, J. Buttet and R. Car, Phys. Rev. B 31, 1804 (1985).

11) W. D. Knight, K. Clemenger, W. A. de Heer, W. A. Saunders, M. Y. Chou and M. L. Cohen, Phys. Rev. Lett., 52, 2141 (1984).

12) B. K. Rao, J. N. Khanna and P. Jena, Chem. Phys. Lett., 121, 202 (1985).

13) B. K. Rao and P. Jena, Phys. Rev. B 32, 2058 (1985).

14) B. K. Rao, S. N. Khanna and P. Jena, Solid State Sommun. 58, 53 (1986).

15) P. Jena, private communication.

16) M. H. McAdon and W. A. Goddard III, Phys. Rev. Lett. 55, 2563 (1985).

17) D. M. Lindsay, Y. Wang and T. F. George, submitted to J. Am. Chem. Soc.

18) I. Boustani, W. Pewestorf, P. Fantucci, V. Bonačić-Koutecký and J. Koutecký, to be published.

19) J. Koutecký and P. Fantucci, Chem. Rev. 86, 539 (1986).

20) E. Schumacher, personal communication.

21) C. Brechignac, Ph. Cahusac and J. Ph. Ray, to be published.

DESORPTION OF LARGE ORGANIC MOLECULES AND CLUSTERS BY FAST ION IMPACT

B U R Sundqvist and P Håkansson
Tandem Accelerator Laboratory, University of Uppsala
Box 533, S-751 21 Uppsala, Sweden

In recent years large efforts have been made to produce gasphase ions of large organic molecules. The reason is the desire to perform mass spectrometric studies on such molecules and possibly in the future to be able to derive structural information from controlled fragmentation of intense ion beams of large molecules. The most successful approach so far has been production of ions of large organic molecules by fast ion impact on a surface.

The field was pioneered by Macfarlane et al [1] and Benninghoven et al [2]. Fast primary ions, i.e. ions with a velocity higher than the Bohr velocity, have been found to be particulary efficient in producing intact ions of large molecules. The Uppsala group has studied a number of biomolecules heavier than 10 000 u by fast ion impact on a sample surface [3, 4]. The largest monomer ion observed so far is that of porcine trypsin with a molecular weight of 23 463 u, see fig. 1 [4]. Macfarlane originally used fission fragments as fast primary ions and the secondary ions were mass analyzed with the so called time-of-flight technigue. The method was given the name Plasma Desorption Mass Spectrometry (PDMS) by Macfarlane and coworkers [1]. The field of PDMS has recently been reviewed by Sundqvist and Macfarlane [5]. In this technique the samples are mainly solid samples. The molecular film is often prepared with the electrospray technique [6]. That means that the film consists of μm-dimension particles and consequently the surface is very rough. A recent and more attractive approach is to use adsorption of biomolecules to a polymer film. A very useful backing is nitrocellulose [7]. The lower part of fig. 1 shows clearly that with such a sample preparation technique more narrow peaks are produced (less internal energy) and multiply charged ion production is enhanced.

In spectra from samples prepared with the electrospray technique cluster ions are often observed. In fig. 2 a spectrum of positive ions of the aminoacid valine is shown. Cluster ions containing up to 25 valine molecules are observed. For larger molecules cluster ions of the phospholipase A2 molecule has been observed up to masses of the order of 70 000 u [8] for positive ions and about 28 000 u for negative ions. Secondary ions produced by fast ion impact often have very large internal energies and the intensity pattern of clusters as a function of cluster number therefore, as observed in time-of-flight mass spectrometers (i.e. at short times after production), are not expected to show any structure as been shown by Standing et al [9]. Using the earlier mentioned technique for sample preparation where the molecules are adsorbed to a polymer backing, the intensity of cluster ions is much less as may be expected.

The understanding of the desorption mechanism involved when a fast primary particle desorbs and ionizes a large thermally labile molecule is still rather poor, although presently it is an active area of research [5]. For fast primary ions the basic ion-solid interaction is electronic and the process is often referred to as electronic sputtering. A basic problem is to understand how the electronic energy is converted into motion of a large molecule. If results for initial axial velocity distributions for light ions like H^+, Na^+ and Cs^+ are extrapolated to high mass molecular ions one may assume that the energy of a mass 10 000 u molecule leaving the surface is of the order of 1 eV. A good description of the process should then explain how momentum is transferred from the electronic system to translational motion of a large molecule. Williams and Sundqvist [10] has recently suggested that the subionization energy electrons in the fast ion-track may efficiently excite low-lying vibrational levels leading to an expansion of the molecular volume followed by desorption. The model addresses the momentum problem mentioned. This model also has the advantage to supply a soft route for desorption and may therefore explain the ejection of large intact molecules.

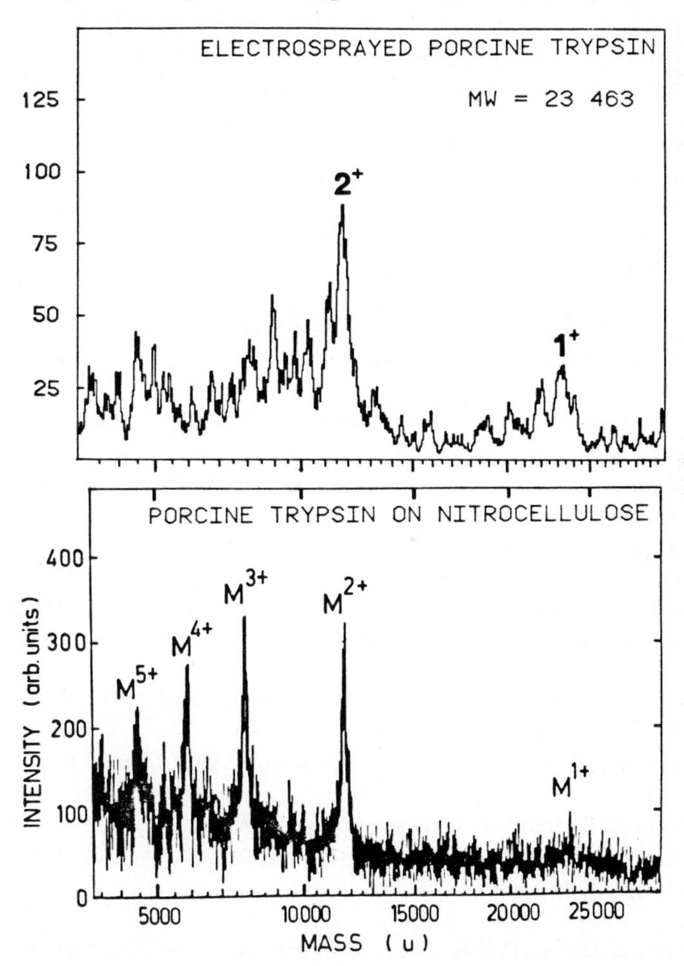

Fig. 1 *Time-of-flight spectrum from a target of porcine trypsin prepared by the electrospray technique (upper) and adsorption on nitrocellulose (lower).*

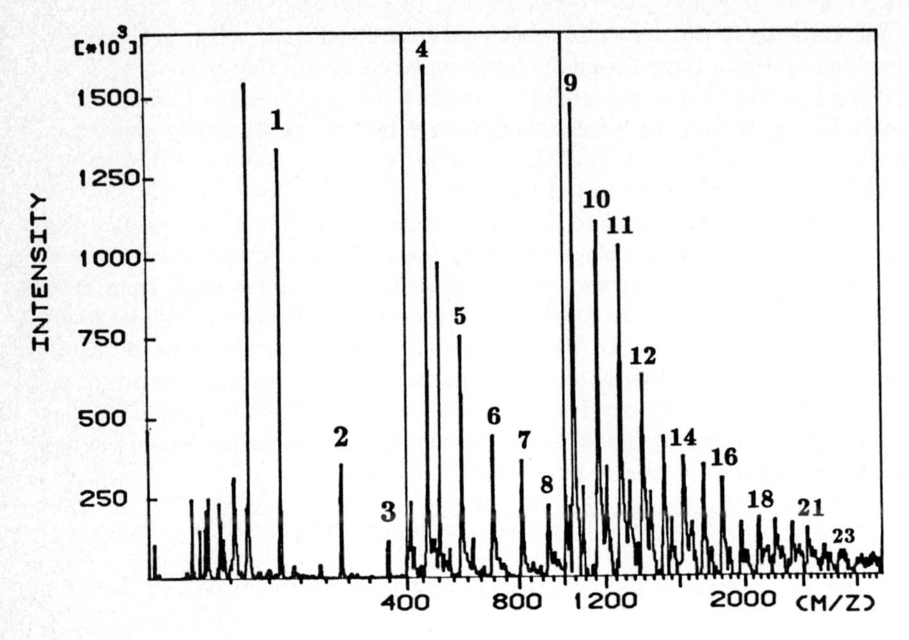

Fig. 2 *Time-of-flight spectrum of valine desorbed by 90 MeV ^{127}I ions.*

REFERENCES

1 D F Torgerson, R P Skowronski and R D Macfarlane *Biochem Biophys Res Commun* **60** (1974) 616-621

2 A Benninghoven, D Jaspers and W Sichtermann, *Appl Phys* **11** (1976) 35-39

3 B Sundqvist, I Kamensky, P Håkansson, J Kjellberg, M Salehpour S Widdiyasekera, J Fohlman, P A Peterson and P Roepstorff *Biomed Mass Spectrom* **11** (1984) 242-257

4 B Sundqvist, P Roepstorff, J Fohlman, A Hedin, P Håkansson, I Kamensky M Lindberg, M Salehpour and G Säve, *Science* **226** (1984) 696-698

5 B Sundqvist and R D Macfarlane, *Mass Spectrom Rev* **4** (1985) 421-460

6 C J McNeal, R D Macfarlane and E L Thurston, *Anal Chem* **51** (1979) 2036-2039

7 G P Jonsson, A B Hedin, P L Håkansson, B U R Sundqvist, B G S Säve P F Nielsen, P Roepstorff, K-E Johansson, I Kamensky and M S L Lindberg *Anal Chem* **58** (1986) 1084

8 B Sundqvist, A Hedin, P Håkansson, I Kamensky, J Kjellberg, M Salehpour G Säve and S Widdiyasekera, *Int J Mass Spectrom Ion Phys* **53** (1983) 167-183

9 W Ens, R Beavis and K G Standing, *Phys Rev Lett* **50** (1983) 27

10 P Williams and B Sundqvist, *Preprint* 1986

Sputtering of Salt Cluster Ions from Liquids by keV Particle Impact

K.P. WIRTH, E. JUNKER AND F.W. RÖLLGEN

Institut für Physikalische Chemie
Universität Bonn
D-5300 Bonn
FRG

1. INTRODUCTION

Secondary ion mass spectrometry (SIMS) applying keV incident particles is a means of generating cluster ions and studying their relative stability. This stability is reflected in the frequency distribution of the cluster ions, showing anomalies i.e. deviations from a pseudoexponential distribution under appropriate conditions. This has been demonstrated for alkali halides /1-9/. Cluster ions of alkali halides have also been generated by a nozzle beam technique and electron impact ionization /10/ providing a broader distribution of cluster sizes than available from the SIMS technique.

Abundant cluster ion signals of alkali halides have been obtained from polycrystalline solids under dynamic bombarding conditions, i.e. by applying high incident particle fluxes and fluences. In constrast, non-volatile complex salts do not provide long lasting cluster ion signals under such conditions. Cluster ions are typically observed for a short time only, due to the accumulation of radiation damage in the solid target. However, long-lasting and abundant cluster ion signals from non-volatile salts can be recorded if the salts are sputtered from a liquid matrix. Sputtering from a liquid matrix is called liquid SIMS or fast atom bombardment mass spectrometry (FAB MS) /11/. Again the frequency distribution of cluster ions can show variations in accordance with the relative stability of the clusters /12-17/.

Cluster ions formed by sputtering are of interest mainly for two reasons: First, they are useful for structural studies, for example for the determination of a relative stability scale of cluster ions, second, they represent a sensitive probe for the investigation and elucidation of the sputtering and ionization mechanism /18/. It is widely assumed that thermally labile and non-volatile molecules are sputtered intact from liquids and under appropriate conditions, even from solids via the ejection of clusters with subsequent fast decomposition and desolvation of molecules and molecular ions. Such a sputtering mechanism forms a natural link between the ejection of organic molecular ions and cluster ions.

In the first part of this paper some effects of a liquid matrix on cluster ion formation are reported. The results are complementary to those of a recent study /16/. The second part of the paper is concerned with the sputtering mechanism leading to the ejection of cluster ions and the question of the origin of the anomalies in the frequency distribution of cluster ions.

2. EXPERIMENTAL

A double focussing mass spectrometer (AEI MS-9) equipped with a home-built atom gun was used in this study. The incident Xe/Xe^+ particle beam was produced in a saddle field discharge source operating at 6 kV anode potential. The atom flux density on the target was about 3×10^{13} particles$/cm^2s$ ($= 5\mu A/cm^2$). The probe surface was in a horizontal position allowing the deposition of a thick liquid layer 0,5 to $1mm$). A potential of +4 kV was applied to the target providing a mass range up to 2300 u. The ion extraction voltage was about 10 V. Accordingly, the residence time of secondary ions of $m/z \sim 1000$ in the source, before passing the counter electrode for acceleration was about 1.5μs. The measured abundance distribution of cluster ions at the detector was essentially determined by the frequency distribution of cluster ions leaving the ion source.

In dissolving salts in glycerol which is hygroscopic, the uptake of some water from the air could not be avoided. Therefore, even in experiments with "pure" glycerol as a matrix water was present contamination.

3. LIQUID MATRIX EFFECTS IN CLUSTER ION FORMATION

Ammonium halides are non-volatile salts. Nevertheless, abundant cluster ions could be obtained under dynamic bombarding conditions from polycrystalline ammonium chloride /19/. This can be explained by very volatile molecules such as NH_3, HCl and Cl_2, produced by bombardment of the chloride, thus avoiding an accumulation of radiation damage on the target surface. In case of the ammonium bromide and iodide, products of radiation damage remain on the surface of solid samples and therefore the formation of larger cluster ions is prevented under the same experimental conditions /16/. However, long-lasting cluster ion signals can be obtained by sputtering of these salts from a liquid matrix such as glycerol. The resulting frequency distribution of ammonium iodide cluster ions is shown in Fig. 1.

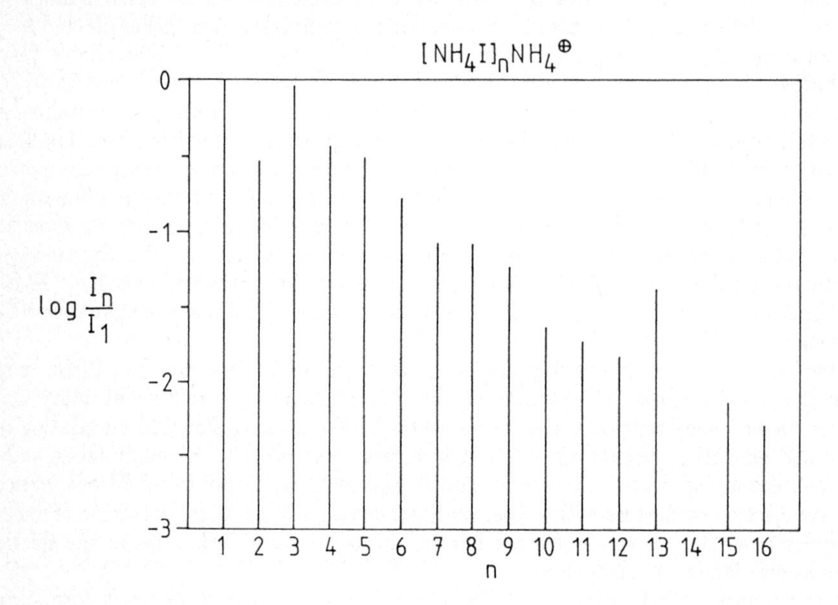

FIG.1: Secondary ion mass spectrum of $(NH_4I)_nNH_4{}^+$ cluster ions obtained from an ammonium iodide/glycerol solution (molar ratio 1:4).

Ammonium halide salts show the same anomalies or magic numbers in the frequency distribution of the cluster ions. Disregarding from the intensity sequence $I_2 < I_3$ the abundance distribution of the ammonium halide cluster ions is very similar to those of alkali halides. This is interesting because the decomposition of alkali halide cluster ions occurs by the removal of salt molecules, while the ammonium halide clusters decompose for energetic reasons by the elimination of NH_3 and HX with $X = Cl$, Br or I. In recent surface ionization experiments we have not found any evidence for the thermal evaporation of intact NH_4X molecules from solid layers /20/. The anomaly $I_2 < I_3$ is indicative for a rather unstable dimer. This instability has been attributed to the directional property of the hydrogen bond between NH_4^+ and X^- /16/. Hydrogen bonding may also have an influence, although not a strong one, one the structure of the most stable larger cluster ions.

Ammonium chloride is less soluble in glycerol than ammonium bromide and iodide. Therefore, the cluster ion signals of ammonium chloride obtained from a glycerol solution are very weak. Typically only a few clusters are above the chemical noise generated by radiation damage of glycerol while mixed clusters of glycerol and the salt are more abundant. This effect of an insufficient solubility of a salt in glycerol for the generation of cluster ions was found for various salts including, for example, KCl. The solubility may be raised by applying appropriate mixtures, in particular those of high dielectric constant. However, this has not been examined yet, except for glycerol/water solvent mixtures as discussed below.

Provided the salt concentration in the liquid matrix is high enough for detecting cluster ion signals, the envelope of the abundance distribution of cluster ions does not change very much from highest (saturation) to lowest concentration. With decreasing concentration the abundance distribution was often found to become even more flat. A further effect of dilution is the reduction of the anomalies in the frequency distribution, which reflects a lowering of the internal energy of the sputtered clusters /15/. This is shown in Fig. 2 for tetramethyl ammonium chloride at three different concentrations. It is also shown in Fig. 2 that the anomaly regarding the intensity ratio I_4/I_3 increases with the time of sputtering of the liquid. This increase reflects the loss of glycerol from the layer by thermal evaporation (vapour pressure of glycerol about $10^{-4} mbar$) and preferential sputtering of glycerol. The anomaly or magic number at $n = 4$ has been discussed before /16/.

The use of a glycerol/water mixture as a solvent for salts provides a means for sputtering of cluster ions from a supersaturated salt solution. To this purpose ammonium chloride was dissolved at high concentration in a glycerol/water mixture (volume ratio about 1 : 1), and in vacuum the evaporation of water gave rise to a precipitation of salt particles. It is reasonable to assume that small solid salt particles are first formed at the surface of the liquid. Sputtering of such a layer (during water evaporation) yielded more abundant cluster ions of the ammonium salt than sputtering of the salt from an unsaturated solution. In addition, abundant mixed clusters of glycerol and the ammonium salt of the general composition $G_n(NH_4Cl)_m NH_4^+$ were found in the spectrum under these conditions. The cluster ion emission can be attributed both to the sputtering of a liquid salt solution and to sputtering of precipitated small solid particles. The latter is indicated by a strong anomaly $I_2 < I_3$ in the frequency distribution of the salt cluster ions while this anomaly is smaller in sputtering of salts from unsaturated solutions.

For tetramethyl ammonium chloride it was found that the precipitation of larger solid particles from a glycerol/water diminished the yield of the cluster ions compared to the application of an unsaturated solution. This observation is consistent with the assumption of an accumulation of radiation damage on the surface of the solid salt particles caused by bombardment at high incident atom fluxes. The question for a similar effect in sputtering of very small particles (up to 10 nm in diameter) precipitated from a less concentrated solution needs to be investigated. First experiments with ammonium bromide and iodide indicate that the radiation damage is not retained in sputtering of very small solid particles formed in solution.

Mixed clusters composed of glycerol or other solute molecules of the matrix solution, and salt cluster ions are particularly abundant with alkali and ammonium halides. It is interesting that the anomaly $I_2 < I_3$ of the ammonium halide series can also be found in mixed cluster ion series. This effect was first reported for stachyose sputtered from a concentrated ammonium halide containing glycerol solution and is shown in Fig. 3 for the M_2 cluster ion series obtained from a mixture of sucrose (M), glycerol and ammonium chloride. The anomaly is also present in the G_1 and G_2 series of glycerol (G) but, surprisingly, not in the sucrose M_1 series. The M_1 and M_2 series could be recorded up to $n = 11$ and $n = 8$, respectively. In contrast to Fig. 3 the anomaly $I_2 < I_3$ was found for stachyose in both cluster ion series. We have no explanation for

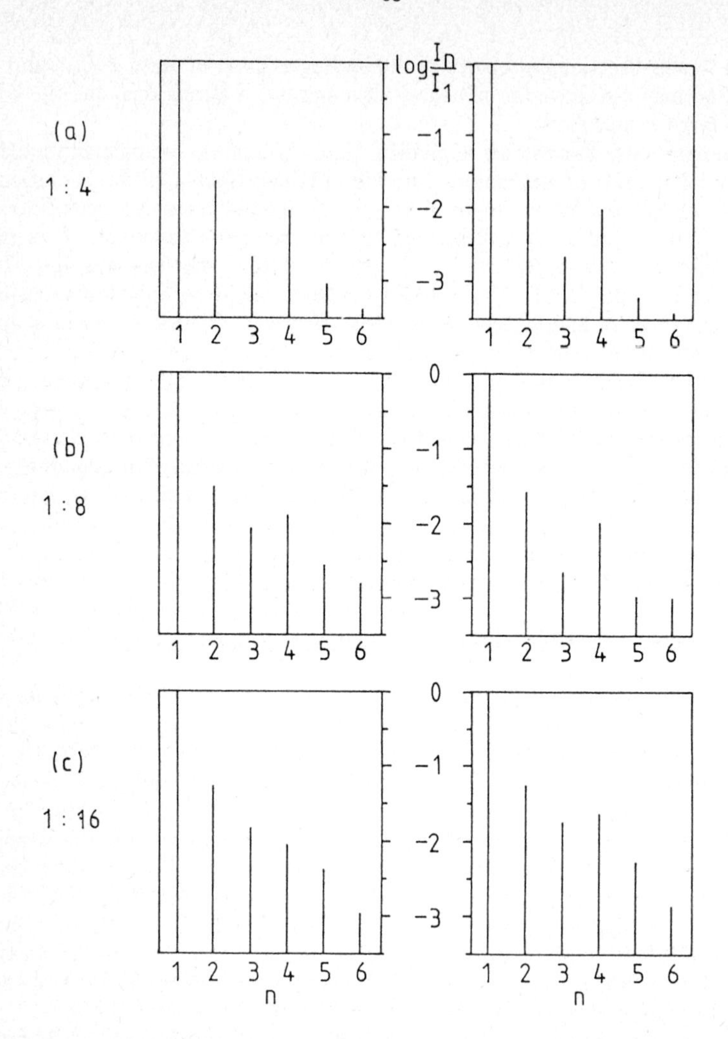

FIG. 2: Frequency distribution of $[(CH_3)_4NCL]_n(CH_3)_4N^+$ cluster ions obtained from tetramethyl ammonium chloride/glycerol solutions at three different concentrations. Molar ratio 1:4 (a), 1:8 (b) and 1:16 (c). The spectra of the first column were obtained immediately after introduction of the mixture, and of the second column after about 10 min bombarding time.

these phenomena but suggest that they are related to a precipitation of the ammonium salt in the surface region of the solution as arising from a supersaturation induced by the evaporation of glycerol and water.

Finally, the results of some experiments shall be reported which were performed to study the effect of a frozen glycerol matrix on the cluster ion emission under dynamic bombarding conditions. For the salts examined (ammonium iodide, tetramethyl ammonium chloride, KI) the formation of cluster ions was found to be prevented by the frozen matrix. This finding collaborates with previous ones of a destructive sputtering of molecules from a frozen glycerol matrix /15/ and the suppression of cluster ion formation from frozen methanol /21/. As already discussed

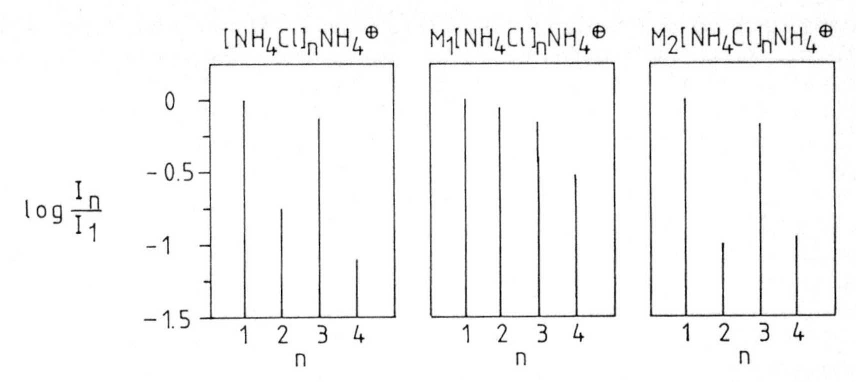

FIG. 3: Frequency distribution of some cluster ions obtained from a mixture of sucrose, ammonium chloride and glycerol (volume ratio 1:3:10). The monomer and dimer of sucrose is denoted by M_1 and M_2, respectively.

before /15/, melting is not involved in sputtering from a frozen matrix and therefore the radiation damage is accumulated under these dynamic bombarding conditions.

4. On the Mechanism of Cluster Ion Formation

The formation of cluster ions by sputtering is related to that of molecular ions because a cooperative sputtering mechanism by a concerted momentum transfer to the atoms of molecules and clusters is required for a non-destructive ejection process. Furthermore, for thermally labile complex salts and molecules the internal energy of sputtered particles must be low enough to avoid fragmentation of molecular ions, and of salt cations and anions by unimolecular decay within the flight time of ions to the detector of a mass spectrometer. Therefore a mechanism of sputtering outside thermal equilibrium leaving the ejected particle vibrationally cold is a prerequiste for the formation of molecular ions of thermally labile compounds and of cluster ions of complex salts. But even for thermally stable compounds it is rather unrealistic to assume that the high temperature needed for a thermal process i.e. for the generation of a vapour pressure of molecules between 1 and more than 100 bar, which corresponds to a keV particle induced sputtering of one or a few intact molecules by a thermal process, can be reached within about 100 ps. For these conditions the generation of a non-equilibrium phase transition process is more likely. The assumption of a thermal sputtering of molecules is also not supported by the time scale of randomization of vibrational energy (vibrational relaxation) within molecules /22/.

In a recent paper /15/ the mechanism of sputtering of molecules and molecular ions from liquids has been discussed in detail. According to this mechanism the ejection of vibrationally cold cluster ions is achieved by a spraying type of sputtering mechanism created by a collision spike /23/ and involves the following steps: 1. The incident keV particle penetrates into the liquid and creates a dense cascade, resulting in destruction of sample and solvent molecules. 2. A high gas pressure is built up by the products of radiation damage. 3. Large charged and uncharged clusters composed of solute and solvent molecules are ejected by the expanding gas. The ejection is most probably assisted by a shock wave. 4. The clusters and cluster ions subsequently decompose preferentially by the loss of solvent molecules and more volatile products of radiation damage. Thus from larger cluster ions smaller ones are formed which finally hit the detector of a mass spectrometer. The unimolecular decay of clusters gives rise to adiabatic (vibrational) cooling.

In sputtering of solids and liquids other effects of collision spikes may dominate such as a complete gasification of the sputtered material /24/. However, the formation of cluster ions of non-volatile salts from liquids or solids, under high fluence or low fluence bombarding conditions, excludes any significant contribution of gas phase collisions to the generation of cluster ions in the sputtering process. Furthermore, the cluster ions obtained by bombarding of non-volatile solids such as the ammonium halides, provide evidence for a non-thermal sputtering mechanism even in the case of solids. Is is very likely that a similar sputtering mechanism as discussed for liquids accounts for the ejection of cluster ions from solid salts i.e. gasification of the salts and shock wave effects should be involved in the sputtering mechanism. The experiments with frozen glycerol /15/ and frozen methanol /21/ show that the formation of a liquid state via melting of the solid does no play a significant role in the sputtering of cluster ions.

Energy distribution measurements of some molecular and cluster ions sputtered from glycerol have been reported /25/. A rather broad energy distribution and a small increase of the width of the distribution with the size of molecular and cluster ions were observed. These results indicate high translational temperatures of sputtered particles and point to a dependence of the size of ions on the energy deposition in the collision spike causing the ejection of these particles. However, measurements of the translational energy do not allow conclusions to be drawn about the vibrational temperature of emitted ions in sputtering by spraying outside thermal equilibrium.

The pseudoexponential decrease of the frequency distribution of cluster ions can be attributed to a similar decrease of the abundance distribution of the size of sputtered particles and the effect of fast unimolecular decomposition of these particles. The anomalies in the frequency distribution of cluster ions arise from the unimolecular decay of larger cluster ions collisionally or vibrationally excited in the sputtering process /16/. They do not originate from sputtering of preformed ions because the kind of anomaly present in a frequency distribution of cluster ions is independent from the physical state of the salt sample exposed to the incident particle beam. Accordingly, there is also no change in the kind of anomaly with time during sputtering of a supersatured salt solution, i.e. during an increase of the rate of precipitation of the salt from solution. Furthermore, a specific solvation of cluster ions of the salts under study is not known in solution chemistry. Solvation of clusters of ionic compounds is only known for surfactants forming micelles above a critical concentration in solution /26/.

Acknowledgement: The authors are grateful to the Deutsche Forschungsgemeinschaft for financial support of this work.

REFERENCES

1. F. Honda, G.M. Lancaster, Y. Fukuda and J.W. Rabelais, J. Chem. Phys. 69 (1978) 4931.
2. T.M. Barlak, J.E. Campana, R.J. Colton, J.J. DeCorpo and J.R. Wyatt, J. Phys. Chem. 85 (1981) 3840.
3. J.E. Campana, T.M. Barlak, R.J. Colton, J.J. DeCorpo, J.R. Wyatt and B.I. Dunlap, Phys. Rev. Lett. 47 (1981) 1046.
4. T.M. Barlak, J.R. Wyatt, R.J. Colton, J.J. DeCorpo and J.E. Campana, J. Am. Chem. Soc. 104 (1982) 1212.
5. T.M. Barlak, J.E. Campana, J.R. Wyatt and R.J. Colton, J. Phys. Chem. 87 (1983) 3441.
6. W. Ens, R. Beavis and K. Standing, Phys. Rev. Lett. 50 (1983) 27.
7. M.A. Baldwin, C.J. Proctor, I.J. Amster and F.W. McLafferty, Int. J. Mass Spectrom. Ion Processes 54 (1983) 97.
8. I. Katakuse, H. Nakabushi, T. Ishihara, T. Sakurai, T. Matsuo and H. Matsuda, Int. J. Mass Spectrom. Ion Processes 57 (1984) 239.
9. T.G. Morgan, M. Rabrenovic, F.M. Harris and J.H. Beynon, Org. Mass Spectrom. 19 (1984) 315.
10. R. Pflaum, P. Pfau, K. Sattler and E. Recknagel, Surface Sci. 156 (1985) 165.

11. M. Barber, R.S. Bordoli, G.J. Elliott, R.D. Sedgwick and A.N. Tyler, Anal. Chem. 54 (1982) 645A
12. S.S. Wong, U. Giessmann and F.W. Röllgen, Proc. 32nd Conf. on Mass Spectrom. and Allied Topics, San Antonio (1984), 186.
13. D.N. Heller, C. Fenselau, J. Yergey, R.J. Cotter and D. Larkin, Anal. Chem. 56 (1984) 2274.
14. J.M. Miller and R. Theberge, Org. Mass Spectrom. 20 (1985) 600.
15. S.S. Wong and F.W. Röllgen, Nucl. Instrum. and Meth., B14 (1986) 436.
16. S.S. Wong and F.W. Röllgen, Int. J. Mass Spectrometry Ion Processes 70 (1986) 135.
17. E. Tolun and J.F.J. Todd, 34th Ann. Conf. Mass Spectrom. and Allied Topics, Cincinnati (1986).
18. B. Schueler, R. Beavis, W. Ens, D.E. Main and K. Standing, Surface Sci. 160 (1985) 571.
19. G. Schmelzeisen-Redeker, S.S. Wong, U. Giessman and F.W. Röllgen, Z. Naturforsch. 40a (1985) 430.
20. U. van Gemmeren and F.W. Röllgen, unpublished result.
21. R.N. Katz, T. Chaudhary and F.H. Field, J. Am. Chem. Soc. 108 (1986) 3897.
22. M. Quack, Bull. ETH Zürich 189 (1984) 19.
23. P. Sigmund in: Sputtering by Particle Bombardment I, ed. R. Behrisch, Springer Heidelberg (1981).
24. D.E. David, Th.F. Magnera, R. Tian, D. Stulik and J. Michl, Nucl. Instr. and Meth. B14 (1986) 378.
25. G.J.Q. van der Peyl, W.J. van der Zande, R. Hoogerbrugge and P.G. Kistemaker, Int. J. Mass Spectrom. Ion Processes 67 (1985) 147.
26. P.C. Hiemenz, Principles of Colloid and Surface Chemistry, Marcel Dekker, New York (1977).

^{252}Cf-desorption of small clusters from Al_2O_3, MgO and SiO_2 surfaces

Bernd Nees, Erwin Nieschler, and H. Voit
Physikalisches Institut
der Universität Erlangen-Nürnberg
D-8520 Erlangen, W. Germany

In this work we investigate the ^{252}Cf induced desorption of clusters from Al_2O_3, MgO and SiO_2 surfaces. The time of flight mass-spectrometer has a flight path of 1 m and a mass resolution $m/\Delta m \approx 750$. The fission-fragments hit the sample perpendicular from the back. The desorbed secondary ions are accelerated (acceleration voltage 10 kV) within a distance of 3 mm, traverse the flight path and impinge on a CsJ coated metal plate. The secondary electrons from this plate are focused into a chevron channel plate. For the time of flight measurement a time to digital converter was used, which is able to accept 30 stop signals correlated to one start signal.

1. Spectra from Al_2O_3 samples

The figures 1 and 2 show mass spectra obtained from a 5 μm Al-foil which had been exposed to normal room atmosphere. The sample surface is therefore covered by a passivating layer consisting of Al_2O_3 and $Al(OH)_3$.
The mass spectrum of positive ions clearly exhibits three cluster series. The most pronounced series (marked by plain numbers in figure 1) consists of $[(Al_2O_3)_nAlO]^+$ clusters, which can be traced up to n = 13 (m = 1368.3) in the spectrum. The second series (marked by ( )) consists of $[(Al_2O_3)_nH]^+$ clusters. It terminates for n = 9 (m = 918.5). $[(Al_2O_3)_nAl_2O_2]^+$ clusters form the third series, which ends for n = 6 (m = 697.6). It is interesting to note, that the intensity of the n = (2) cluster (m = 204.9) in the second series is

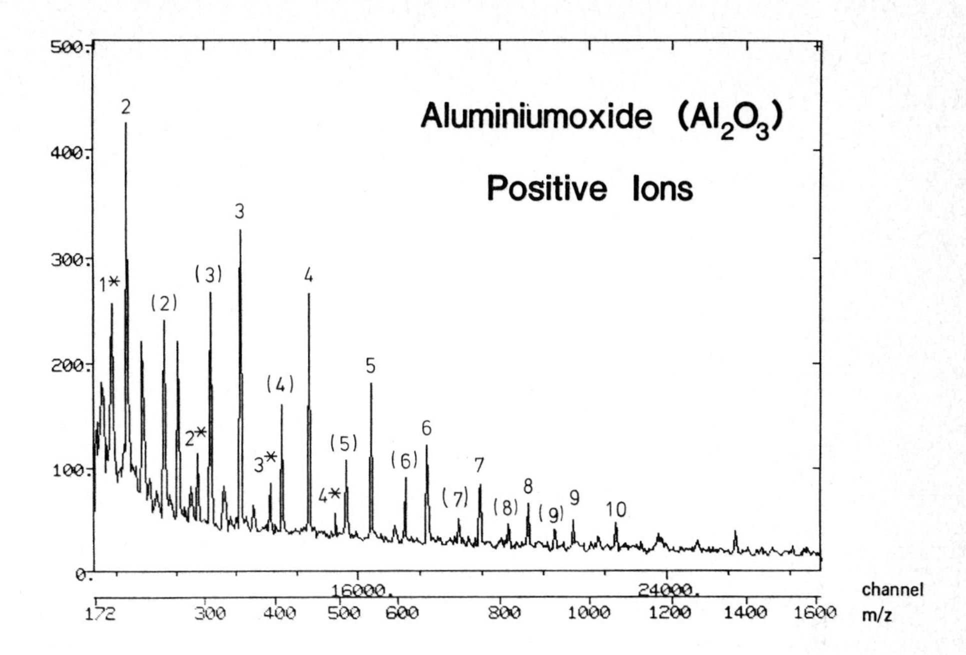

Figure 1: Mass spectrum of positive ions desorbed from a Al_2O_3 sample

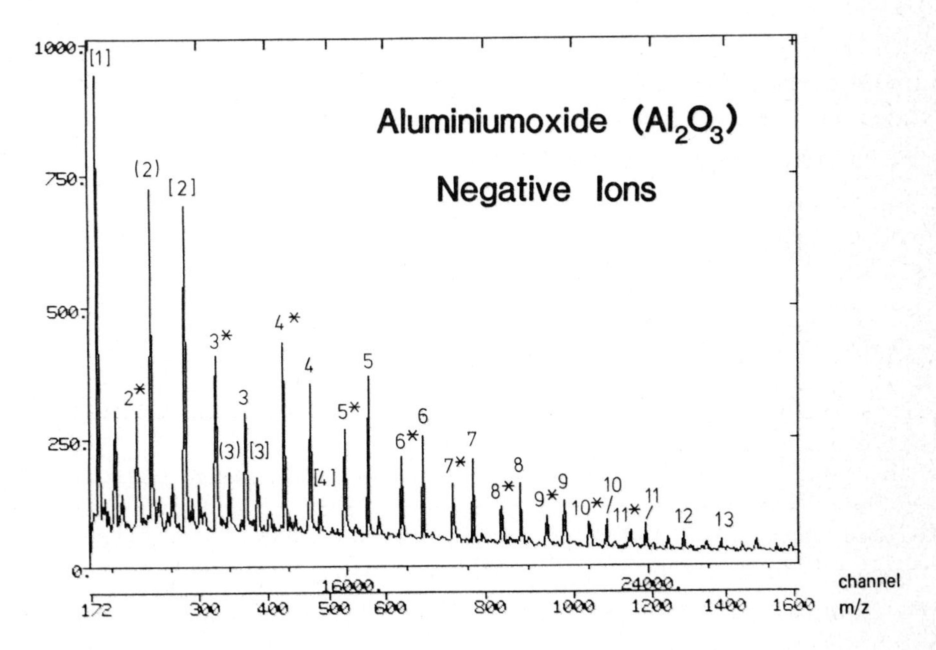

Figure 2: Mass spectrum of negative ions desorbed from a Al_2O_3 sample (lower mass region)

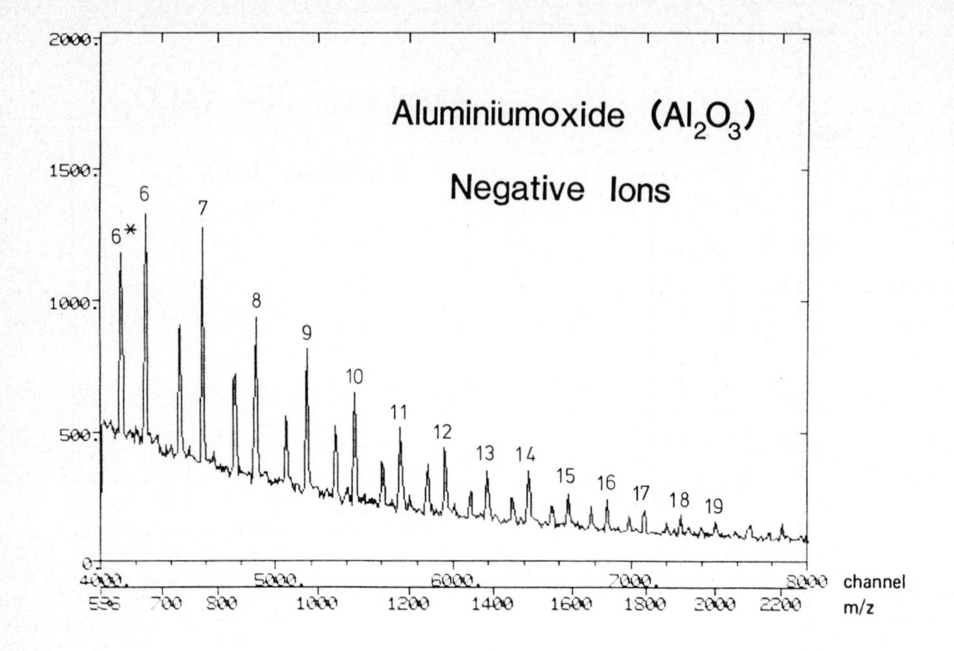

Figure 3: Mass spectrum of negative secondary ions from Al_2O_3 (higher mass region)

smaller than that of the n = (3) cluster whereas quite generally a decrease with increasing n is observed (partially due to the decreasing detection efficiency).

The mass spectra of the negative ions (see figures 2 and 3) show four series, which are based on the following cluster configurations: $[(Al_2O_3)_nAlO_2]^-$, $[(Al_2O_3)_nOH]^-$, $[(Al_2O_3)_n(AlO_2)H_2O]^-$, $[(Al_2O_3)_nH(OH)_2]^-$. The series are marked in figures 2 and 3 by n, n*, [n] and (n), respectively. They can be followed up to n = 21 (m = 2199.9), n* = 17* (m = 1751.1), [n] = [5] (m = 586.7) and (n) = (4) (m = 442.8). It should be noted, that the intensity has a maximum for n = 4 in case of the first two series.

Very similar cluster series can be observed in the mass spectra obtained from aluminized polyester foils, which are often used as support for organic samples. Inhomogeneous covering of this foil by the sample may lead to background peaks due to the Al_2O_3 clusters.

2. Spectra from MgO samples

The MgO sample was prepared by evaporating MgO from a W-boat onto a gold covered (40 $\mu g/cm^2$) aluminized polyester foil. The sample thickness was approximately 30 $\mu g/cm^2$. The figures 4 and 5 show portions of the mass spectrum for the positive ions. This spectrum contains four series based on the following clusters: $[(MgO)_nH]^+$, $[(MgO)_nO_2H]^+$, $[(MgO)_nMg]^+$ and $[(MgO)_nOH]^+$. Only the series containing the first two cluster configurations are explicitly labeled in the figures 4 and 5 (by n and n*, respectively).

The relative large widths of the peaks is due to the fact, that three stable isotopes are contributing. In case of the $[(MgO)_nH]^+$ series an additional effect, already observed in ref.[1], is of importance: with increasing n an increasing number of H atoms is attached to the clusters. The different contributing isotopes are clearly resolved in case of the n = 1 member of the $[(MgO)_nH]^+$ series as can be seen from figure 6.

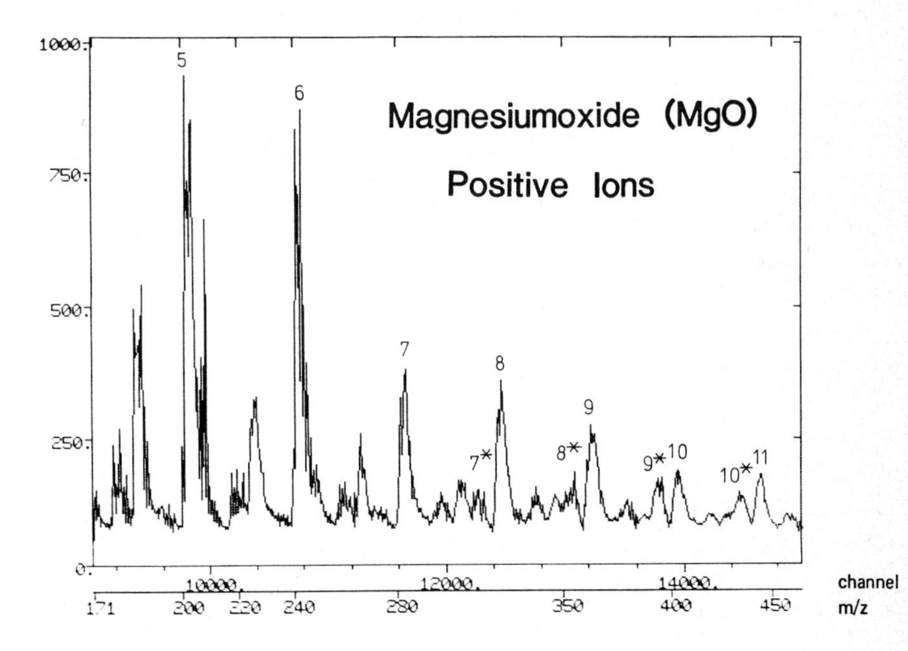

Figure 4: Section of the mass spectrum from a MgO sample (m = 170-450, positive ions)

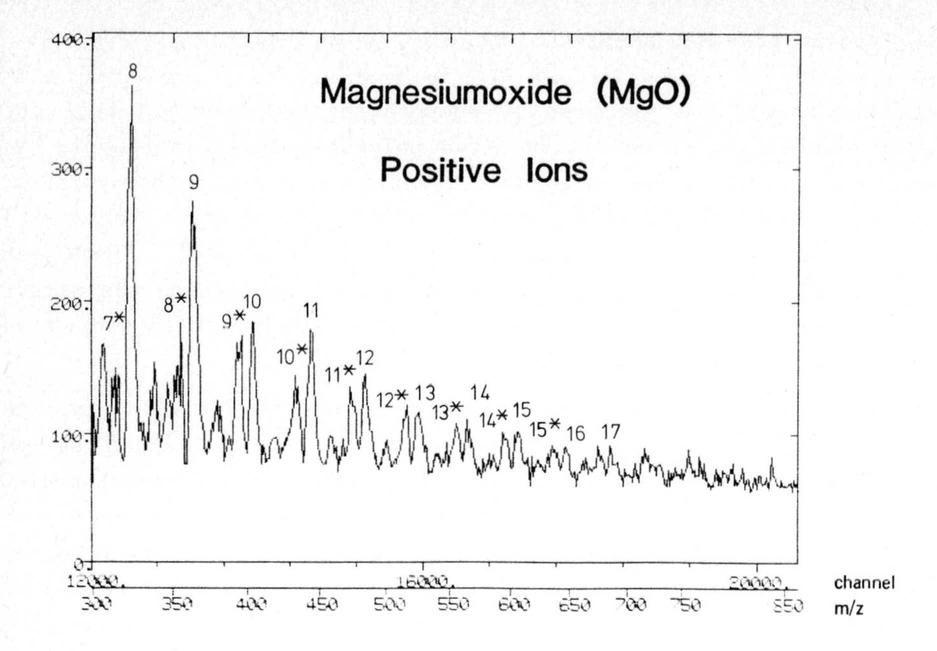

Figure 5: Section of the mass spectrum from a MgO sample (m = 300-850, positive ions)

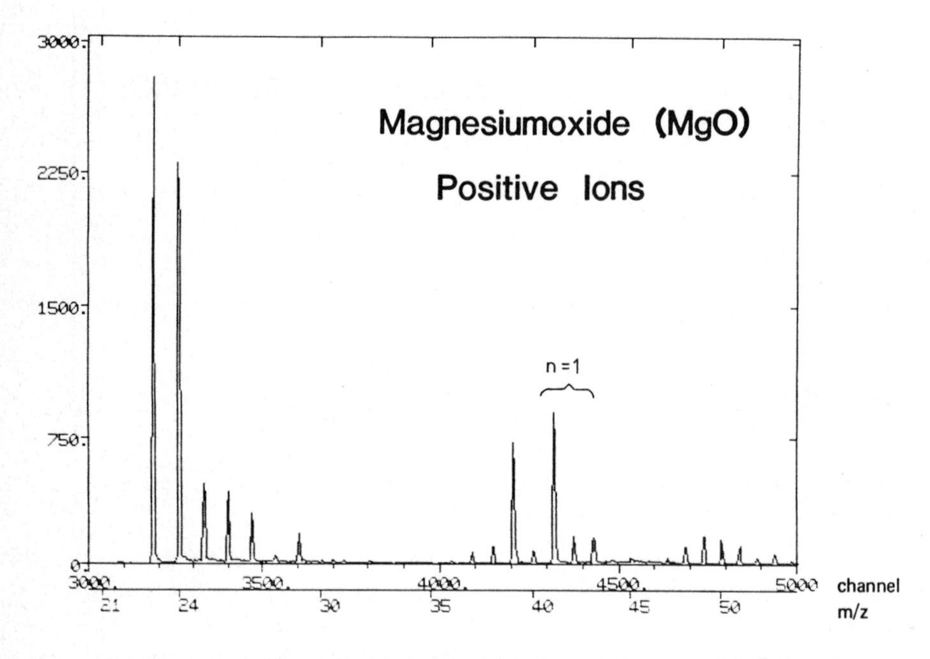

Figure 6: Section of the mass spectrum desorbed from a MgO sample (m = 20-55, positive ions)

This figure shows in addition, that Mg^+ ions (with the correct isotope ratio) can be desorbed with large yield. This is in contrast to the Al_2O_3 case where no Al^+ ions are observed.

The mass spectra of the negative ions show no sample specific peaks (see also ref. [1]). However, in the mass region 200-500, groups of peaks can be observed, which belong to different tungsten clusters reproducing the natural isotope distribution (see figure 7).

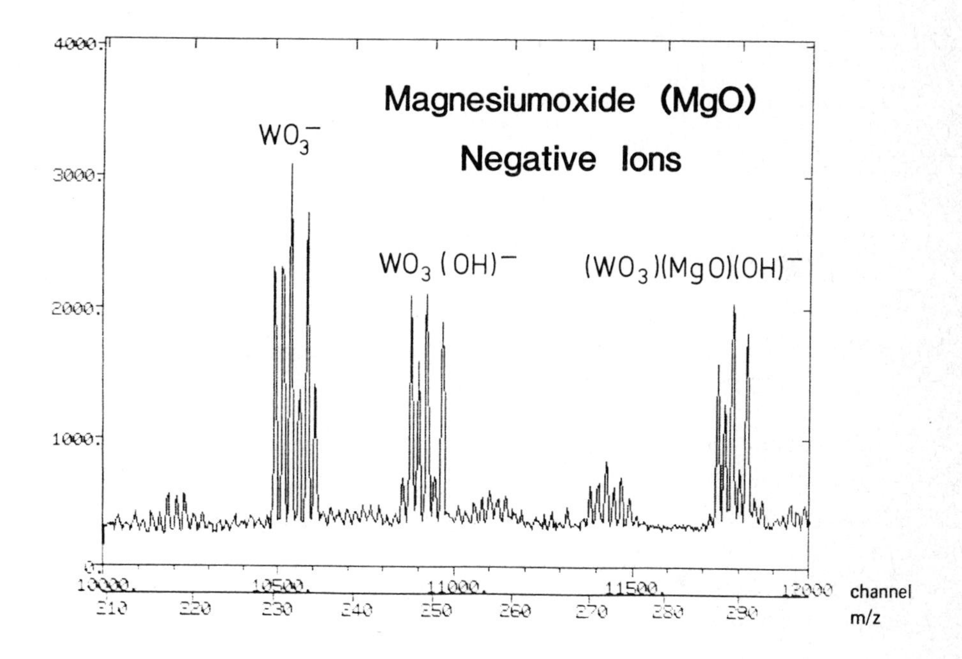

Figure 7: Mass spectrum for negative ions desorbed from a MgO sample

3. Spectra from a SiO_2 sample

The SiO_2 sample was prepared in the same way as the MgO sample. The mass spectra of the positive ions do not contain sample specific masses.

The mass spectra for negative ions (see figure 8) contain a series based on a $[(SiO_2)_nOH]^-$ cluster (labeled by n in figure 8). The series starts at n = 2 (m = 136.9) and ends at n = 9 (m = 556.6). Besides this a series based on a $[(WO_3)(SiO_2)_nOH]^-$ cluster is observed (labeled with n* in figure 8).

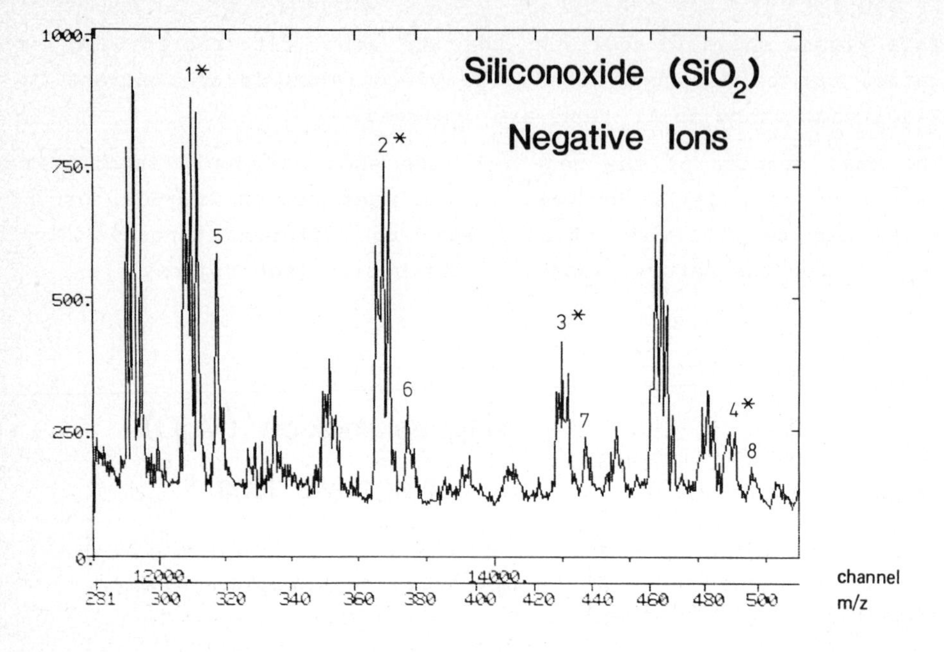

Figure 8: Mass spectrum from a SiO_2 sample: negative secondary ions

In conclusion we have investigated the desorption of clusters from Al_2O_3, MgO and SiO_2 samples. The largest yield was obtained for the desorption of negative ions from a Al_2O_3 sample. In contrast to this the MgO sample did not yield sample specific negative ions. The yield of positive ions from the MgO sample is, however, rather large, it exceeds the positive yield desorbed from the Al_2O_3 sample. The SiO_2 sample does not give a sample specific yield for positive ions.

Clusters could be followed up to n-values of 21 (m = 2199.9, see figure 3). For some cluster series a maximum yield was observed for a particular n value.

[1] O. Becker, thesis, Darmstadt 1985, unpublished

This work was supported by the Bundesministerium für Forschung und Technologie (BMFT).

ION TRACK ASPECTS OF ELECTRONIC SPUTTERING

A. Hedin P. Håkansson and B.U.R. Sundqvist
Tandem Accelerator Laboratory,
University of Uppsala,
Box 533, S-751 21 Uppsala, Sweden

An ion track model for fast heavy ion induced desorption of molecules has been developed [1]. Fast heavy primary ions (MeV/amu) impinging on a solid insulating target will interact primarily with the electronic structure of the target material in a limited region around the ion path, the infra track. From the infra track, a distribution of secondary electrons is ejected into the so called ultra track. In the ion track model, the flux of secondary electrons in the ultra track is related to the desorption yield of molecular ions by the requirement that m electrons must hit a target molecule in order to desorb it as an ion. The likelihood that a molecule is not damaged by the secondary electron energy deposition is described by the survival probability, p_s. m and p_s are the two free parameters in the model.

The flux of secondary electrons in the ultra track will vary linearly with the electronic stopping power of different primary ions if the primaries have the same velocity. The model has been fitted [1] to yield data from such stopping power experiments [2] for several bio-molecules. The agreements are good and m is shown to increase with molecular size, whereas p_s is smaller for a larger molecule.

For molecules of a certain size, m is an approximate measure of the energy deposited in a molecule and hence a measure of the energy required for ionization and desorption of the molecule. Recently, it has been demonstrated that sample molecules adsorbed to a nitro cellulose backing give high positive molecular ion yields of massive ($\geq$ 5000 amu) bio-molecules compared to samples prepared with the electro spray method which is the standard technique. By applying the described stopping power experiment to samples prepared with the two techniques, it has been demonstrated, using the ion track model [3], that desorption of molecular ions from nitro cellulose samples require less energy for ionization-desorption than from electro sprayed samples, see fig. 1.

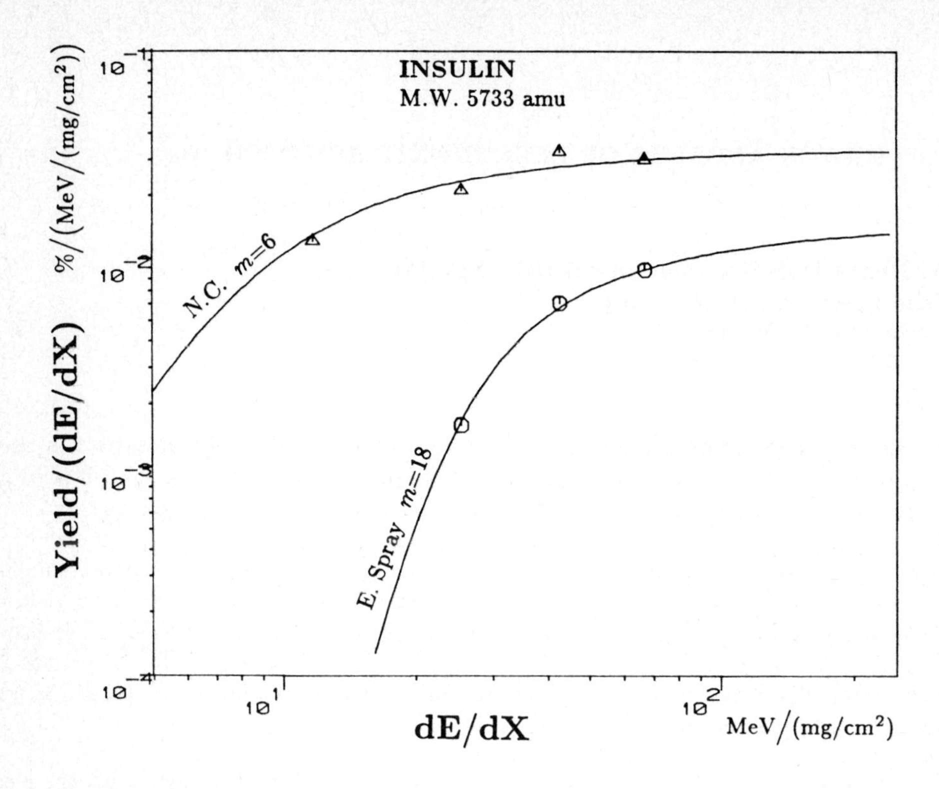

Fig. 1 *Yield divided by the electronic stopping power as a function of the electronic stopping power for nitrocellulose-adsorbed and electro sprayed insulin samples. The solid lines represent fits of the ion track model to the data. The lower m value for the N.C. sample indicate that less energy is required to ionize and desorb the sample molecules in this case.*

REFERENCES

1. A Hedin, P Håkansson, B Sundqvist and R E Johnson
   Phys Rev B **31** (1985) 1780

2. P Håkansson, I Kamensky, M Salehpour,
   B Sundqvist and S Widdiyasekera
   Radiat Eff **80** (1984) 141

3. A Hedin, P Håkansson and B U R Sundqvist
   Tandem Accelerator Report TLU 145/86, Uppsala 1986.
   To be published.

THERMAL SPIKE MODEL FOR HEAVY ION INDUCED DESORPTION

I. NoorBatcha and Robert R. Lucchese
Department of Chemistry, Texas A&M University
College Station, Texas 77843

1. Introduction

The passage of a high linear energy transfer (LET) ion through a solid is know to induce the desorption of large molecules and molecular ions from the surface of the solid [1]. This process, often referred to as heavy ion induced desorption (HIID), has been observed using both heavy ions from nuclear fission [1] and using ions from accelerators [2]. One important application of the process is to mass spectrometry of large molecules of biological interest [3], in this context the method has been referred to as particle desorption mass spectrometry (PDMS). A number of theories have been proposed [4-15] to explain how the energy which is deposited by the passage of the heavy ion, mostly as electronic excitation [16], is transformed into nuclear motion of the adsorbate away from the surface.

In this paper, we will briefly discuss the basic processes which occur during the HIID desorption event, then we will describe a model for understanding this process which is based on a local thermal equilibrium picture of the vibrational excitation of the surface [10]. The parameters of the model were obtained by fitting to the experimental data of Hakansson *et al*. [17] which has been given in tabular form by Hedin *et al*. [13] This experimental data [13,17] gives the yield of HIID molecular ions ranging from valine to insulin, where the molecules were desorbed from an aluminum surface onto which they had been electrosprayed. The experimental data were obtained for a variety of primary ions in their equilibrium charge state all with the same velocity. The use of the equilibrium charge state assures that the energy deposition of the primary ion is uniform in the direction of motion of the ion. The constant velocity of the primary ions implies that the diameter of the initial electronically excited region is the same for all of the ions considered.

We have used our thermal model [10] which fit the data of Hakansson *et al*. [17] to extrapolate the experimental data to higher and lower LET and to different initial track widths. We have also considered the distribution of desorbing particle kinetic energies which would be predicted by the thermal model. We have assumed that the initial distribution of kinetic energies of the molecules leaving the surface is given by the temperature of the surface at the time of desorption. Then we have considered the effects of collisions in the gas phase which would occur after the desorption process. To do this we have employed a Monte Carlo simulation of the gas-dynamics of the molecules leaving the surface [18] and we have made a variety of assumptions concerning the mass distribution and number of particles leaving the surface. We have found that if there is a lighter mass component then there can be a significant acceleration of the heavier mass particles which is proportional to their mass.

2. Microscopic Model of HIID

When an ion passes through a solid it interacts with both the electrons and nuclei of the solid. At the ion energies of interest in HIID, ~1 MeV/amu [17], most of the energy which is transferred to the solid is transferred to the electronic degrees of freedom of the system and not to the nuclear motion [16]. Thus immediately after the primary ion has passed through the solid, the state of the solid can be characterized by a high degree of electronic excitation and relatively low level of excitation of the nuclear motion [8]. The excited electronic states of the constituent atoms can relax through various Auger processes and the excited electrons can also loose energy by scattering with other electrons producing more excitations in the solid. The nuclei also begin to move under the influence of the potential energy surfaces which have been modified by the electronic excitations. The vibrational excitations then begin to propagate through the solid. Thus both the electronic and nuclear degrees of freedom become excited due to the passage of the primary ion through the solid, although the radial and time dependence of these excitations has not been well characterized to date. It is believed that the electronic excitations have in large part relaxed in a time on the order of 10^{-11} secs [8]. In insulating materials the half life of the nuclear excitation can be estimated to be on the order of 10^{-10} secs by using the thermal diffusivity of the material [10].

The desorption of molecules and molecular ions from the surface of the solid is then induced by the excitation of the nuclear degrees of freedom of the solid. There are several models for how the desorption occurs, including thermal spike models [4-10], the "popcorn" model [11], RRKM models [12], secondary electron bond disruption [13], high frequency polarization of the electron plasma [14], and solid continuum mechanics models [15]. These models can be divided into two classes, the first being models which assume that the vibrational excitation of the solid is to some extent randomized and can be described by statistical distribution functions [4-10,12], and the second class being models which assume that the desorption occurs by the preferential excitation of the nuclear vibrational mode which leads to desorption [11,13-15]. In the present study we will consider a thermal model which makes the statistical assumption [10].

The final microscopic step which must be considered for the PDMS method is the mechanism for the production of molecular ions during the desorption. It is known that the number of ions produced is a very small fraction of the total mass desorbed from the surface of an ion [19]. Thus the ionization process may either be a rare event or it may be such a hard ionization that only a few of the ionized molecules have a lifetime long enough with respect to fragmentation for the molecules to remain intact during the acceleration phase in the PDMS spectrometer [20]. There are several possible mechanism which can lead to the production of molecular ions including the existence of preformed ions on the surface, non-adiabatic desorption dynamics leading to ion pair formation, electron-molecule impact ionization, and ion-molecule reactions in the selvedge [21]. In the present paper, we will not further consider the ionization process and we will make the simplifying assumption that the probability of ionization of a given desorbed molecule is a uniform function of the distance from center of the primary ion track.

3. Thermal Model of HIID

In order to model the excitation of the nuclear degrees of freedom of a solid due to the passage of a fast heavy ion, we will consider a primary ion track which is normal to the surface of the solid. We will then assume that the level of vibrational excitation can be described by a function of r, the distance from the center of the primary ion track, and t, the time since the passage of the primary ion. The value of this function will be taken to be the mean vibrational excitation at a given r and t and will be written as a temperature $T_s(r,t)$ using the vibrational excitation profile suggested by Mozumder [22]

$$T_s(r,t) = T_o(1+4\delta t/r_o^2)^{-1} \exp[-r^2/(r_o^2+4\delta t)] \tag{1}$$

where the width of the initial excitation is characterized by the radius r_o, δ is the thermal diffusivity of the solid, and T_o is proportional to the LET which we will estimate using [22]

$$T_o = (dE/dX)/\pi\rho C_v r_o^2 \tag{2}$$

where dE/dX is the LET of the primary ion, ρ is the density of the solid and C_v is the heat capacity of the solid. For high energy ions, dE/dX can be as large as 900 eV/Å which leads to T_o on the order of 10^4 to 10^5 K with r_o ranging from 10-25Å.

For many systems of which one would like to obtain mass spectra the simple vaporization of a sample is not feasible. For these systems, slow heating rates lead to molecular decomposition before vaporization [23]. If the thermal model is correct, then the only difference between the vaporization of molecules using standard heating sources and using a fast heavy ion, is the difference in the heating rate. For standard heating procedures the peak rate is on the order of up to 10^3 K sec^{-1}, using a laser source one can obtain heating rates on the order of 10^{12} K sec^{-1} with nanosecond laser pulses [24] and up to 10^{15} K sec^{-1} for sub-picosecond laser pulses [25], and with the HIID the heating rate at $r = 2r_o$ is also on the order of 10^{15}K sec^{-1}[10].

There are two main effects which are well known to occur when a system is rapidly heated. The first effect is that the relative rates of competing pathways can be altered leading to end products which differ from those one would have obtained if the system had been heated at a slower rate [26]. In particular consider a system which can react by two pathways having rates given by the two Arrhenius rate laws

$$R_1 = A_1 \exp(-\Delta E_1/k_B T) \tag{3}$$

and

$$R_2 = A_2 \exp(-\Delta E_2/k_B T) \tag{4}$$

Now, if $\Delta E_1 < \Delta E_2$ and $A_1 < A_2$ then at low heating rates the lower energy pathway, governed by rate R_1, will predominate, whereas at high heating rates the relative reaction rates will be proportional to the ratio of the A factors so that the high energy pathway, governed by rate R_2, will predominate. This behavior has been seen experimentally by Hall and Bares [26] in the competition between surface reactions and desorption from a surface.

Another effect of rapid heating is that all modes of nuclear motion of a system will not become excited at the same rate. This disequilibrium in vibrational modes is well known in the study of solid systems [27]. The disequilibrium between different modes of nuclear motion has also been theoretically predicted to occur in rapid heating desorption from surfaces [28]. In simulations of NO desorption from an LiF(100) surface, the vibrational modes of the desorbing molecule were found to be colder than the rotational modes which were in turn colder than the translational modes. Extending this observation to the case of HIID of large molecular systems would suggest that the internal modes of the desorbing molecule may not be in thermal equilibrium with the temperature of the surface at the time the molecule is desorbed from the surface. Both of the effects of rapid temperature jumps mentioned here are probably to some degree responsible for the sucess of HIID in producing large molecular ions which do not immediately fragment after desorption.

4. Model for Desorption Process

The surface temperature profile given in Eq. (1) gives an assumed functional form for the vibrational excitation of the surface. We must then connect this vibrational excitation of the surface to motion away from the surface by gas-phase molecular ions. One essential point in making this connection is that there is experimental evidence that the molecular ions which are directly formed in the HIID process come from the surface of the solid [29]. We will then assume that as long as we are trying to describe the direct secondary ion yield, we can assume that only molecules desorbing form the top layer of the system will contribute. Once we have made this assumption we can then use transition state theory to obtain a rate expression for desorption from the surface of the Arrhenius form [30]

$$k(r,t) = A \exp[-\Delta E/k_B T_s(r,t)]. \tag{5}$$

Using transition state theory, A is the frequency characteristic of the adsorbate-surface vibrational mode and ΔE is the adsorbate-surface interaction energy. The probability for finding a molecule on the surface at a given r and t is then governed by the differential equation

$$\frac{dP(r,t)}{dt} = - k(r,t)\ P(r,t) \tag{6}$$

with

$$P(r,t{=}0) = 1. \tag{7}$$

The yield of the molecular ions in the HIID desorption process is then given by

$$Y = \eta n \int_0^\infty 2\pi r[1-P(r,t{=}\infty)]dr, \tag{8}$$

where the efficiency η is the product of the probability that a molecule which is desorbed from the surface does so in an ionized state and the efficiency with which the ions are detected and n is the surface number density.

In addition to the one-site models of the type outlined in Eqs. (5)-(8), we have also considered multi-site models for desorption form surfaces [10]. In a multi-site model, it is assumed that for a molecule to desorb from a surface a number of bonds must be broken in sequence. Between a molecular state with i bonds broken and i+1 bonds broken we have assumed that there is a rate expression for bond cleavage of the form given in Eq. (5) and a rate for bond reformation which is related to the rate for bond cleavage by assuming that there is no barrier to bond cleavage and by using the equilibrium populations of the i bond broken and i+i bond broken molecules [10]. We have found that the multi-site model for desorption from surfaces gives molecular ion versus LET curves equivalent to those which can be obtained from one-site models. When comparing the A factors and ΔEs for a one-site model and a N-site model we have found that the bond energy found in the one-site model is nearly the same as the total bond energy in the N-site model. In a similar fashion, the A factors obtained from a one-site models are N time smaller than the A factors for the individual bond cleavage reactions found in the N-site model. Thus we have found [10] that the is no loss of generality by assuming that the desorption process can be written as a one-site model for desorption with a suitable interpretation of the resulting model parameters.

In addition to the desorption process, the rapid temperature jump also leads to fragmentation of the adsorbates. Thus in general the fragmentation channel is competing with the desorption channel. We believe that during the rapid heating the intramolecular vibrational modes are excited at a rate which is less than the rate of excitation of the intermolecular vibrational modes. This heating effect is incorporated into our one-site model for desorption by defining an internal temperature T_i which is coupled to the surface temperature T_s by a heat transfer equation

$$\frac{dT_i(r,t)}{dt} = k_T \, [T_s(r,t) - T_i(r,t)]. \tag{9}$$

We then assume that $T_i(r,t=0) = T_s(r,t=0)$ and that the desorbing flux at a given time t and position r can only contribute to the observed unfragmented molecular ion peak if $T_i(r,t) < T_{frag}$ where T_{frag} is a temperature which characterizes when the probability of fragmentation before desorption or during the acceleration phase of PDMS is high.

5. Fit of Experimental Data

We have fit the model for HIID outlined in Eqs. (5)-(9) above to the experimental data for molecular ion yields of Hakansson *et al*. [17] which has been reported in tabular form by Hedin *et al*. [13]. The parameters found in these fits for valine, insulin$^+$, and insulin$^-$ were $100\eta n r_o^2$ = 0.16, 0.15, 0.16, $(\pi C_v r_o^2/k_B)\Delta E$ = 3.8, 11., 13. MeV cm^{2}mg^{-1}, respectively. The remaining parameters were assumed to be the same for all of the molecular systems in the fit and they were $(r_o^2/4\delta)A$ = 11., $(r_o^2/4\delta)k_T$ = 0.086, and $(\pi C_v r_o^2)T_{frag}$ = 0.92 MeV cm^2 mg^{-1}. The best fit model is compared to the experimental data in Fig. 1.

For these systems the yield as a function of LET can be seen to have a highly nonlinear dependence on dE/dX at low LET and to be linearly proportional to dE/dX at high LET. This behavior is not observed for all molecular systems. Hakansson *et al*. [17] have found that for ergosterol at high LET the yield scaled as $(dE/dX)^2$.

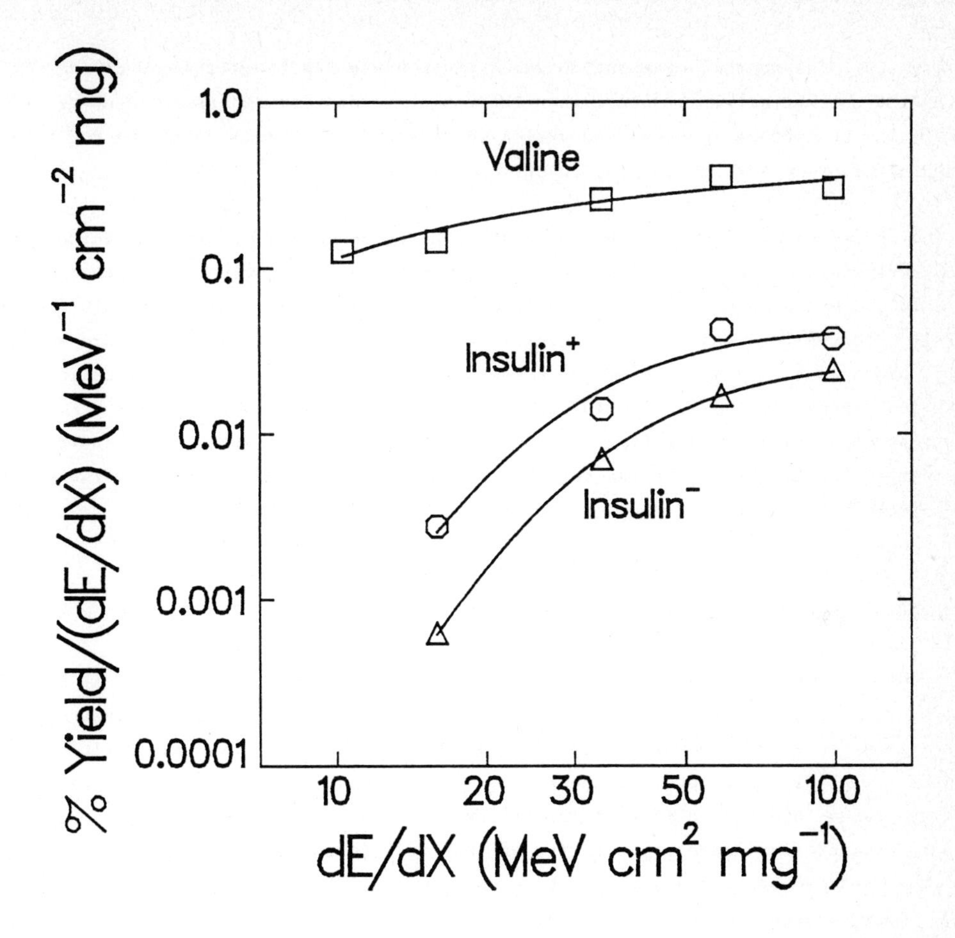

Fig. 1. %Yield/(dE/dX) vs dE/dX for valine, insulin$^+$, and insulin$^-$, data points are from Ref. 13 and the curves are from the thermal spike model described in Eqs. (5)-(9) of the text.

The $(dE/dX)^2$ dependence of yield has also been seen in the HIID yields of systems such as UF_4 [31] and ice [32] and also for the total neutral molecule yields of large molecular systems [19]. To obtain the linear dependence at high LET we needed to assume that the ions were only coming from the surface of the system. If, on the other hand, the ion yield could also come from subsurface layers then the yield would be given by [4-7,9]

$$Y = \eta n \int_0^\infty 2\pi r dr \int_0^\infty dt \; A \exp[-\Delta E/k_B T_s(r,t)]. \tag{10}$$

At high LET, the thermal evaporation yield given by Eq. (10) would scale as $(dE/dX)^2$. Using the thermal model, we would then interpret yields which scale as dE/dX as coming from a surface process and yields which scale as $(dE/dX)^2$ as being due to an evaporative process which include subsurface layers.

The parameters given above can be given definite values by assuming values for the parameters r_o, n, C_v, and δ appropriate for the experimental conditions. Using r_o = 10Å, C_v= 2.0 J $gm^{-1}K^{-1}$, and δ = $5.4\times10^{-4} cm^2 sec^{-1}$, and n = 3.4×10^{-2}, 2.7×10^{-3}, and 2.7×10^{-3} for valine, insulin$^+$, and insulin$^-$, respectively, the model parameters become [10] ΔE = 0.83, 2.4, 2.8 eV and η = 0.05, 0.55, 0.59 % for valine, insulin$^+$, and insulin$^-$, respectively, A = $2.4\times10^{12} sec^{-1}$, k_T= $1.9\times10^{10} sec^{-1}$, and T_{frag}= 2,300 K. Due to the uncertainties in estimating the physical constants of the system and due to the experimental errors, these best fit parameters have uncertainties on the order of a factor of 2-4.

6. Predictions of the Model

The thermal model of the desorption ionization process in HIID discussed in Sec. 5 can be viewed as a semiempirical method for extrapolating the measured values of the ion yield as a function of LET to other values of LET and initial excitation track widths. In Fig. 2 we present the results of the model given above for an extended energy range and with different initial track widths. We have varied the value of r_o in the model up (down) by a factor of two, this in turn changes the value of all of the reduced parameters given above up (down) by a factor of four. For a given LET, this change in r_o corresponds to decreasing (increasing) the initial energy density in the track by a factor of four. For the lower LET the reduction of the radius of energy deposition leads to a greater than an order of magnitude increase in the yield, whereas at high LET the increased yields are much smaller. This shows that thermal models are very sensitive to the initial radial distribution of the excitation energy for low LET and much less sensitive at high LET. This observation is in agreement with the insensitivity noted [13] to the use of differing distribution profiles when only the limit of high LET is considered. The plateaus in the r_o, and 2.0 r_o curves in Fig. 2 at low LET are due to the inclusion in the model of T_{frag}. At the very low LET, the surface never gets above T_{frag}, then as the LET is increased, the surface reaches T_{frag} and the molecules begin to fragment at the point where there is an inflection in the yield curves. As the LET is further increased, the molecules can desorb before T_i reaches T_{frag} so that the yield begin to increase again. At high LET the yields are seen to increase slightly less rapidly than the first power of dE/dX leading to the slightly negative slope in the yield curves in Fig. 2 at high LET.

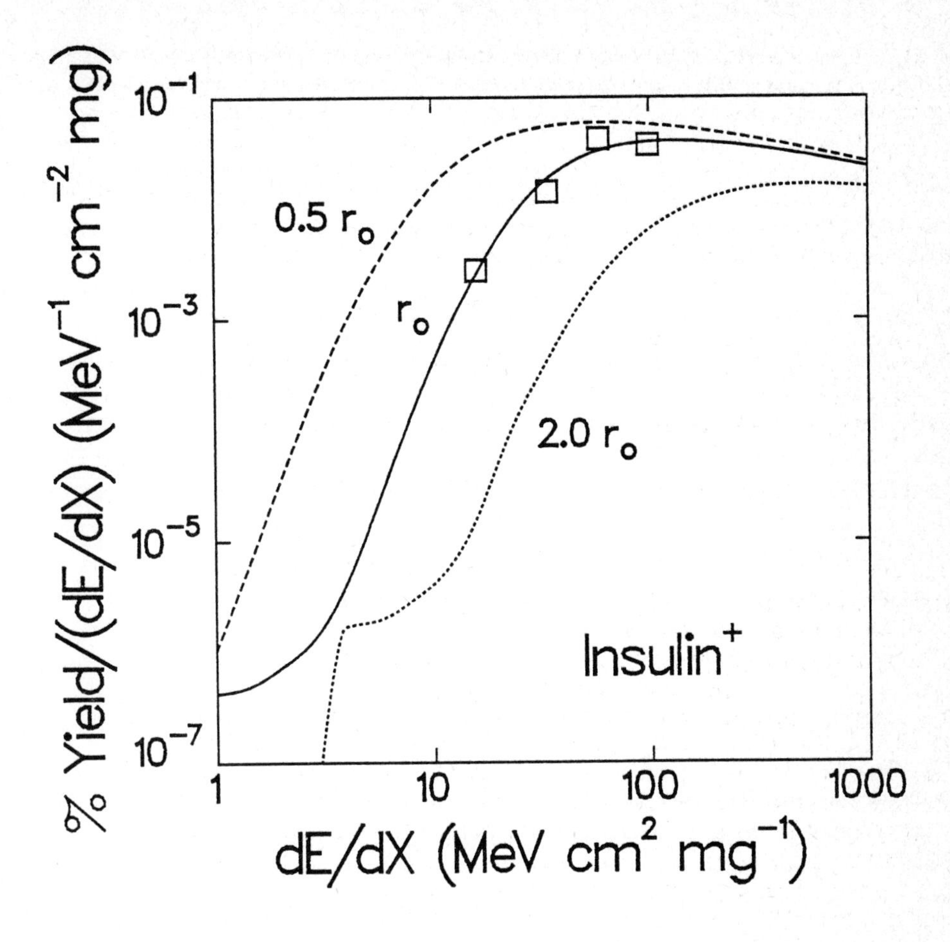

Fig. 2. %Yield/(dE/dX) vs dE/dX for insulin$^+$ over an extended range of dE/dX using $r'_o = 0.5\ r_o$, r_o, and $2.0\ r_o$ predicted by the thermal spike model described in Eqs. (5)-(9) of the text with the data points from Ref. 13.

Another experimentally observable quantity is the kinetic energy and angular distributions of the desorbing molecules and molecular ions. We have not yet made direct simulations of the dynamics of the desorption process for large molecular systems, however simulations of rapid laser desorption of small molecules have shown [28] that the translational degrees of freedom of the molecules can be expected to have a kinetic energy distribution which is characteristic of the temperature of the surface at the time of desorption. Thus we would expect that the initial energy distribution of the particles desorbed in an HIID event would have a energy and angular distributions which are related to the temperature in the thermal spike at the time of desorption.

The initial velocity and angular distributions are not directly observable since they will be modified by collisions which occur in the gas phase between desorbed molecules. Given the initial mass, energy, and angular distributions, it is possible to obtain the final distributions of the molecules which are observable using the Monte Carlo simulation method for gas dynamics [33]. We have implemented this method for the study of post-desorption collisions [18] and have applied the method to the HIID desorption problem. We have performed this simulation using a one dimensional representation of the distributions of the system. Thus we followed all three components of the velocities of the particles but we assumed that the distributions of the particles were uniform in the two directions parallel to the surface and only depended on the distance in the direction normal to the surface. We have not simulated the internal degrees of freedom and only considered elastic collisions between molecules. We have assumed that the initial flux of desorbing molecules was that given by the model discussed in Sec. 4 for insulin$^+$ at $r = 2.5\ r_o$. The initial velocity distributions were taken to be thermal distributions with the temperature of the surface at the time of desorption given in Eq. (1). We have performed the simulations in three stages using the parameters given in Table I. For the purposes of performing collisions between the particles, we divided the simulated region into boxes in the z direction with dimensions as indicated in Table I. We used 10,000 boxes which was enough boxes so that no particles left the system during the simulation. We have made a variety of assumptions about the distributions of masses leaving the surface and the relative abundance of the masses as indicated in Table II. The coverages used for particle 1 were equivalent to one monolayer coverage with the surface number density given by

$$n = \left(\frac{\rho}{m} \right)^{2/3} \tag{11}$$

where $\rho = 0.785$ amu/Å^3. The coverages used for particle 2 corresponded to multilayers each of which had a number density given by Eq. (11) such that the mass per unit area of particle 2 divided by the mass per unit area of particle 1 was the mass ratio given in Table II. The collisional cross sections, σ_{AB}, between two particles, A and B, were taken to be

$$\sigma_{AB} = \left[\frac{1}{2} \left(n_A^{-1/2} + n_B^{-1/2} \right) \right]^2 \tag{12}$$

where n_A and n_B are the monolayer number densities for the two particles. The results of these simulations are given in Table II.

Table I. Parameters for multistage Monte Carlo simulation of postdesorption collision dynamics.

Time Range (sec)	Time Step Size (sec)	Box Size (Å)
$0 - 3\times10^{-11}$	2×10^{-13}	10
$3\times10^{-11} - 5\times10^{-10}$	1×10^{-12}	50
$5\times10^{-10} - 1\times10^{-8}$	1×10^{-11}	100
$1\times10^{-8} - 1\times10^{-7}$	1×10^{-10}	1000

Table II. Results of Monte Carlo simulations of postdesorption collisions in HIID desorption from surfaces.

Mass Particle 1 (amu)	Mass Particle 2 (amu)	Mass Ratio	Mean Kinetic Energy Particle 1 (eV)	$\langle\cos\theta\rangle$ Particle 1
6000	500	1.0	2.9	0.916
6000	500	3.0	5.2	0.968
6000	500	6.0	7.5	0.981
3000	500	3.0	3.6	0.952
6000	500	3.0	5.2	0.968
12000	500	3.0	8.1	0.986
6000	500	3.0	5.2	0.968
6000	250	3.0	8.2	0.981
Initial Distributions Particles 1 and 2			1.8	0.667

From the first set of numbers given in Table II we can see that with increasing mass ratio the acceleration effect on particle 1 is enhanced. The second set of data given in Table II show that given the same mass ratio of light to heavy particles and using the same light mass, the heavier the mass of particle 1 the higher the final kinetic energy of particle 1. The final set of data in Table II show that the final kinetic energy of particle 1 is very sensitive to the mass of the light particle. All of these simulations show a significant focusing effect on the angular distributions of particle 1 towards the surface normal. We can see from these results that there is probably a significant acceleration effect due to postdesorption collisions, which could produce final kinetic energies of heavy molecular ions

3-4 times their initial values. It is also clear that to give a quantitative prediction of the final kinetic energies of desorbed molecular ions a precise knowledge of the mass distributions of all particles desorbing from the system would be needed.

7. Conclusions

We have seen that the surface thermal spike model of HIID/PDMS can reproduce the linear dependence of ion yield versus LET found by Hakansson et al. [17] at high LET. We have also shown that postdesorption collisions can significantly effect the angular and kinetic energy distributions of the desorbing molecular ions. We might also speculate that the postdesorption collisions could be an important ionization mechanism and could lead to cooling of internal modes of the molecular ions, thus leading to enhanced molecular ions yields compared to those which would be obtained without postdesorption collisions.

At present we are extending these studies by developing a dynamical model for the desorption process so that we can study the competition between energy transfer and desorption in larger molecular systems. In addition we are refining the Monte Carlo simulations by performing two dimensional simulations which consider the dependence of the velocity and density distributions in both the direction normal to the surface and in radial direction normal to the primary ion track.

Acknowledgments

Acknowledgment is made to the Donors of the Petroleum Research Fund, administered by the American Chemical Society, to the Dow Chemical Company Foundation, to the Monsanto Company, and to the Celanese Chemical Company for partial support of this research. In addition, this material is based upon work in part supported by the National Science foundation under Grant CHE-8351414. Further support for this work was provided by the Office for International Coordination of Texas A&M University. I.N. would like to thank the Government of Tamilnadu (India) for granting him leave to pursue this research.

References

[1] R.D. Macfarlane and D.F. Torgerson, Science 191, 920 (1976).
[2] P. Duck, W. Treu, H. Frohlich, W. Galster, and H. Voit, Surf. Sci. 95, 603 (1980).
[3] R.D. Macfarlane, Physica Scripta T6, 110 (1983).
[4] G.H. Vineyard, Radiat. Eff. 29, 245 (1976).

[5] R.W. Ollerhead, J. Bottiger, J.A. Davies, J. L'Ecuyer, H.K. Haugen, and N. Matsunami, Radiat. Eff. 49, 203 (1980).

[6] R.E. Johnson and R. Evatt, Radiat. Eff. 52, 187 (1980).

[7] L.E. Seiberling, J.E. Griffith, and T.A. Tombrello, Radiat. Eff. 52, 201 (1980).

[8] R.H. Ritchie and C. Claussen, Nucl. Instrum. Methods 198, 133 (1982).

[9] M. Urbassek and P. Sigmund, Appl. Phys. A 35, 19 (1984).

[10] R.R. Lucchese, J. Chem. Phys., submitted for publication.

[11] P. Williams and B. Sundqvist, Phys. Rev. Lett., submitted for publication.

[12] B.V. King, A.R. Ziv, S.H. Lin, and I.S.T. Tsong, J. Chem. Phys. 82, 3641 (1985).

[13] A. Hedin, P. Hakansson, B. Sundqvist, and R.E. Johnson, Phys. Rev. B 31, 1780 (1985).

[14] F.R. Krueger, Surf. Sci. 86, 246 (1979).

[15] H.F. Kammer and E.R. Hilf, Solid State Comm. 58, 465 (1986).

[16] J.F. Ziegler, Stopping Cross-Sections for Energetic Ions in All Elements (Pergamon, New York, 1980) p. 19.

[17] P. Hakansson, I. Kamensky, M. Salehpour, B. Sundqvist, and S. Widdiyasekera, Radiat. Eff. 80, 141 (1984).

[18] I. NoorBatcha, R.R. Lucchese, and Y. Zeiri, J. Chem. Phys., submitted for publication.

[19] P. Hakansson, in Proceedings of the First International Workshop on Physics of Small Systems (Springer, Berlin, 1986).

[20] B.T. Chait, Int. J. Mass Spectrom. Ion Phys. 53, 227 (1983).

[21] R.D. Macfarlane, Acc. Chem. Res. 15, 268 (1982).

[22] A. Mozumder, Adv. Radiat. Chem. 1, 1 (1969).

[23] R.J. Beuhler, E. Flanigan, L.J. Greene, L. Friedman, J. Am. Chem. Soc. 96, 3990 (1974).

[24] J.P. Cowin, D.J. Auerbach, C. Becker, and L. Wharton, Surf. Sci. 78, 545 (1978).

[25] C.V. Shank, R. Yen, and C. Hirlimann, Phys. Rev. Lett. 50, 454 (1983).

[26] R.B. Hall and S.J. Bares, in Chemistry and Structure at Interfaces, edited by R.B. Hall and A.B. Ellis (VCH Publishers, Deerfield Beach, 1986) p. 85.

[27] W.E. Bron, in Nonequilibrium Phonon Dynamics, edited by W.E. Bron (Plenum Press, New York, 1985) p.1.

[28] R.R. Lucchese and J.C. Tully, J. Chem. Phys. 81, 6313 (1984).

[29] R.D. Macfarlane, C.J. McNeal, and C.R. Martin, Anal. Chem. 58, 1091 (1986).

[30] A. Redondo, Y. Zeiri, and W.A. Goddard III, Phys. Rev. Lett. 49, 1847 (1982).

[31] J.E. Griffith, R.A. Weller, L.E. Seiberling, and T.A. Tombrello, Radiat. Eff. 51, 223 (1980).

[32] W.L. Brown, W.M. Augustyniak, E. Brody, B. Cooper, L.J. Lanzerotti, A. Ramirez, R. Evatt, and R.E. Johnson, Nucl. Instrum. Methods 170, 321 (1980).

[33] G.A. Bird, Molecular Gas Dynamics (Clarendon Press, Oxford, 1976).

CHARGE OF FAST HEAVY IONS IN SOLID: APPLICATION TO DESORPTION PROCESS

G. MAYNARD, C. DEUTSCH

Laboratoire de Physique des Gaz et des Plasmas
Bat 212, 91405. Orsay
France

ABSTRACT

A model describing the charge of fast heavy ions in a solid is presented here. Dynamical and density effects are taken into account. Relaxation near the surface to an equilibrium charge is calculated and the application to the desorption process is pointed out.

INTRODUCTION

The problem of energy losses suffered by energetic ions moving in condensed matter has been investigated for more than fifty years mainly for nuclear physics applications. Since the mid-seventies physicists have looked into energy production by nuclear fusion induced by irradiation of intense ion beams [1] and calculations of stopping power of fast ions in plasma have been developed. We present in this paper a model related to this last problem with a plasma which may be a gas, a liquid or a solid, ionized or not. To cover the full range of charge and velocity of the incident ion, and of the density and temperature for the plasma, is a formidable task and one has to make some reliable simplifications. In this paper we present the main approximations of our model, details of calculations are reported elsewhere [2]. We show then results for ions in a solid which are compared with some experimental results. Finally inferences of charge variation in desorption process from a solid surface are emphasized.

DESCRIPTION OF THE MODEL

Description of the plasma (solid):

We use Relativistic Hartree-Fock-Slater Calculations as described by B.F. Rozsnyai [3] for a mean atom of the plasma. If R_o is half the mean distance between two atoms, each atom is supposed to be enclosed in a neutral spherical cell of radius R_o. The potential inside the cell is given by the Poisson equation and a local approximation for exchange and correlation. The Dirac equation for this potential gives the energy levels and the orbital wave functions. Thomas Fermi approximation is used to calculate the free electron-density, and Fermi statistics on the energy levels to find the bound electron-density. The chemical potential acts as a normalisation factor to assure the neutrality of the cell. All the calculation is self-consistent. The crucial fact is that all the characteristics of atoms of the plasma that have to be used in stopping power or in charge-exchange calculations are evaluated in situ and not for an isolated ion.

Description of incident ions:

As it can be seen from beam foil spectroscopy experiments, fast heavy ions in a solid are in highly excited ionized states. Exitation can change the ionization cross section by more than a factor of two and cannot be neglected. To describe the incident ions we have chosen the screened hydrogenic ionization model as used by R.M. More [4]. Each ion has ten shells defined by principal quantum number n and occupied by P_n electrons.

The energy eigenvalue En for electrons of the nth shell is taken to be (atomic unities are used)

$$P_n = E_{n0} - \frac{Q_n^2}{2n^2}$$

with Q_n and E_{n0} being

$$Q_n = Z_0 - \sum_{m<n} S(n,m) \cdot P_m - \frac{1}{2} S(n,n) \cdot P_n \quad ,$$

$$E_{n0} = \frac{S(n,n) \cdot P_n}{2r_n} + \sum_{m>n} P_m \frac{S(m,n)}{r_m} \quad ,$$

r_n is the orbit radius $= n^2/Q_n$ and Z_0 is the atomic number. $S(n,n)$ are taken from [4].

Energie loss calculation:

Collisions of plasma electrons or ions with the projectile change either its energy or its charge. In a first approximation we can separate the two effects and calculate the electronic stopping power of a point- like particle using the formula [5]:

$$\frac{dE}{dx} = K \frac{Z^2}{v^2} \{L_0 + L_1 \cdot \frac{Z}{v} + f(\frac{Z^2}{v^2}) \quad . \tag{1}$$

Z is the charge of the projectile and v its velocity. The right hand side of (1) has to be calculated for free and bound electrons.

L_0 is the Bohr-Bethe term which corresponds to the Born I approximation. For bound electrons it is calculated as in [6], and for free ones as in [5].

L_1 is the Barkes [7] term which is the Born II approximation for distant collisions,but only on bound electrons; it is calculated as in [8].

Last term on the right hand side of (1) is the Bloch [9] term which gives the good classical limit for close collisions with high charge. Velocity distribution of the plasma electrons is taken into account in this term [2]. The charge Z in (1) is a parameter which has to be calculated separately and at the same time as E because, as it has already been noticed [10], E can change more or less rapidly than Z depending on the ionization of the plasma and on the value of Z.

Determination of Z:

We present here the different collision processes that change the charge. Details on the calculation of cross sections can be found in [2].

- Electronic excitation, deexcitation and ionisation by free electrons are calculated by quantum theory
- Ionic excitation, deexcitation and ionisation by ions using classical binary encounter approximation
- Dielectronic recombination and autoionisation: The dielectronic recombination is calculated as a threshold process of excitation
- Radiative decay using hydrogenic oscillator strength
- Radiative recombination from free electrons
- Electrons exchange from a bound level of plasma ions to a bound level of the projectile; it is calculated using a classical model.

We can sum up all these effects and write:

* $P_n L(n,n)$ for the rate of ionisation from level n
* $P_n(2m^2 - P_m)L(n,m)$ for the rate of excitation from level n to level m

* $(2n^2 - P_n)G(n,n)$ for the rate of recombination on level n
* $(2n^2 - P_n)P_mG(m,n)$ for the rate of deexcitation from level m to level n.

To reduce the number of variables that we have to determine at each time of the slowing down we use the mean atom model as in [11]; this mean ion has a charge given by

$$Z = Z_0 - \sum_n P_n \tag{2}$$

which changes with time due to

$$\frac{dz}{dt} = -\sum_n \frac{dP_n}{dt} \tag{3}$$

with

$$\frac{dP_n}{dt} = -P_n\{L(n,n) + \sum_{m \neq n}(2m^2 - P_m)L(n,m)\} + (2n^2 - P_n)\cdot\{G(n,n) + \sum_{m \neq n} P_mG(m,n)\} \quad . \tag{4}$$

Equations (1), (3) and (4) can be used to describe the evolution of charge and energy of an incident ion, initially - for instance, at a solid surface - with energy E and charge Z all along its path.

RESULTS

In Figures 1a, 1b, and 1c we present results of our calculation for incident ions Ne, Ar and Kr of $1.16MeV/a.m.u.$ for different initial charges impinging a solid of mass density $1.371g.cm^{-3}$, atomic number 6 and atomic mass 8.333 - it has the same mass density as coronene $C_{24}H_{12}$ but the same ion density as carbon -.

Some comparisons with experimental results can be made: The theoretical equilibrium charge agrees very well with experiments of [12]. The biggest error on stopping power is in the Kr case. Table data give $47MeV.cm^{-1}$ in Carbone, and our result is 56.9 which is not too bad due to the large Z/v of Kr at $1.16MeV$.

We can parametrize our results using a formula suggested by Bohr [13]

$$Z = Z_{\text{eq}} + (Z_0 - Z_{\text{eq}}) \cdot \exp(-x/d) \quad . \tag{5}$$

Z_0 is the initial charge, Z_{eq} the equilibrium charge, x the distance from the surface in Ångstrom and d is the distance of relaxation to equilibrium.

$Z_{\text{eq}} = 7.62$ for Ne, 11.75 for Ar and 17.9 for Kr; values of d are given in Table 1. In Fig. 2 we can compare the parametrization result with the theoretical one for a Kr projectile. The importance of excitation can be seen in this case which makes sensitive differences with parametrization near the surface.

APPLICATIONS TO DESORPTION PROCESS

In experiments described in [12] variation of relative yield for desorption of different ions with the charge of incident projectile at given velocity has been measured. In many cases - see figures in [12] - the relative yield depends rapidly with the incident charge and it varies also from one ion to another. For instance, the yield of Kr^{10+} is greater than the yield of Ar^{10+} and yet these two ions have nearly the same stopping power at the surface. One can explain this fact taking into account a certain depth D_s inside the solid which interacts with the surface for the desorption process.

In the depth D_s is small compared with d defined in (5), desorption is seen as a surface process and the yield depends only on the incident charge for a given velocity. In the opposite

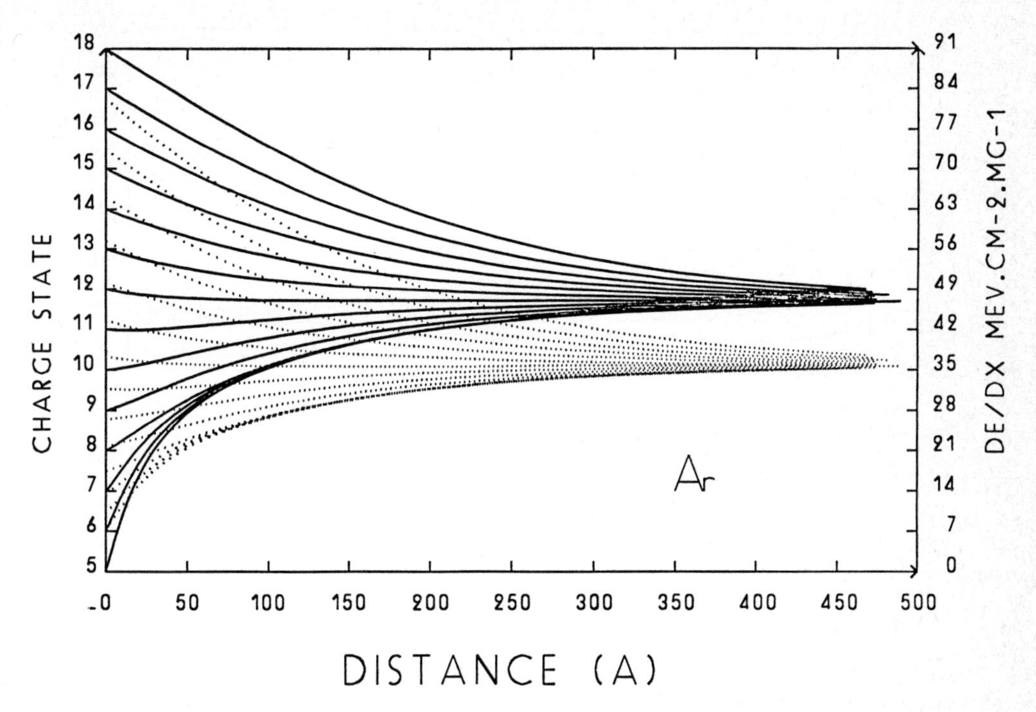
CHARGE STATE
DE/DX MEV.CM-2.MG-1
Ar
18 17 16 15 14 13 12 11 10 9 8 7 6 5
91 84 77 70 63 56 49 42 35 28 21 14 7 0
0 50 100 150 200 250 300 350 400 450 500
DISTANCE (A)

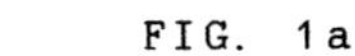
FIG. 1a

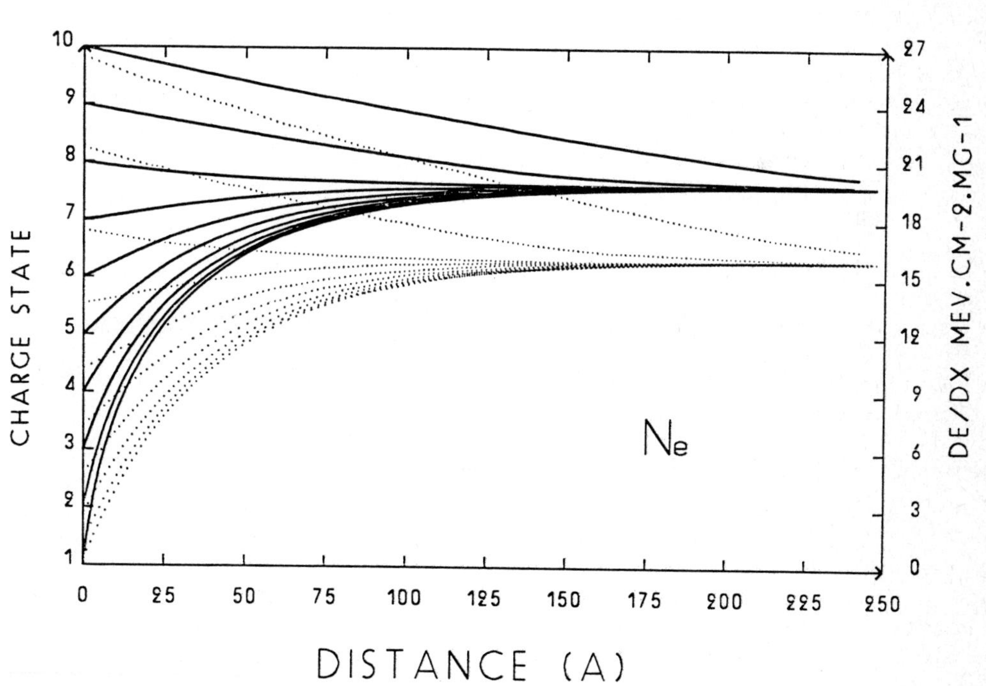
CHARGE STATE
DE/DX MEV.CM-2.MG-1
Ne
10 9 8 7 6 5 4 3 2 1
27 24 21 18 15 12 9 6 3 0
0 25 50 75 100 125 150 175 200 225 250
DISTANCE (A)

FIG. 1b

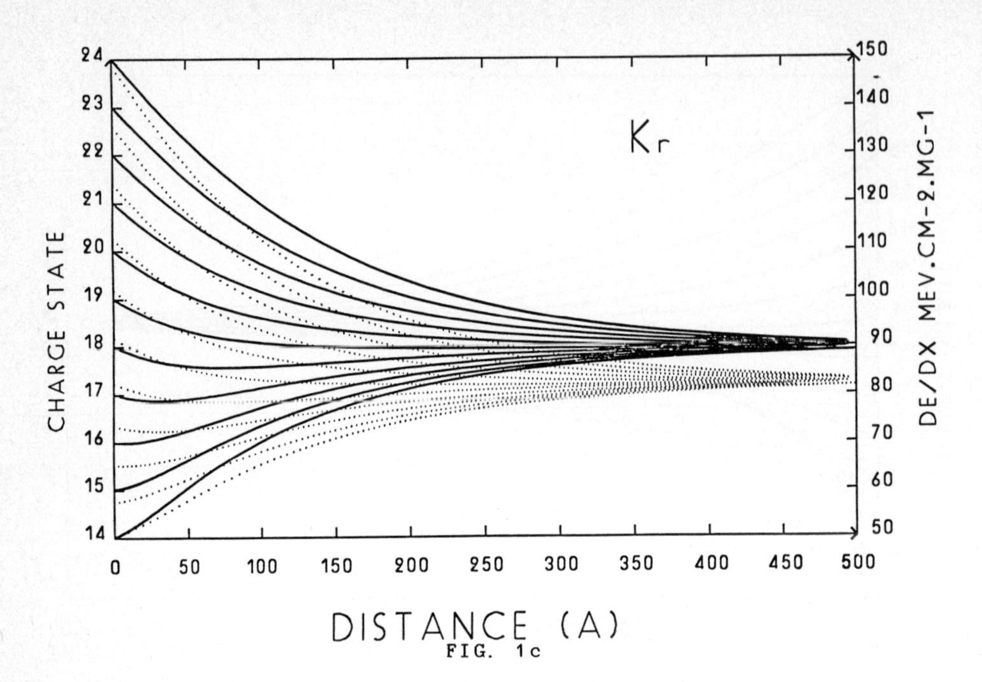

FIG. 1c

FIG. 1a, 1b, 1c : Variation of charge and stopping power of Ne, Ar and Kr with 1.16 Mev/a.m.u on Coronene. Staight line is the charge with the left scale and Dotted line is stopping power with the right scale

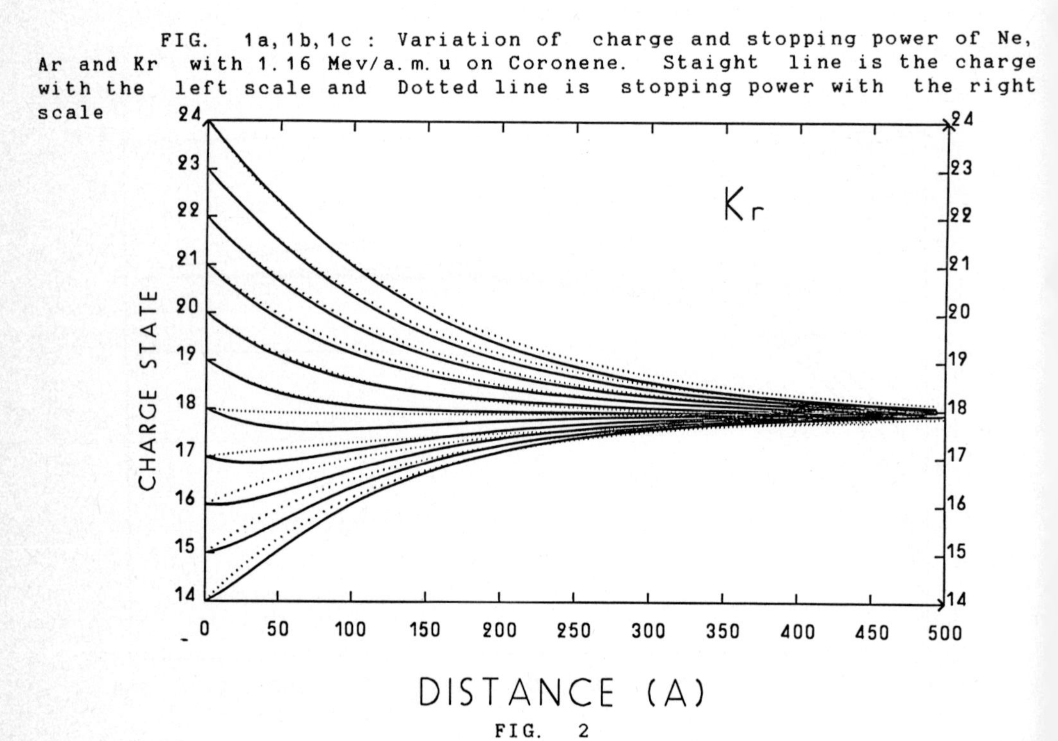

FIG. 2

FIG. 2 : Same as FIG. 1 for the charge. Dotted line is the parametrisation (5)

TABLE 1

Relaxation distance d, define in (5), for Ne, Ar and Kr at different initial charges.

Ne		Ar		Kr	
ZO	d	ZO	d	ZO	d
1	25.0	5	64.5	14	125.0
2	28.6	6	70.9	15	133.3
3	33.3	7	87.0	16	142.9
4	36.6	8	119.0	17	250.0
5	40.0	9	135.5	18	100.0
6	44.4	10	149.3	19	57.1
7	53.5	11	333.3	20	94.3
8	37.0	12	100.0	21	124.2
9	107.5	13	106.4	22	133.3
10	161.3	14	135.5	23	142.9
		15	149.3	24	149.3
		16	158.7		
		17	168.1		
		18	181.8		

limit of D_s much larger than d the relative yield depends only on the equilibrium charge. More often, D_s is comparable with d, and as d varies from Kr to Ne, D_s varies also and this variation has an influence on the relative yield so that any theory which wants to describe the desorption process has to take into account a depth of interaction in the solid and the variation of this depth with the incident ion. Moreover, if the incident charge is not in the equilibrium charge state one has to consider the variation of this charge all along the path of the projectile.

CONCLUSION

We have presented a model that can describe the variation of charge of fast heavy ions near a surface. These calculations can give information on the depth inside the solid which interacts with the surface in the desorption process. Our results show that this depth varies with the incident ion and then, even when the projectile is in the equilibrium charge state one has to consider the variation of this depth with the stopping power and to include it in any desorption theory. Desorption cannot been seen only as a surface effect.

ACKNOWLEDGEMENTS

We thank S. Della-Negra and Y. LeBeyec who gave us experimental data prior to publication and have initiated us in PDMS physics, we thank also K. Wien for stimulating discussions.

REFERENCES

1. Deutsch, C.: Ann. Phys. Fr. 11 (1986) 1
2. Maynard, G., Deutsch, C.: to be published
3. Rozsnyai, B.F.: Phys. Rev. A5 (1972) 1137
4. More, R.M.: J. Quant. Spectros. Radiat. Transfer 27 (1982) 345
5. Maynard, G., Deutsch, C.: J. Physics 46 (1985) 1113
6. Garbet, X., Deutsch, C.: Euro. Letters (1986)
7. Barkas, W.H., Dyer, J.W., Heckmann, H.H.: Phys. Rev. Lett. 11 (1963) 26
8. Maynard, G., Deutsch, C.: J. Physique Lett. 43 (1982) 223
9. Bloch, F.: Ann. Phys. 16 (1933) 285
10. Nardi, E., Zinamon, Z.: J. Phys. (Paris) 44 (1983) C8-93
11. Bailey, D., Lee, Y.T., More, R.M.: J. Phys. (Paris) 44 (1983) C8
12. Della-Negra, S., Le Beyec, Y., Monart, B., Standing, K., Wien, K.: this volume
13. Bohr, N.: Phys. Rev. 59 (1941) 270

DESORPTION MECHANISM OF AMINOACID COMPOUNDS ANALYSED BY PDMS

Y. Hoppilliard
D.C.M.T., Ecole Polytechnique
91128 Palaiseau (France)

Y. Lebeyec and C. Deprun
Institut de Physique Nucl.
B.P.n°1, 91406 Orsay (France)

A. Delfour
Groupe de neurobiologie, U.A. 554
96, bd Raspail, 75006 Paris (France)

INTRODUCTION

Recently, it has been shown that organic matrix and in particular the reduced Glutathione, mixed in equimolecular parts with peptides of high weight (>5000 daltons), enhances the intensity of the pseudomolecular M^+ and M^{2+} ions, analysed by 252 Cf PDMS (1).

It seems interesting, in function of the compound or of the compound family to be desorbed, to find the best matrix. A solution could be found understanding why, how and in what conditions a matrix is helping the desorption-ionisation of a sample. In others words, what kind of interaction is favourable to the formation of molecular ions in 252 Cf PDMS.

STUDIED COMPOUNDS

Most of the time, the organic molecular ions observed in PDMS correspond to the protonation (or to the deprotonation) of neutral molecules. The reaction leading to the formation of these ions is a bimolecular reaction needing a proton donor molecule and a proton acceptor molecule.

The best compounds to study the formation of the $(MH)^+$ and $(M-H)^-$ ions are the aminoacids. Each compound having a proton donor site, the carboxylic acid function and a proton acceptor site, the amine function.

Six compounds have been choosen to PLAY ALTERNATIVELY the part of the matrix and that of the sample : four aminoacids, one dipeptide and the reduced glutathione. The four aminoacids are : the valine VAL, the glutamic acid GLU, the cysteine CYS and the arginine ARG.

Each aminoacid has been choosen in function of the reactivity of its lateral chain : the valine has no reactive lateral chain ; the glutamic acid has a second carboxylic function ; the cysteine has a terminal thiol function able to give an another proton ; the arginine is substituted with a very basic function : the guanidine group. The glutathione has been choosen as reference matrix. The dipeptide is the argynylglycine, selected for its great basicity.
Two organic acid has been used as matrix only : the crotonic acid and the 3 chloropropanoic acid.

EXPERIMENTAL CONDITIONS

The samples have been analysed by 252 Cf PDMS. Each compound has been first analysed alone and then in EQUIMOLECULAR BINARY MIXTURES.

Mass spectrometry

Mass spectrometry measurements were made with the 252 Cf plasma desorption T.O.F. mass spectrometer of the "I.P.N." located in Orsay. The system was operating at an acceleration voltage of 15 kV. The ions were detected after a flight of 40 cm. The lenght of the measurement was fixed at 30'.

Sample preparation

Multilayer deposits were prepared by the electrospray method. Isopropanol/water 1/1 solutions containing 2.10-7 mole of solute were sprayed, on the aluminium side of a mylar round spot. This preparation gives amino acid samples in the solid phase.

PD MASS SPECTRA OF THE DIFFERENT COMPOUNDS ANALYSED ALONE

MH+ : an intense formation of the $(MH)^+$ is observed for the different aminoacid compounds. On the contrary, the organic acids according with their weak basicity give few $(MH)^+$ ions.
Fragment ions : the classical elimination of HCOON from $(MH)^+$ is observed followed by secondary fragmentations.

ENERGY ASSOCIATED WITH THE FORMATION OF MOLECULAR IONS

What is the energy necessary to transform the neutral aminoacid molecules in the solid phase into protonated gaseous ions ?
It is well known that in solution or in the solid state the aminoacid compounds have a zwiterionic form : $[\,^+NH_3 - CH(R) - COO^-]$ sol. - On the contrary in the gas phase, it has been demonstrated that the aminoacids keep a classical form NH_2 - CH(R) - COOH (2).
- The stabilisation energy ΔH_1 associated with the solidification of the amino-acid :

$$\text{M gaseous} \xrightarrow{\Delta H_1} \text{M solid}$$

is of the same order of magnitude as the solvation energy of this aminoacid in the water. For example this energy has been calculated to be -10 $kcal.mol^{-1}$ for the valine.

- Starting again from neutral valine molecules, in the gas phase, to form one mole of protonated valine, two moles of gaseous valine are necessary :

$$HOOC\ VAL\ NH_2\ g\ +\ HOOC\ VAL\ NH_2\ g\ \overset{H_2}{}\ HOOC\ VAL\ NH_3^{\ +}\ +\ NH_2\ VAL\ COO^{\ -}$$

It is easy to demonstrate that the energy associated to this reaction is equal to the difference between the acidity H_a of the valine in the gas phase minus the proton affinity of the valine in the gas phase also.

$$H_2 = H_{a(VAL)} - P.A._{(VAL)}$$

The proton affinities of the aminoacids have been measured by J.F. Gal using I.C.R. experiments (3). The P.A of the valine is 218 $kcal.\ mol^{-1}$.

Concerning the acidities of aminoacids in the gas phase, any value has been determined. Taking as Ha(VAL) an average value between the extreme values of the known aciditie (4) one may evaluate H_2 to be 123 $kcal.mol^{-1} \pm 7\ kcal.mol^{-1}$.

- Under an accelerating voltage of + 15 kV only the protonated valines are detected in the gas phase. The deprotonated valines stay in the solid phase. The energy H_3 associated with the solidification of the anion can be estimated to be around - 30 $kcal.mol^{-1}$(5).

Eventually the energy H necessary to desorb and ionise one mole of valine is :

$$H = -2\ H_1 + H_2 + H_3 = 113\ kcal.mol^{-1} \pm 7\ kcal.mol^{-1}$$
$$= 4.9\ eV \pm 0.3\ eV$$

(Scheme 1)

Scheme 1

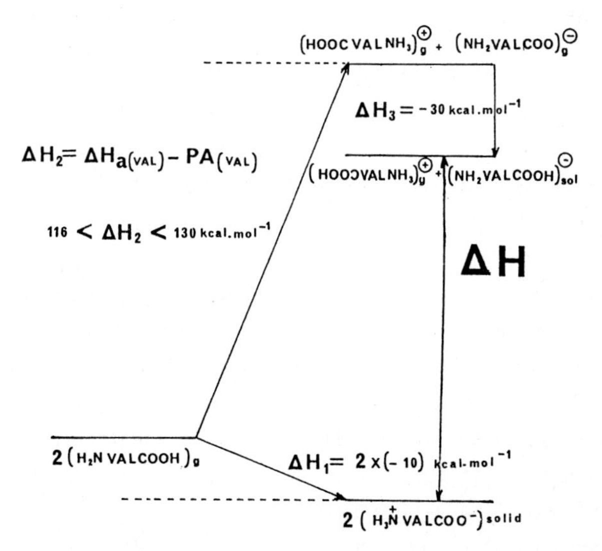

The desorption-ionisation of a solid compound will increase in proportion with the diminution of the ΔH value.

The ΔH_3 value being nearly constant from one compound to another, the difference in the H energy will be the result of either ΔH_1 and ΔH_2 variation.

So the difficulties to desorb and ionise both organic acids and peptides can be explained as follow :

For the organic acids the stabilisation energy associated with the solidification depends on the formation of the two hydrogen bonds linking one molecule of acid to two other ones. Comparing with scheme 1, this ΔH_1 value is low, but the proton affinity associated with the carbonyl of the carboxylic group is much lower than that of the valine, consequently ΔH_2 is high. Eventually the ΔH necessary to desorb an organic is about 40 $kcal.mol^{-1}$ higher than that necessary to desorb the valine.

The case of the polypeptide is different. Each lateral chain of each aminoacid residue forming the polypeptide is susceptible to give an intermolecular interaction with a lateral chain of another polypeptide molecule. Each polypeptide allowing several intermolecular bonding, the ΔH_1 is significantly increased. On the contrary the ΔH_2 value is of the same order of magnitude than that an aminoacide. In short the energy necessary to desorb and ionise a polypeptide is some 30 $kcal.mol^{-1}$ higher than that necessary to desorb the valine.

HOW TO CHOOSE A MATRIX HELPING THE FORMATION OF THE PROTONATED VALINE (VAL)

If H is the energy associated to the reaction :

$$VAL_{sol} + VAL_{sol} \xrightarrow{\Delta H} [VALH]^+ g + [VAL - H]^-_{sol} .$$

In order to favour the formation of $VAL\ H^+_g$ we have to find a matrix MAT such as the energy H' associated with the reaction.

$$MAT_{sol} + VAL_{sol} \xrightarrow{\Delta H'} [VAL\ H]^+ g + [MAT - H]^-_{sol}$$

should be $\Delta H' < \Delta H$. i.e. $\Delta H'_1 > \Delta H_1$ or $\Delta H'_2 < \Delta H_2$ or "a fortiori" both together.

1) Influence of an aminoacid (choosen as matrix) on the D.I. (desorption-ionisation) of the valine (choosen as sample).

Each aminoacid being on the zwiterionic form the ΔH_1 value associated to the solidification of the pure valine is of the same order of magnitude as that associated with the solidification of a mixture valine - matrix - the only one way to decrease the $\Delta H'_2$ value is to choose a matrix having an acidity and a proton affinity lower than those of the valine. Because of the electronic repartition, the acidity and the proton affinity values of an aminoacid increase or decrease together. Consequently to find a good matrix for the valine, it is sufficient to find an aminoacid having a P.A lower than that of the valine. The cysteine (P.A =

215 kcal.mol^{-1}) has been choosen as matrix. Comparing the spectra 1 and 2, we observe that for the same quantity of valine initially sprayed, and in the same experimental conditions, the intensity of the analyzed protonated valine ions is multiplied by two in presence of cysteine : the cysteine is a good matrix for the valine.

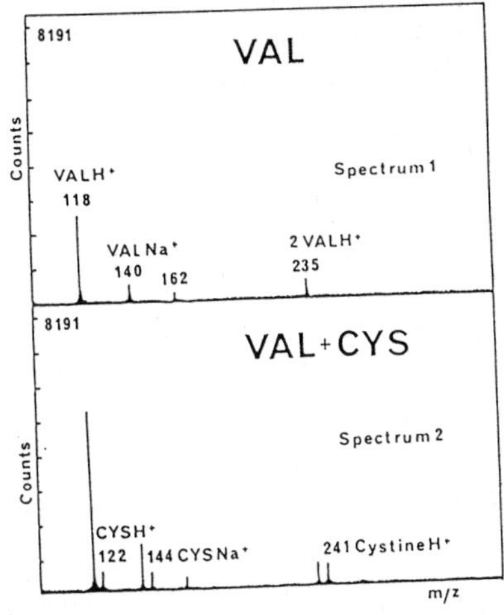

This analysis has been extended to each possible binary mixture between our six reference compounds. The results are :

- the best matrix is the cysteine, having the lower P.A, followed by the glutanic acid.
- two compounds have an ambivalent behaviour : the reduced glutathione and the valine.For example the reduced glutathione mixed with the argine helps the D.I. of the protonated argine (spectra 3 and 4). On the contrary the reduced glutathione mixed with the cysteine prevents the formation of protonated cysteine (spectra 5 and 6).

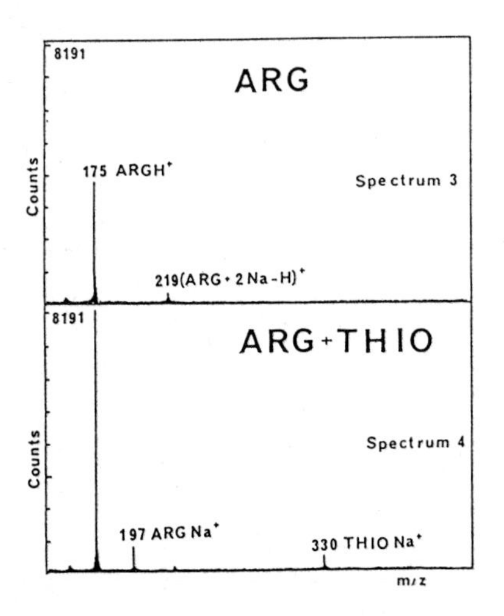

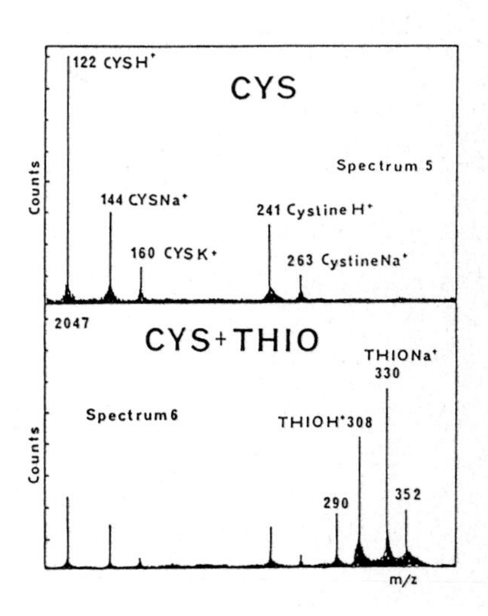

- the argine and the argynylglynine are very bad matrix but good samples.

2) Influence of an organic acid on the D.I of the valine.

In presence of crotonic acid the intensity of the protonated valine ions is two times the intensity of the $VAL\text{-}H^+$ ions when the valine is sprayed alone. In presence of 3 chloropropanoic acid the reference intentity is multiplied by four. The question is : why the organic acids are better matrix than the aminoacids ? Two reasons can be suggested :

- their intrinsic acidity
- their interactions with aminoacids. They are certainly lower than those binding together aminoacids in zwiterionic form. The organic acid breaks the structuration of the solid aminoacid.

CONCLUSION

1) What is important to help the formation of any MH^+ ion in PDMS ?

- on the sample molecule the intrinsic basicity of the function fixing the proton
- on the matrix molecule the acidity of the function giving the proton.

2) What is important to thelp the formation of a protonated peptide ?

To break the intermolecular bonds between the different molecules of peptides using a matrix giving small intermolecular interactions sample-matrix and separating the peptide molelcules ones from the others.

BIBLIOGRAPHY

(1) M. ALAI, P. DEMIREV, C. FENSELAU and P.J. COTTER, Analytical Chemistry 58, 1302 (1986)
(2) M.J. LOCKE and R.Mc IVER, J. Amer. Chem. Soc. 105, 4226 (1983)
(3) J.F. GAL, unpublished results
(4) J.E. BARTMESS and R.T. Mc IVER, Gas phase ion chemistry, vol. 2 Academic Press, 1979, p. 87-121
(5) P. HABERFIELD and A. KRAKSHIT J. Am. Chem. Soc. 98, 4393 (1976).

Continuum Mechanical Model for Heavy Ion Induced Desorption

HANS FRIEDRICH KAMMER*
Fachbereich Physik, Universität Oldenburg
Postfach 2503, D-2900 Oldenburg, West Germany

Abstract

The concept of a "hot core track region" (limited region in the track of the projectile ion, where the electron temperature is large compared to the lattice temperature before equilibration), as used in other works, is investigated. The results (lifetime, dimension and temperature of the HCTR) show that in insulator targets the formation of this HCTR can play an important role for the excitation of molecular motion. One possible mechanism for this excitation is investigated and discussed. In the macroscopic limit molecular neutral secondary yields are calculated by an estimate of the energy available for cluster formation.

Introduction

In spite of the large number of theoretical models for HIID in the electronic stopping regime, there is up to now no satisactory explanation for the basic mechanism. The aim of this work is therefore not the establishment of a new model isolated from and besides many others, than a more serious investigation of concepts, which have been introduced and used in earlier works only in a quite brief way.

Despite striking similarities in the desorption behaviour of different target types there must be more than one mechanism contributing to the observed secondary ion- and neutral yields. The different yield characteristics of positive and negative ions as well as the different orders of magnitude of charged and neutral [1] yields indicate that either neutral and charged particle desorption are two principal distinct processes, or particle release from the surface and possible ionization can occur independently. Nanosecond-delayed emission on the one hand can be seen as an argument for thermal evaporation mechanisms, while the occurrence of desorbed protonized Insuline-molecules [2] on the other hand can hardly be explained by equilibrium thermostatic approaches.

This work will concentrate on the last aspect: the release of clusters or large fragile molecules, which obviously must occur *before* equilibration of electrons and molecular motion. Recent measurements of LeBeyec et al. [3] indicate that electronical excita-

* Supported by Bundesministerium für Forschung und Technologie

tion in deeper lying layers contributes essentially to desorption from the surface. The investigation of desorption from a very clean CaF_2 surface by Wien et al. [4] shows that even more than the first layer must contribute to the secondary ion yield. For this reason collective motion effects are assumed to contribute to those observations in HIID which are of great and even practical interest.

I. The concept of the hot core track region (HCTR)

The concept of a HCTR which is produced by energy deposition to the electron system, and its possible influence to molecular motion was used in a more than only implicit way first by Watson and Tombrello in connection with their "modified lattice potential model" [5]. However, the band structure of the electron system enters only very incomplete into the model, and excitation of lattice motion is described in terms of equilibrium thermostatics.

The concept itself deserves interest, because after all any explicit investigation of energy transfer from the electron to the lattice requires information about the range and intensity of electronical excitation and can be quite simplified if the assumption of a locally honogenous thermalized "hot" electron gas makes sense. The crucial point is the relation between internal equilibration and external interaction: If it can be assumed that a locally high excited electron–hole–plasma (in an insulator) thermalizes internally fast, compared to external relaxation (i.e. energy transfer to molecular motion and spreading–out of electronical energy), then the maybe very complicated initial excitation state is less important than an approximate equilibrium configuration of the electron excitation.

Therefore the existence of the HCTR in an insulator solid is assumed at the moment and some of its properties are calculated. The conditions were assumed as simple as possible: two energy bands for electrons with identical effective mass and separated by a gap G; the Fermi–level E_F lies in the middle between the band edges. The description of solids with a more complicated electronic structure by such a model is frequently used in solid physics [6]. The parameter G is not necessarily identical with the lowest distance between the two bands because dispersion within the bands is neglected. Furthermore, within a cylindrical region (radius R) an electron temperature T_e was assumed; R, T_e and the energy loss of the projectile $(dE/dx)_e$ are interconnected by energy conservation:

$$(dE/dx)_e = \pi R^2 \varepsilon(T_e).$$

In calculations for targets formed by heavier elements this turned out to be problematic: obviously one cannot assume that the entire energy loss is spent into higher–band excitations only; the deviations become the more important, the more inner shells are present for being excited.

ε is the energy density of the excited electrons. ε and the particle density of electrons in the conduction band n_e are calculated with the Fermi–distribution. The most other temperature–dependent functions are calculated for convenience with the approxima–

tion for $k_B T_e \ll G$, i.e. for the Fermi-distribution function $[\exp\{(E-E_F)/k_B T\}+1]^{-1}$ the lowest order term of an expansion in powers of $\exp\{(E-E_F)/k_B T\}$ was used for $E<E_F$, and for $E>E_F$ $\exp\{(E_F-E)k_B T_e\}$ was used. This should give qualitatively plausible results as long as $k_B T_e/G<1$.

For a given T_e the density n_e and average energy $\varepsilon/n_e=2(E_F+m_e\langle v^2\rangle/2)$ of the excited electrons and holes are obtained. From both a Debye-screening constant

$$\lambda = [4\pi n_e e^2/m_e\langle v^2\rangle]^{1/2}$$

is calculated, which modifies the Coulomb-interaction by an exponential factor: the two-body potential becomes

$$U(\boldsymbol{x}-\boldsymbol{x}') = \frac{e^2}{|\boldsymbol{x}-\boldsymbol{x}'|} \exp\{-\lambda|\boldsymbol{x}-\boldsymbol{x}'|\}.$$

With these ingredients the mean free path l_f for e-e-interaction can be calculated (see fig.1):

$$l_f = (\pi/12)^{1/2} (\hbar^2/m_e e^2) [1-\Lambda\exp\{\Lambda\}Ei(\Lambda)]^{-1} \quad \text{with } \Lambda=\hbar^2\lambda^2/4m_e k_B T_e.$$

Ei is the exponential integral function. This result is obtained from the transition rate for scattering of particles interacting by a screened Coulomb-potential. Integrating once and then averaging the transition rate for two particles with wavenumbers $\boldsymbol{k}_1$ and $\boldsymbol{k}_2$ over the phase space with the approximated Fermi-distribution (see above) yields an average transition rate η. The mean free path is then taken as $l_f=\langle v^2\rangle^{1/2}/\eta$. For materials where more than one band can be assumed to contribute, l_f must be divided through the number of conduction bands.

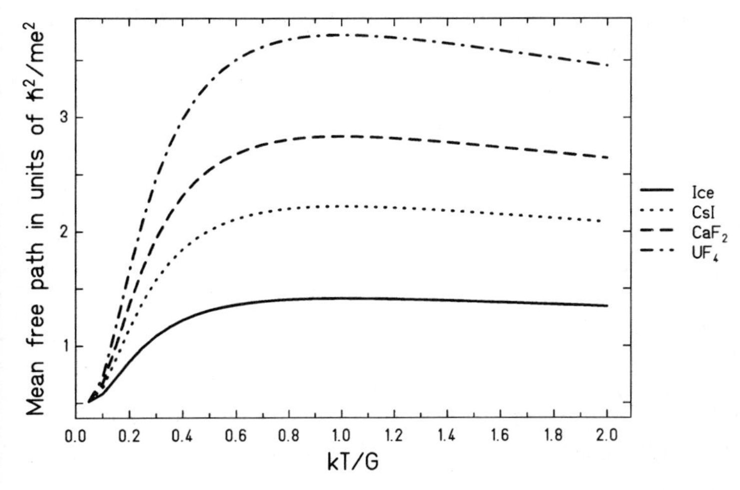

Fig.1: Mean free path l_f of an electron within the HCTR in units of $\hbar^2/m_e e^2$

It turns out that l_f increases in the temperature range of particular interest here ($kT_e<G$) from a finite value for $T_e=0$ with increasing temperature. The reason of this surprising result (one should assume that l_f decreases with the density of potential

scattering partners) is Debye–screening, reducing the e–e–interaction even for larger densities. This picture, however, is realistic only as long as there is a considerable amount of excited electrons; otherwise other effects limiting the interaction will become more important. The calculations here were performed in a temperature range, where this condition still holds.

Since vibrational electron scattering can be neglected on a time scale below 10^{-11}s [7], a cooling off (and at the same time enlargement) of the HCTR is caused mainly by Auger–type transitions [5]: scattering of a conduction electron by shifting a valence electron over the band gap. The inverse process (pair annihilation) is in balance with pair production under equilibrium conditions, which are assumed within the HCTR. In pair production reactions only electrons can be involved, which carry a kinetic energy larger than G, because they must shift another electron over the gap. The average transition rate γ was calculated in the following way:

- The wavefunctions of electrons in the conduction band were assumed as plane waves, the wavenumber $\boldsymbol{k}_i$ of the incident electron large compared to those of the outcoming two ($\boldsymbol{k}_1$ and $\boldsymbol{k}_2$). Let ψ_n be the wavefunction of an electron in the valence band (the Bloch–function properties do not enter here). Then the transition matrix element of this reaction for screened Coulomb interaction is approximately dependent only on $\boldsymbol{k}_i$ (the volume norm factor $V^{-1/2}$ applies to the plane wave functions and to ψ; $\boldsymbol{k}_1$ and $\boldsymbol{k}_2$ are wavenumbers of the final states):

$$U_{if} = V^{-2} \frac{4\pi e^2}{|\boldsymbol{k}_i - \boldsymbol{k}_1|^2} \quad \psi_n(\boldsymbol{k}_i - \boldsymbol{k}_1 + \boldsymbol{k}_2)$$

- Integration of the transition rate for the single reaction over the phase space of the finite states ($\boldsymbol{k}_1$, $\boldsymbol{k}_2$) yields then the total transition rate for the incident electron being scattered by the "bound" electron in a state near the valence band upper edge and with a wave function $\psi_n(\boldsymbol{x})$ as

$$\gamma_{if} = (m_e e^4/\hbar^3)\ |\psi_n(\boldsymbol{k}_i)/V|^2\ \Theta(E_i - G)\ [(E_i - G)/(E_i + G)]^2$$

  where $\psi_n(\boldsymbol{k}_i)$ is the spatial Fourier–transform of $\psi_n(\boldsymbol{x})$.

- Summation over all populated states ψ_n yields the total transition rate dependent of E_i. For convenience and since only an upper limit for the transition rate was to be found, in this summation the step function was neglected and

$$\sum_n \psi_n^\star(\boldsymbol{x})\ \psi_n(\boldsymbol{x}') = \delta^3(\boldsymbol{x} - \boldsymbol{x}')$$

  was assumed. Then the sum of $|\psi_n(\boldsymbol{k}_i)/V|^2$ over n yields 1 independently of $\boldsymbol{k}_i$, if the valence electrons are treated as a Fermi–gas.

- Averaging with the Fermi–distribution over the phase space for conduction electrons yields an average transition rate

$$\gamma = (m_e e^4/\hbar^3)\ \exp\{-\Gamma\}\ [(15/8)\Gamma^2 + (3/4)\Gamma], \qquad \Gamma \text{ being } G/k_B T_e.$$

This result must be multiplied by the number of valence bands contributing. The main difference to the transition rate for e–e–scattering consists of the Boltzmann–factor. This difference justifies the assumption of homogenity within the HCTR. One can assume that only electrons with a distance shorter than l_f from the boundary and propagating towards this boundary contribute to the Auger–transition rate. The decrease of ν, the portion of electrons excited into the conduction band, can be described by

$$d\nu/dt = -(l_f/R)\ \gamma\ \nu$$

Fig.2 shows the dependence of γ from T_e for a given R in different materials (in calculations the material constants of water ice H_2O, CsI, CaF_2 and UF_4 were used because these targets have been frequently tested in experiments). For a given $(dE/dx)_e$ R increases with decreasing T. The dependence of γ from T_e for a given $(dE/dx)_e$ is shown in fig.3. Since here an upper limit for γ was searched, the lifetime of the HCTR is probably underestimated. One finds, however, that an highly excited electronic state can exist within the time scale of molecular motion (10^{-13}s), out of that of thermal equilibration of electrons and molecules. γ is plotted in units of $m_e e^4/\hbar^3 \approx 4.1\times10^{18} s^{-1}$.

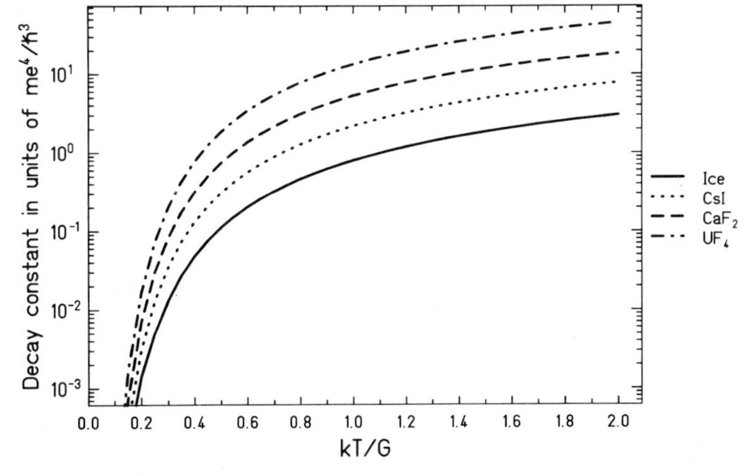

Fig.2: Average decay constant γ for the HCTR in units of $m_e e^4/\hbar^3$ for constant $R=r_0$

There are two other points which may become interesting in a further study of the properties of the HCTR. The first is the effect of Debye–screening to the Auger–transition rate. In above calculations it was assumed that Auger–transitions take place only in a region where screening effects are negligible, i.e. in the cold region outside of the HCTR. This assumption is not very reasonable because there must be a transition zone with a dimension at least of l_f, where the most Auger–transitions occur. In fig.4 one can see the effect of Debye–screening to γ for different materials within the HCTR: the Coulomb–interaction is reduced by the screening factor λ as above, which results in an additional factor $[(1-\Lambda)^2+\Lambda^4]^{-1}$ (Λ as above) to γ. This factor results from the replacement of $|\boldsymbol{k}_i-\boldsymbol{k}_1|^{-2}$ by $[|\boldsymbol{k}_i-\boldsymbol{k}_1|^2+\lambda^2]^{-1}$ in U_{if}. Since the pro–

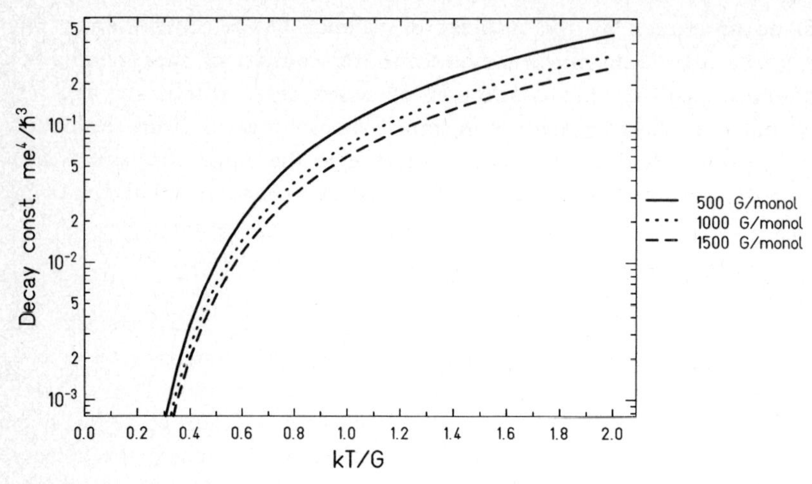

Fig.3: Average decay constant γ for the HCTR in units of $m_e e^4/\hbar^3$ for given values of $(dE/dx)_e$ in gap units per monolayer

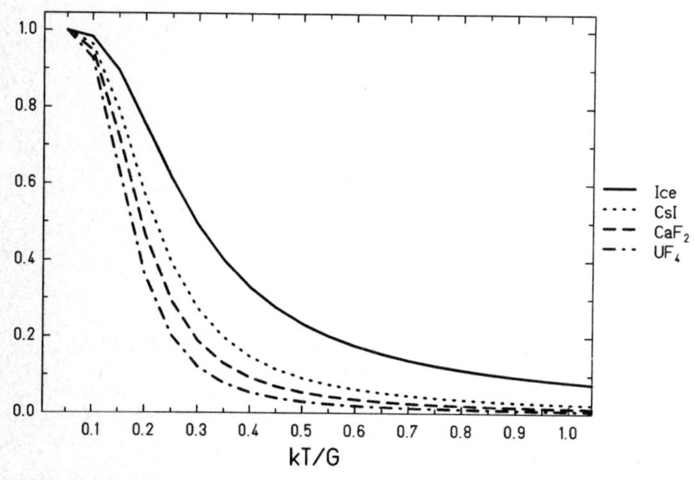

Fig.4: Reduction factor to the decay constant γ due to Debye–screening for different materials in dependence of T_e

perties of the boundary of the HCTR were not studied further, this effect was neglected in the following calculations. In a more realistic calculation Debye–screening in the transition region should be included and could then be expected to stabilize the HCTR.

Another stabilizing effect can be expected from electron–hole–interaction. It is well known that the electronical structure will deviate from the band structure, if there are more than one particles or holes interacting attractively. In the case investigated here the many particle limit applies, i.e. the main contribution to single–particle–energies will not come from single–exciton formation but from a contribution of the whole plasma environment. The attractive interaction of (conduction–) electrons with holes is not cancelled entirely by the repulsive e–e–interaction, because in the last case antisymmetrization between spin–equal electrons applies. The result is that the exchange energy contributes with a negative sign. This exchange energy ΔE was calculated here pertubationally by the assumption that electron– and hole–wavefunctions are plane waves and for electrons with a kinetical energy equal to G. This selection was taken because one can expect that electrons leaving the HCTR must carry this additional energy for being involved in Auger–processes outside of the HCTR: the change of the environment at the boundaries plays the role of a potential barrier. ΔE is plotted in units of G in fig.5. Obviously the contribution is negligible in the temperature region of interest here, but even for high temperatures one can expect visible stabilization effects. Unfortunately the approximations made here do not allow an exact interpretation for this temperature range.

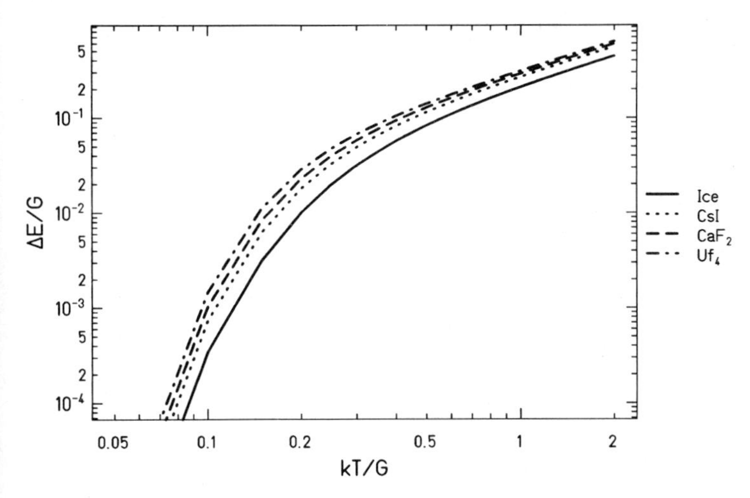

Fig.5: Energy shift for conduction band electrons with a kinetic energy E_k=G in units of G and in dependence of T_e

II. Electronical Heat and Lattice Motion

Possible mechanisms for excitation of molecular motion by interaction with an electronical plasma, as it is described in I, have been discussed especially by Tombrello and coworkers: a pressure produced by the excited electrons was proposed [5], or it was assumed that at atomic locations, where attractively interacting holes meet, some bond–breaking occurs [8]. Both models are from a solid–physical point of view somewhat unsatisfying: pressure of electrons moving within the conduction band says little about their interaction with the lattice; in the case of holes with an attractive interaction also no explicit information is given about the forces on atoms of molecules. The "popcorn"–model of Sundqvist and Williams [9] assumes excitation of long wave thermal vibrations within large molecules, which shall lead to an expansion. Here the critical point lies in the assumed selectivity of long–wave excitation by electron–phonon–scattering, which is required in order to explain the desorption of biomolecules.

In this work a simple approach shall be investigated for the cases of ionically or covalently bound solids, where one can assume that attractive forces between atoms or molecules are caused essentially by the electrons in the valence band. In the case of covalent bonds, attractive forces are mainly exchange forces; in the case of ion bonds attractive Coulomb–interaction is caused by the localization of valence electrons of the cations in orbits close to the anions. In the first case it is reasonable to assume that the exchange forces are reduced proportional to the part of electrons leaving the valence band; in the second case it is even qualitatively clear that the localization of wavefunctions to a binding potential is reduced by an energy shift. In a strict calculation the contributions of conduction electrons would have to be taken into account, too. But since here binding states as well as anti–binding ones will be produced, this contribution is simply neglected. For this reason the following calculations cannot hold for Van der Waals–solids, where it is well known that electronical excitation causes for lower energies (single particle excitation) attractive interaction.

The microscopic picture resulting from this argumentation is that interatomic or –molecular bonds consist of a repulsive and an attractive part, the latter one contributing proportinal to the number of electrons remaining in the valence band. By fitting three parameters for the potentials of the form $V(r)=V_r(r)-V_a(r)$ in a way that they reproduce known material data (Debye–frequency ω_0 as a measure for molecular oscillation frequency, lattice–spacing r_0 and sublimation heat U) one can then make conclusions from the electronic excitation state to lattice forces. Let ν be the portion of valence electrons shifted into the conduction band, calculated as mentioned in I. Then from the modified potential $V_r(r)-(1-\nu)V_a(r)$ ω_0 modified by a factor β, a changed r_0 and U can be calculated. In the Born–Oppenheimer–approximation the atomic locations remain unchanged; so an intermolecular force $-\nu V_a'(r)$ results from the potential modification, if the atoms/molecules are in balanced positions for t=0. Actually not the most common empirical potentials were used here (Born–Mayer, Lennard–Jones, Morse), because they usually describe single bonds, but not macroscopically averaged material properties. The different results which came out by using different types of parametrization show, that this is a crucial point of the model. The two potential types used here were

$$V(r) = [Q/r - (1-\nu)W] \exp\{-\lambda r\} \quad \text{and}$$

$$V(r) = W\ [(r_0/r)^n - (1-\nu)(n/m)(r_0/r)^m].$$

The three parameters of both potentials were adjusted so that $V(r_0)=-U_0$, $V'(r_0)=0$ and $V''(r_0)=M(r_0\omega_0)^2$. Then $\omega/\omega_0=\beta$, U and r were recalculated for $\nu>0$. The effective pressure was taken as $p=(1-\nu)V'(r_0)/r^2$ and an effective compression as $pr/(M\omega^2)$.

III. Continuum Mechanical Model

In most works concerning electronic sputtering an estimation of interatomic or intermolecular forces is followed by a yield calculation with the more or less implicit assumption of individual motion of desorbed atoms/molecules. This would be rather unconsequent in an argumentation, which claims to give an alternative to thermal equilibrium models. Therefore especially in cases where one can expect large yields due to a large range of interaction and strong forces driving desorption at the same time, a description in terms of the classical theory of continuum motion should be more appropriate. This is just the antipodal to the approach of independent molecular desorption and *not* necesarily connected to the driving mechanism proposed in II.

In the macroscopic limit the results of II – modification of material properties and a nonzero pressure within the HCTR – define a continuum mechanical problem. Due to the nonharmonic potentials an analytical solution giving a detailed description of the time evolution of the system can not be calculated even in the interesting case of large amplitude motion. A use of the harmonic approximation can nontheless give qualitative information. Although effects near the surface are to be found here, transversal motion is neglected, thus treating the target substance like a fluid. A sound velocity reduced by a factor β within a radius R defines two types of degrees of freedom: "axial" modes (soundwaves being totally reflected at the boundary) and "radial" ones (transmitted also radially). The radial modes are neglected in the following calculations because they are assumed to give no rise to transport of compressional energy to the surface (the material reacts to the pressure within the HCTR partially by a cylindrical sound wave without serious effect to the surface). This neglect causes an underestimation of the effects contributing possibly to desorption, because even in the case of homogenous sound velocity (β=1), when there is no total reflection at all, one can expect that the surface will oscillate.

The displacement field vector $\boldsymbol{S}(t,\boldsymbol{x})$ is derivable from a potential Φ. For the axial modes the Fourier-transformed potential $\Phi(\omega,\boldsymbol{x})$ is proportional to $\sin(kz)J_0(Kr)$ for $r<R$ and to $\sin(kz)K_0(\kappa r)$ else (free boundary at the surface). Boundary conditions (Φ smooth at r=R) and the conditions $k^2-\kappa^2=\omega^2/v^2$ and $k^2+K^2=\omega^2/(\beta v)^2$ define the modes for any ω; κ and K can be found easily by Newton-iteration. The initial compression distribution $\mathrm{div}\boldsymbol{S}(t=0,r,z)=\mu\Theta(R-r)$ defines the coefficients of an expansion of the axial part of Φ into the eigenmodes. It turns out that about 90% of the axial energy is spent into the zero modes (smallest K for any ω). The axial Fourier-transform of the initial compression, $\mathrm{div}\boldsymbol{S}_{a0}(t=0,r=0,k)$ of the zero modes is quite similar to a Lorentz-curve; so the qualitative behaviour of the zero-axial contribution can be described by an initial condition $\mathrm{div}\boldsymbol{S}_{a0}(t=0,r=0,z)=A\exp\{-k_m z\}$; A and k_m are found as

$$k_m = \frac{\beta R^{-1}}{[1-\beta^2]^{1/2}}(0.704496+1.18687\beta) \quad \text{and} \quad A = 8\mu k_m R\,(0.372168-0.299306\beta)/\pi\beta$$

The time evolution resulting from this condition is quite trivial. It represents a rarefaction wave front running into the inner of the material:

$$\mathrm{div}\boldsymbol{S}_{a0}(t,r=0,z) = \begin{array}{ll} -A\exp\{\beta k_m vt\}\sinh(k_m z) & \text{for } z<\beta vt \\ +A\exp\{k_m z\}\cosh(\beta k_m vt) & \text{for } z>\beta vt \end{array}$$

This means: The peak rarefaction is reached at least at some distance from the surface which scales with $R\sim 1/k_m$.

One could now try to find some tear–off point by this approximation, but the actual three–dimensional motion must be expected to give a more complicated behaviour in the extremely nonlinear case than a one–dimensional harmonical calculation can give. Therefore in order to obtain an estimate for the total yield to be expected by continuum mechanical motion the energy required for surface formation in the "desorption" of one large cluster drop was compared to the energy available in the axial modes. The harmonical picture yields that even for infinite t half of the axial energy is involved in the wave running into the material, and the half of the remaining energy will be spent into kinetical energy of the desorbed "cluster". The axial energy was calculated by numerical integration from the spectral expansion.

So the total yield was estimated by the formula

$$Y = (\pi/3)\,(2E_{ax}/U\pi)^{2/3} \qquad \text{with} \qquad E_{ax} = \pi R^2 A^2 M \omega^2 / 4rk_m$$

IV. Yield Results

Yield calculations using the results of I–III were done in the following way:

1. For a given $(dE/dx)_e$ a T_e and R are calculated so that a decay rate γ equal to the Debye–frequency used in II results. The reason is that this selection on the one hand justifies the Born–Oppenheimer–approximation; on the other hand one can be sure that there is enough time before the further decay of the electronical HCTR by heating the lattice.
2. The T_e found that way defines ν; β and the modification of material parameters are calculated from ν as shown in II.
3. A total yield is calculated as shown in III from the total axial energy.

The result for different target materials is shown in fig.6. The proportionality of Y to $(dE/dx)^{9/4}$ is obvious; this characteristic is quite similar to the square dependence

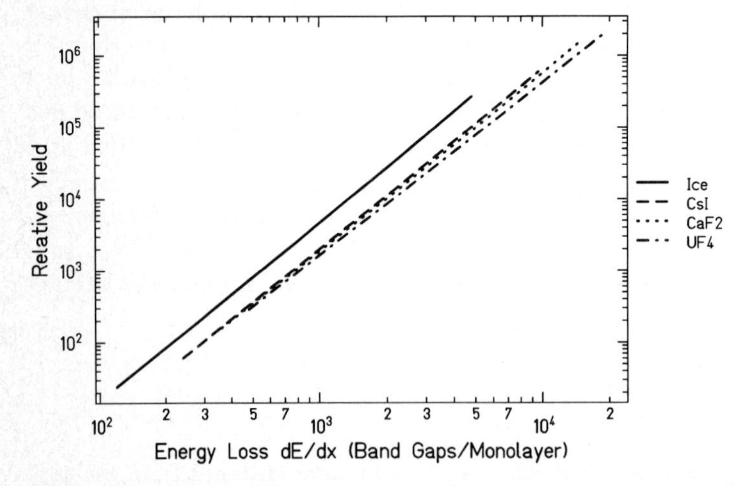

Fig.6: Neutral secondary particle yields as a function of $(dE/dx)_e$

found in many experiments for the neutral yield. Maybe this is only by chance: the assumption that only surface formation energy is required for desorption and an asymptotically constant electron temperature for different values of $(dE/dx)_e$ are mainly responsible for the result.

There are, of course, still important problems to be solved for this model. The one is the dependence of yield results on the way of parametrization of the potentials. It is not to be solved in the trivial, but technically complicated way of replacing one single potential by "realistic" intermolecular and interatomic potentials under inclusion of geometry etc.. One reason is that even in well proven cases attractive and repulsive terms cannot be assigned to valence electron– or screened Coulomb–repulsion–contributions. The other reason is that for larger deformations the coordination of atoms becomes important, so that the neglect of transversal and optical modes becomes invalid. An explicit and mor basic calculation of lattice forces under the conditions of a hot electron gas will therefore be necessary.

Another problem can be seen on fig.7: The ratio of the yield for a Si target and organic carbon (graphite) targets, bombarded by the same projectile ($(dE/dx)_e$ is calculated by the Ziegler–formula [10] and the Bragg–rule) is obviously too large. The reason is that in I the contribution to inner shell excitation from $(dE/dx)_e$ is neglected, i.e. the calculations are done as if all the energy loss is spent only into valence band electron excitation. This becomes problematical even for heavier targets.

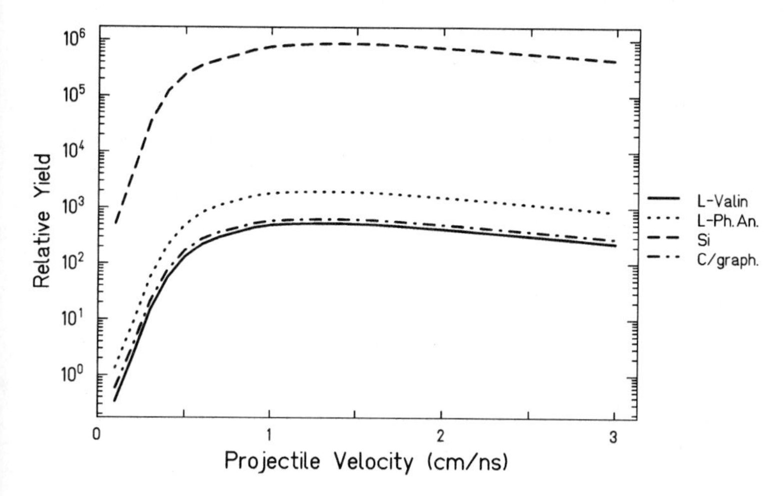

Fig.7: Secondary particle yield as a function of the projectile velocity for different target materials and for an effective charge Z_{eff}= 20

There is one qualitative prediction which can be concluded from the model already at this stage: desorbed material must be released in clusters (the "one big" cluster is an artefact of the very simple ansatz) which are still "cold" in the sense of molecular motion but have a hot electron system. The observed spectra can then be expected to be controlled by the decay mechanisms which take place for this sort of excitation:

ionization by Auger–transitions, thermal electron evaporation and following Coulomb–fragmentation on the one hand, and electron–phonon heating on the other hand. The observations of Schmidt and Jungclas [11] seem to support this result: coincidence of different ionized molecular fragments indicate that fragmentation *after* desorption plays an essential role for observeable secondary ion spectra. A further investigation will be necessary here too; it is possible that HIID is an effective tools for the preparation and investigation especially of highly exited clusters.

References

[1] Teilchendesorption von Nichtleiteroberflächen durch Schwerionen hoher Energie Guthier, W.; *Thesis*, TH Darmstadt **(1986)**

[2] 127–I Plasma Desorption Mass Spectrometry of Insuline Håkansson, P.; Kamensky, I.; Sundqvist, B.; Fohlmann, J.; Peterson, P.; McNeal, C.J.; MacFarlane, R.D., *J.Am Chem. Soc.* **(1982),** *104,* 2948–1949

[3] Della Negra, S.;Becker, O.; Cotter, R.; LeBeyec, Y.; Monoret, B.; Standing, K.; Wien, K.; IPNO–DRE–86–09 Orsay (France), to be published in NIM B

[4] Wien, K.; priv. comm., presented to the Workshop on Physics of small Systems, Isle of Wangerooge, Sept. 1986

[5] A Modified Lattice Potential Model of Electronically Mediated Sputtering Watson, C.C.; Tombrello, T.A.; *Rad.Eff.* **(1985)** *89,* 263–283

[6] Phillips, C.J.; *Revs.Mod.Phys. 42* **(1970)** 317

[7] Seiberling, L.E.; Griffith, J.E.; Tombrello, T.A.; *Rad.Eff.* **(1980)** *52,* 201

[8] Itoh, N.; Nakayama, T.; Tombrello, T.A.; *Phys.Lett. 108A* **(1985)** *9,* 480–484

[9] Sundqvist, B.; priv. comm., presented to the Workshop on Physics of small Systems, Isle of Wangerooge, Sept. 1986

[10] Ziegler, J.F.; Biersack, J.P.; Littmark, U.; Handbook Vol.5 of "The Stopping and Ranges of Ions in Solids", Perg.Press **(1980)**. The formula was calculated using a numerical program from the Darmstadt group (O.Becker, K.Wien).

[11] Correlation Effects of Secondary Ion Emission Schmidt, L.; Jungclas, H.; see article in this book

Adatoms and Adclusters: On Imaging Studies by Scanning Tunneling Microscopy

KLAUS SATTLER

Department of Physics, University of California
Berkeley, California 94720

ABSTRACT

The scanning tunneling microscope (STM) as a new tool for cluster research is discussed. The imaging of *Au* and *Ag* clusters on graphite in air is described and cluster pictures are shown in two operation modes: the constant current and the variable current mode. Images of large (~ 350Å) and small clusters (a few atoms) are shown. Movement and growth of adatoms and adclusters is observed with the STM on the atomic scale.

In recent years it has been demonstrated that the scanning tunneling microscope (STM)/1/ is a powerful new technique for investigations of surface structures. It provides images in real space, in three dimensions, with atomic resolution. It can be applied both for periodic and for non-periodic structures.

The STM has been used to measure the topographical and electronic properties of crystalline semiconductors /2/ and evaporated metal films /3/ and it has been applied to the study of surface roughness /4/, superconductivity /5/ and charge density waves /6/. Further studies showed that the instrument operates (besides in UHV) in air, liquid nitrogen, water and paraffinoil /7/.

The STM works by the quantum mechanical phenomenon of tunneling. A metal wire with a finely prepared ground tip is mechanically rastered across the surface of the sample. If the tip is close enough, a few angstroms, the wave functions of the electronic quantum states of an atom at the end of the tip and of those in or on top of the substrate overlap slightly. The overlap allows electrons to jump across the vacuum gap between the tip and the surface, when a voltage is applied between them. Thereby a tunneling current is produced, which increases exponentially as the gap distance decreases.

The tunneling current is, according to STM theory, directly proportional to the surface electronic charge density at the position of the center of curvature of the tip. The charge density is normally concentrated where the atoms are. This accounts for the interpretation of bumps in STM images as atoms. However, it is possible that an enhanced local charge density that gives such a bump is not associated with an atom at all. For example, it has been shown that variation of the tunneling current can also be generated by charge density waves /6/. Furthermore, results from *Ag* cluster studies shown in this work indicate tunneling into bonds instead of tunneling into atoms.

Within a few angstroms apart from the surface the tunneling current varies typically by one order of magnitude if the width of the vacuum gap is changed by one angstrom. Therefore electrons tunnel into the outermost part of the spatial charge density distribution of the object to be imaged. S-states usually are more diffuse than p- or d-states. Thus, tunneling into s-states can be preferred even if the density of states within the adjusted energy window (typically 10 meV) is higher for p- or d-states.

In principle every metal or semiconductor could be used as a substrate for clusters to be imaged. However, in order to see intrinsic cluster properties the substrate should be inert and reasonably flat. Graphite seems to be a suitable material. Both, its electronic and atomic structures are well known. The localized σ and π states, as well as fairly delocalized interlayer states /8/ and recently discovered empty surface states /9,10/ coexist within a range of a few eV around the Fermi energy. Graphite is conductive (it is a semimetal) and it has no reconstruction problems. Therefore the graphite surface structure is expected to remain unchanged after the deposition of adlayers. The atoms on a cleaved graphite surface (0001 plane) are located at the vertices of the hexagons of a two-dimensional honeycomb pattern with diameter of 2.4Å. The six carbon atoms making up a ring are not equivalent. The difference stems from the fact that three of the carbon atoms have neighboring atoms in the layer immediately below, whereas the other three do not.

The STM works by two modes of operation. In the **topographic mode** the tip is maintained at a fixed distance above the surface during scanning by keeping a constant current with an electronic feedback circuit. In this way the trajectory of the tip traces out a profile of the surface including the bumps due to clusters or single atoms on top of the substrate. In the **variable current mode** a mean distance between tip and surface is held constant and the modulation of the tunneling current is imaged.

With the topographic mode 3D-clusters can be imaged on top of the substrate. The clusters are not destroyed by contact with the tip because the feedback circuit provides a constant distance. In this mode a picture is taken in a few minutes and therefore it is also called the **slow-mode**. It can only be used for rigid surface structures or adlayers with very small diffusion times.

With the variable current mode only 2D-clusters can be studied because the substrate lattice should be imaged, simultaneously. A tunneling current of a few nA, i.e., a distance to the substrate of a few angstroms is adjusted. When the tip passes an adatom or a group of adatoms the tunneling current increases by several orders of magnitude. 3D-clusters are destroyed by the tip passing by. Additionally during this process the tip can pick up atoms from the cluster. The atomic structure at the end of the tip therefore can be changed during the experiment and in consequence the contaminated tip has to be replaced by a clean one. The advantage of the variable current mode is that a picture is taken in a few seconds and therefore it is also called the **fast mode**. Fast imaging is required for clusters which move on the substrate. The jump frequencies have to be small enough in order to get reproducible pictures.

We now describe STM studies of *Au* and *Ag* clusters deposited on the surface of highly oriented pyrolitic graphite /11/. In the topographic mode 500 × 500Å areas have been adjusted and islands of various sizes and shapes have been imaged. In the variable current mode 24 × 24Å areas have been adjusted and individual adatoms and adclusters have been observed with atomic resolution. Additionally, diffusion and growth of atoms and clusters on the substrate were observed.

An STM of conventional design has been used. The x-, y- and z- directions of the tungsten tip are varied by piezoelectric elements. The graphite substrate was prepared by peeling off some top layers followed by immediate transfer into a vacuum chamber for the adlayer preparation. The preparation technique of the graphite substrate is standard in surface science /12/. It guarantees quite clean surfaces because of the extremely low gas adsorption efficiency of graphite. This is supported by STM images obtained in air which reveal a perfect lattice extending over thousands of angstroms /13/. The *Ag* clusters have been generated by inert gas condensation (2.2 torr of *Ar*) before deposition. The *Au*-clusters have been formed on the substrate after vapour deposition of single atoms. The covered substrates have been transferred to the STM where pictures in air at atmospheric pressure were obtained. Images from hundreds of different areas have been taken, which showed a wide variety of features, ranging in size from a few atoms to many thousands of atoms.

Figure 1 shows an image of a large *Ag* cluster obtained in the topographical mode at a voltage of 16 mV (tunneling current: several nA). The cluster is roughly cylindrical with a diameter of 350Å and a height of about 30Å. It is composed of several smaller clusters each 30 to 100Å in diameter. The shape of the substructures was stable during the 10 minutes time of observation

Fig. 1: Top view of a 350Å Ag cluster on graphite taken in the topographic mode (height ≈ 30Å) /11/

(one single image required 3 minutes). Obviously the deposited small particles with 30 to 100A diameters moved at the substrate and the observed island is the result of coalescence. It is interesting that the particles retain (at least partly) their individuality after coalescence.

Fig. 2 shows a series of five 24 × 24Å images taken in the variable current mode. Fig. 2(a) displays the clean graphite surface with individual carbon atoms being imaged. As mentioned before the structure has not the expected honeycomb pattern.

Fig. 2(b) shows a Ag cluster roughly 15Å long and 5Å wide. The substructure can be explained by an array of dimers. The electrons tunnel into the covalent bonds and consequently the individual atoms of the dimers are not imaged separately. The individual dimers, however, are resolved which can be explained by van der Waals coupling between the dimers.

The remaining pictures in Fig. 2 were selected from a long series of scans within 37 minutes: $t = 0$ (b), $t = 15min$ (c), $t = 17min$ (d), $t = 22 - 37min$ (e). Within this time period more than 500 pictures have been observed, one picture every 4 seconds. Ag atoms diffused into the region to join with the original cluster. The structure of the cluster changed several times. Finally the cluster has taken on a Y shape which remained stable for 15 minutes.

Fig. 3 shows a Au cluster of about seven atoms surrounded by clean graphite. Different from Ag, for the Au cluster single atoms are displayed. Furthermore, the atoms are close packed with one atom in the center surrounded by one shell. The Ag-dimers are commensurate with the graphite lattice but the Au-atoms in Fig. 3 are not. This can be explained by the different bond energies for Ag and Au clusters (Ag_2 : $159 \pm 6kJmol^{-1}$, Au_2 : $221 \pm 1kJmol^{-1}$, Ag_3 : $253 \pm 13kJmol^{-1}$, Au_3 : $367 \pm 13kJmol^{-1}$, Ref. 14). For lower cluster bond energies the cluster substrate interactions become more important.

Now we consider possible influences of the instrument on the initial cluster distributions and shapes. Both, for Ag and Au, clusters in a broad size range and islands with all kinds of shapes were imaged. In the fast mode, many features, however, did not show atomic resolution because obviously the tip crashed into the uppermost layers of 3D-clusters brushing the atoms aside. The areas shown in Fig. 2 and 3 were selected for presentation because the structure of the displayed clusters was stable over many scans and underwent relatively slow, progressive changes.

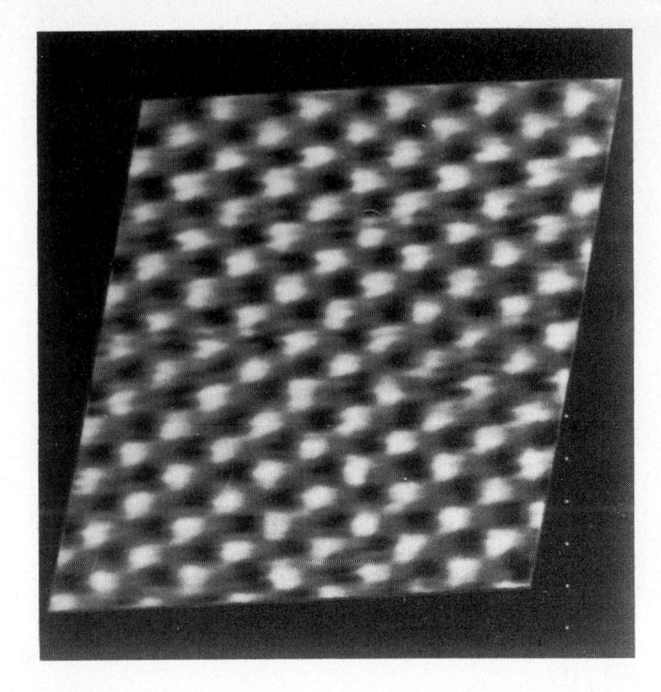

Fig.2a

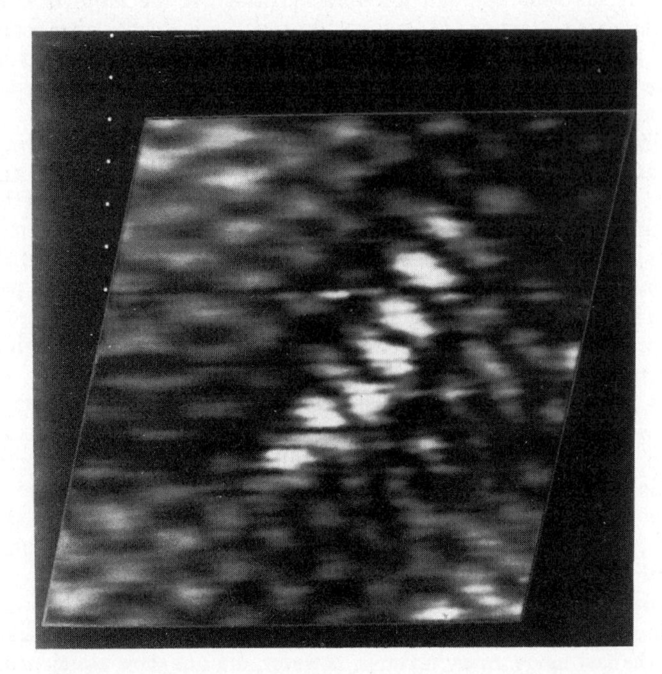

Fig.2b

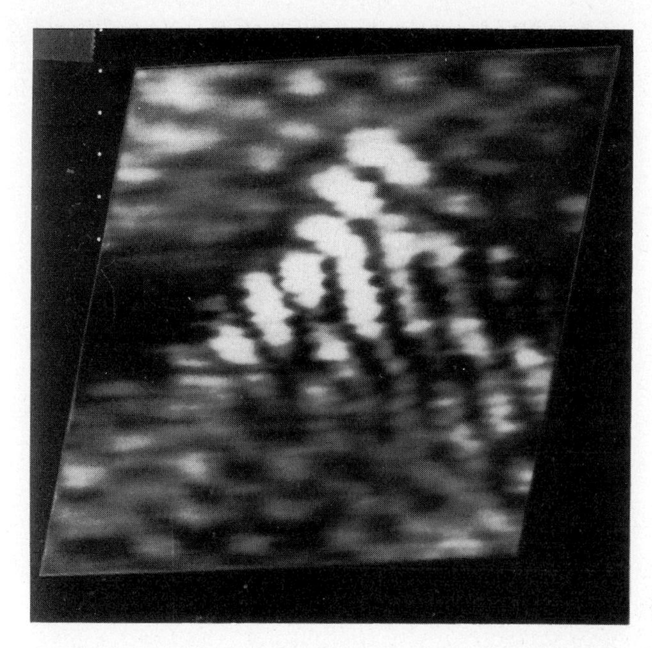

Fig.2c

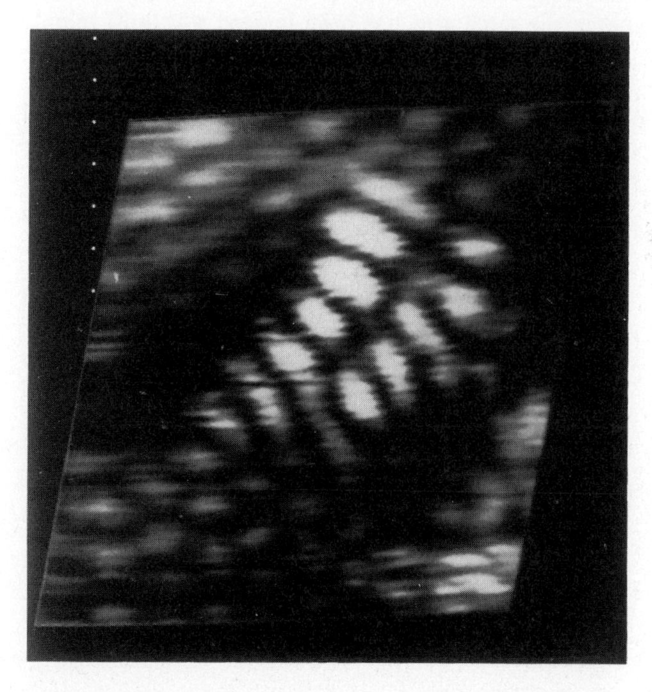

Fig.2d

Fig.2e

Fig. 2: Top view of clean (a) and covered (*Ag*, (b)-(e), top to bottom) graphite obtained in variable-current mode (24Å × 24Å frame). The evolution of the features in the same area has been imaged within 37 minutes /11/

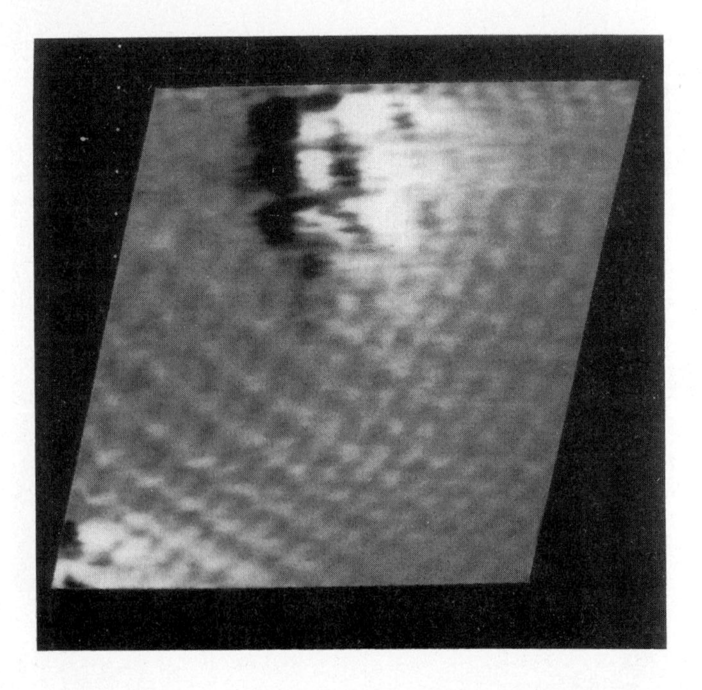

Fig. 3: Top view of a *Au* cluster on graphite in the variable current mode (24Å × 24Å frame)

For Ag_3 and Au_3 predominantly s-orbitals have been found to contribute to the bonds (80 – 90%) /15/. Likely these s-states of adatoms and adclusters couple to the conduction band of the substrate building up a joint density of state. Otherwise, in the case of pure physisorption, adatoms and very small adclusters would not be observed with the STM because the probability for the sharp eigenstates of the adparticles to lie within the 10 meV energy window (given by the bias voltage) would be considerably small.

Theoretical morphology studies yield different cluster structures depending on the model approaches and on the parameters being used. For free *Ag*-clusters theoretical calculations predict either linear /16,17/ or three-dimensional structure /18/. As the STM studies show *Ag* cluster structures being commensurate with the substrate lattice the geometry of the adsorbed particles cannot be compared with the predictions for free clusters. The graphite substrate seems not to be inert concerning *Ag* coverage.

As shown in Fig. 2, movement and growth of single atoms and clusters on the substrate is observed. Dynamics of adatoms can be described by relatively simple statistical mechanical methods /19/. For clusters, however, the description of the random walk on a corrugated lattice network is complicated since each link in the unit mesh may correspond to a different jump probability, reflecting different activation energies for the various configuration changes of the cluster. Direct observations of mean square displacements of adatoms and adclusters yielding tracer diffusion coefficients, elementary jump distances and jump frequencies would be extremely helpful to understand nucleation and growth of submonolayer thin films. For mean displacements of a few angstroms per second STM studies in the fast mode should give quantitative results concerning the quantities mentioned above. Studies of this kind are in progress.

In summary, the STM has been applied to image *Au* and *Ag* clusters on graphite. Single adatoms and clusters in a wide size range as well as diffusion and growth have been observed on the atomic scale.

The author gratefully acknowledges a Heisenberg fellowship from the Deutsche Forschungsgemeinschaft. He thanks his colleagues D. W. Abraham, E. Ganz, H. J. Mamin, R. E. Thomson and J. Clarke for helpful discussions.

References

1. G. Binnig and H. Rohrer, Surf. Sci. 126, 236 (1983)
2. R. S. Becker, J. A. Golovchenko, E. G. McRae, and B. S. Swartzentruber, Phys. Rev. Lett. 55, 2028 (1985);
   R. J. Hamers, R. M. Tromp, and J. E. Demuth, Phys. Rev. Lett. 56, 1972 (1986)
3. G. Binnig, H. Rohrer, Ch. Gerber, and E. Weibel, Phys. Rev. Lett. 49, 57 (1982)
4. R. Miranda, N. Garcia, A. M. Baró, R. Gracia, J. L. Peña, H Rohrer, Appl. Phys. Lett. 47, 367 (1985)
5. S. A. Elrod, A. L. de Lozanne and C. R. Quate, Appl. Phys. Lett. 45, 1240 (1984)
6. R. V. Coleman, B. Drake, P. K. Hansma and G. Slough, Phys. Rev. Lett. 55, 394 (1985)
7. R. Sonnenfeld and P. K. Hansma, Science, 232, 211 (1986)
8. M. Posternak, A. Baldereschi, A. J. Freeman, E. Wimmer, and M. Weinert, Phys. Rev. Lett. 50, 761 (1983)
9. Th. Fauster, F. J. Himpsel, J. E. Fischer and E. W. Plummer, Phys. Rev. Lett. 51, 430 (1983)
10. M. Posternak, A. Baldereschi, A. J. Freeman and E. Wimmer, Phys. Rev. Lett. 52, 863 (1982)
11. D. W. Abraham, K. Sattler, E. Ganz, H. G. Mamin, R. E. Thomson and J. Clarke, Appl. Phys. Lett., in print

12. M. F. Toney and S. C. Fain jr., Phys. Rev. B30, 1115 (1984)
13. G. Binnig, H. Fuchs, Ch. Gerber, H. Rohrer, E. Stoll and E. Tosatti, Europhys. Lett. 1, 31 (1986)
14. K. A. Gingerich, I. Shim, S. K. Gupta and J. E. Kingcade jr., Surf. Sci. 156, 495 (1985)
15. J. A. Howard and R. Sutcliffe, Surf. Sci. 156, 214 (1985) and references therein
16. R. C. Baetzold, J. Chem. Phys. 55, 4363 (1971)
17. A. B. Anderson, J. Chem. Phys. 68, 1744 (1978)
18. C. Bachman, J. Demuynck and A. Veillard, Faraday Symp. Chem. Soc. 14, 170 (1980)
19. A. N. Berker, S. Ostlund and F. A. Putnan, Phys. Rev. B17, 3650 (1979)

MODEL CALCULATIONS OF SINGLY AND DOUBLY

CHARGED CLUSTERS OF (S^2) METALS.

by G. DURAND, J.P. DAUDEY and J.P. MALRIEU

Laboratoire de Physique Quantique

Unité Associée au C.N.R.S. n°505

Université Paul Sabatier

118, route de Narbonne

31062 TOULOUSE CEDEX (FRANCE)

Abstract

Starting from a Valence Bond description of singly and doubly charged clusters of (S^2) metal atoms, treated as one hole or two holes in a completely full s band, a model Hamiltonian is derived which includes instantaneous polarization energies and Coulombic repulsion of the holes in doubly charged clusters. This model is applied to Mg_n^+ and Mg_n^{++} clusters, for which complete explorations of the energy hypersurfaces are reported for $n \leqslant 5$ (for Mg_n^+) and $n \leqslant 8$ (for Mg_n^{++}). Paradoxical stabilities, recently observed for small Hg_n^{++} clusters is also theoretically predicted for the Mg_n^{++} case for $n \geqslant 7$).

Introduction

The present study of doubly charged clusters of metal atoms has been motivated by a recent observation of an Hg_5^{++} cluster from electron impact on a beam of mercury clusters.[1] The existence of such a small doubly charged cluster is surprising from simple electrostatic considerations which should induce its breaking into two singly charged fragments. Bréchignac et al. [1] proposed that polarization forces created by a central Hg^{++} atom on surrounding neutral atoms might explain this apparent contradiction. A theoretical approach of this problem requires full explorations of energy surfaces for the different singly and doubly charged clusters. Ab-initio quantum methods have not yet reached the point where such studies would be feasible at a reasonable computing cost. The most recent extensive studies (including electron correlation which is a determinant factor in the determination of equilibrium geometries and energetics of small clusters [2]) were limited either to very small species or restricted to some special symmetric clusters.

For atoms of column II of the periodic table, it may be considered that in small clusters they keep their ground state s^2 (^1S) configuration. (Hybridization with the sp ($^{1,3}P$) configuration would appear for larger systems as shown recently for Be clusters by Bauschlicher et al. [3] in the compact configurations). A model Hamiltonian has been recently proposed [4] to treat the positive ions as a single hole or two holes in a completely full s band. In the first section, the derivation of the model and the extraction of the parameters from accurate ab-initio calculations on Mg_2, Mg_2^+ and Mg_2^{++} dimers is briefly recalled. The second part deals with the results obtained with small clusters and the last one presents larger clusters for which doubly charged species are expected to be stable with respect to singly charged decomposition products.

I. Model Hamiltonian for Mg_n^+ and Mg_n^{++} clusters

The model Hamiltonian is derived in the frame of the Valence Bond description, in which the total wavefunction is a linear combination of localized functions corresponding to different atomic situations. For the singly charged clusters, these different atomic situations correspond to the n different positions of the hole on the various atoms. If the hole is on atom i, the corresponding valence bond structure is

$$\phi_i^+ = \mathcal{A}\left[\prod_{\substack{j=1\\ j\neq i}}^{n} \varphi_j\,(2j-1, 2j)\right] \cdot S_i\,(2i-1)$$

where φ_j is the wavefunction of the neutral (s^2) configuration. If we call ψ_0 the wavefunction of the neutral cluster

$$\psi_0 = \mathcal{A}\left[\prod_{i=1}^{n} \varphi_i\,(2i-1,\ 2i)\right],$$

ϕ_i^+ is obtained by the annihilation of a β spin electron on site i ; $\phi_i^+ = a_{\bar{i}}\,\psi_0$.
The matrix elements of the model Hamiltonian $\bar{H}$ in the ϕ_i^+ basis are easily derived using the same hypothesis than the DIM [5-6] method. For the diagonal element,

$$\langle\phi_i^+|\bar{H}|\phi_i^+\rangle = \sum_{j>k\neq i} R_{jk} + \sum_{j\neq i} R_{ij}^+ + IP_1$$

R_{jk} corresponds to the interaction between neutral centres and is derived from the knowledge of the neutral Mg_2 dimer. R^+_{ij} corresponds to the interaction between positive site i and the neutral site j. IP_1 is the lowest ionization potential of the atom.

The non-diagonal element, $\langle\phi^+_i|\overline{H}|\phi^+_j\rangle = F^+_{ij}$ represents the hopping integral between the positive sites i and j. Both R^+_{ij} and F^+_{ij} are supposed to be identical to their values in the positive Mg_2^+ dimer, according to the DIM hypothesis. For Mg_2^+, the 2 x 2 matrix of $\overline{H}$

$$\begin{bmatrix} R^+_{12} & -F^+_{12} \\ & R^+_{12} \end{bmatrix}$$

generates two roots $^2\Sigma^+_u$ and $^2\Sigma^+_g$ corresponding to the energies

$$E(^2\Sigma^+_u) = R^+_{12} + F^+_{12}$$

$$E(^2\Sigma^+_g) = R^+_{12} - F^+_{12}$$

The knowledge of these potential curves for the diatomic cation therefore leads to the values of R^+_{12} and F^+_{12}.

In the doubly charged case, two different VB situations are possible

1. two holes on the same site, corresponding to the wavefunction

$$\phi^{++}_{i\bar{i}} = \mathcal{A} \prod_{j\neq i} \varphi_j = a_{\bar{i}}\, a_i\, \psi_0$$

2. two holes on different atoms, with two possible wavefunctions (with $S_z = 0$)

$$\phi^{++}_{i\bar{j}} = \mathcal{A}\left[\prod_{k\neq i,j} \varphi_k \cdot S_i\, S_{\bar{j}}\right] = a_{\bar{i}}\, a_j\, \psi_0$$

$$\phi^{++}_{\bar{i}j} = \mathcal{A}\left[\prod_{k\neq i,j} \varphi_k \cdot S_i\, S_{\bar{j}}\right] = a_i\, a_{\bar{j}}\, \psi_0$$

The derivation of the corresponding matrix elements is detailed in ref. 4, and the results are only given here for sake of completeness.

$$\langle \phi^{++}_{i\bar{i}}|\overline{H}|\phi^{++}_{i\bar{i}} \rangle = IP_2 + \sum_{(k>\ell)\neq i} R_{k\ell} + \sum_{k\neq i} R^{2+}_{ki}$$

where IP_2 : double-ionization potential

R^{2+}_{ki} : effective interaction between Mg and Mg^{++}

$$< \phi^{++}_{\bar{i}\bar{j}} |\bar{H}| \phi^{++}_{\bar{i}\bar{j}} > = 2IP_1 + \sum_{k>\ell\neq i,j} R_{k\ell} + \sum_{k\neq i,j} (R^{+}_{ki} + R^{+}_{kj}) + R^{++}_{ij} + \delta_{pol}$$

R^{++}_{ij} : Coulombic repulsion between holes on i and j

δ_{pol} : correction to the polarization energy of the neutral atoms in the field created by the positive ions i and j.

The total polarization energy of the system, with holes i and j, is

$$E_{pol} = \sum_{k\neq i,j} \left[-\frac{\alpha_1}{2} \left(\frac{\underline{r}_{ki}}{r^3_{ki}} + \frac{\underline{r}_{kj}}{r^3_{kj}}\right)^2 \right]$$

Since each R^{+}_{ki} term involves a polarization contribution $-\frac{\alpha_1}{2}\left(\frac{\underline{r}_{ki}}{r^3_{ki}}\right)^2$, the non-additive δ correction is

$$\delta_{pol} = \sum_{k\neq i,j} -\alpha_1 \left(\frac{\underline{r}_{ki}}{r^3_{ki}} \cdot \frac{\underline{r}_{kj}}{r^3_{kj}}\right)$$

$< \phi_{\bar{i}\bar{j}} |\bar{H}| \phi_{\bar{i}\bar{k}} > = F^{++}_{jk}$ is the hopping integral between sites j and k with another hole on site i and is related to the hopping integral F^{+}_{jk} introduced for the singly positive cluster : $F^{++}_{jk} = F^{+}_{jk} - \langle j|J_i|k\rangle$, where J_i is the Coulomb operator relative to the S_i orbital on site i. This three-center integral should be small and neglected with respect to F^{+}_{jk} .

$< \phi^{++}_{\bar{i}\bar{i}} |\bar{H}| \phi^{++}_{\bar{i}\bar{j}} > = - F^{2+}_{ij}$ is the hopping integral between the doubly charged site i and the two singly charged sites i and j situations.

Other matrix elements correspond to transfers of two electrons and the corresponding two-electron integrals can be neglected. In the doubly charged cluster hamiltonian matrix three supplementary parameters (R^{2+}_{ij} , F^{2+}_{ij} and R^{++}_{ij}) appear which are deduced from the knowledge of the first potential curves for Mg_2^{++}. Details of the ab-initio calculations for Mg_2^{+} and Mg_2^{++} as well as the parameters fitting as analytic functions of the internuclear distance are reported elsewhere.[7]

It is worth to point out here that

i) our model Hamiltonian fully includes correlation effects because it is based on a VB description,

ii) the dimension of the matrix, n for Mg_n^+ and n^2 for Mg_n^{++} (n situations with the two holes on the same center and n(n-1) situations with the two holes on different centers) is not prohibitive and extensive explorations of the potential hypersurface are still feasible for n ≅ 10 atoms.

II. Results for Mg_n^+ and Mg_n^{++}, n = 3-5.

These results have been extensively reported and discussed in ref. 4. Table I concentrates the principal points concerning conformations and energies.

Table I : Conformations, energies (E in eV) and nearest-neighbour distances (a in Å) for Mg_n^+ and Mg_n^{++}, n = 2-5.

n	conformation	E (+)	a (+)	E (++)	a (++)
2	linear	6.35	2.99	17.82	2.93
3	linear	5.81	3.10	16.21	2.99
4	square	5.23	3.20	14.39	3.12
	centred triangle	5.30	3.15	14.99	2.94
5	centred tetrahedron	4.77	3.16	13.87	2.93
	centred square	4.75	3.15	13.96	2.95

The most interesting feature is the near degeneracy between very different structures found for the largest clusters (n = 4 and 5). For Mg_4^+ and Mg_4^{++}, the square structure is more stable than the centred triangle by 0.07 eV (Mg_4^+) and 0.6 eV (Mg_4^{++}). For Mg_5^+ and Mg_5^{++}, near degeneracy occurs between the centred square (more stable by 0.02 eV for Mg_5^+) and the centred tetrahedron (more stable by 0.09 eV for Mg_5^{++}). Ab-initio Hartree-Fock calculations have confirmed that all these structures were real local minima but the relative ordering of these minima was different (see ref. 4). In recent test calculations including correlation performed for Mg_4^{++}, the square and centred triangle structures were found almost degenerated in agreement with the present model. This tendency of ab-initio Hartree-Fock to favour compact structures could be explained by the neglect of instantaneous polarization effects which can be recovered only with a correlated multi-determinant wavefunction and it will be discussed in a forthcoming contribution[8].

III. Results for Mg_n^{++}, n = 6-8

In this region of cluster size we expect that doubly charged clusters might be more stable than their decomposition products. The topology of the energy hypersurface is now complicated as illustrated by the results in Table II, in which we have reported absolute minima and some different local minima which are close in energy. For n=6, the most stable cluster corresponds to a compact structure (a square bipyramid) and the most compact (the perfect octahedron) lies very close in energy, although they correspond to different spin states. For n=7 and n=8, very stable structures can be found simply by adding a neutral Mg atom to the most stable cluster with n-1 atoms (see 7b and 8c) but competing structures are obtained by surrounding a stretched Mg_2^{++} dimer (see $\overline{16}$ distance in case 7a or $\overline{12}$ in case 8b with respect to 2.93 in the dimer) by a crown of quasi neutral Mg atoms. For n=8, the most stable (8a) is obtained by adding two Mg atoms to the relaxed structure 6a, but this process does not keep the original D_{4h} symmetry and the absolute minima is found in C_{2v} symmetry. In many cases presented in Table II there exist local minimas both for the singlet and for the triplet with very similar geometrical parameters.

A more detailed account of the topology of the energy hypersurface will be reported later on.

Table II : Conformations, energies (E in eV), geometrical parameters (in Å) and net charges for n = 6-8.

	Conformations	E	distances	net charges
n=6				
a) square bipyramid (singlet)	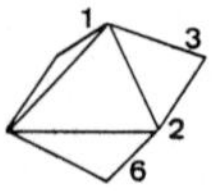	11.87	$\overline{23}$(3.69) $\overline{12}$(3.20) $\overline{16}$(3.69)	1(.64), 2(.18)
b) octahedron (triplet)		12.08	$\overline{12}$(3.27)	1(.33)
c) relaxed bicapped tetrahedra (singlet-triplet)	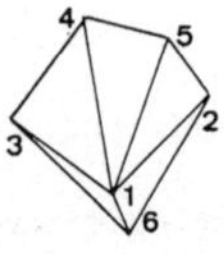	12.41	$\overline{12}$(3.00) $\overline{34}$(3.01) $\overline{14}$(3.57)	1(.50), 3(.43), 5(.06)

Table II (continued)

d) pentagonal pyramid			
(triplet)	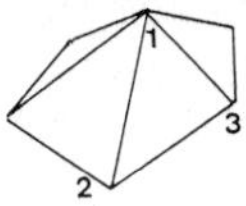 12.76	$\overline{12}$(3.67),$\overline{23}$(3.13)	1(.00),2(.40)
(singlet)	12.96	$\overline{12}$(3.65),$\overline{23}$(3.10)	1(.00),2(.40)
e) centred trigonal bipyramid (singlet)	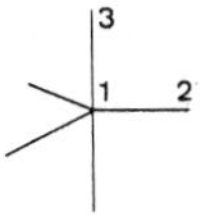 12.81	$\overline{12}$(2.93) $\overline{13}$(2.93)	1(1.04), 2(.21) 3(.16)
n=7			
a) pentagonal bipyramid			
(singlet)	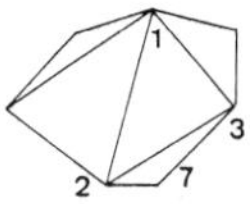 10.91	$\overline{12}$(3.29),$\overline{23}$(3.35)	1(.77),2(.09)
(triplet)	10.93	$\overline{12}$(3.29),$\overline{23}$(3.36)	1(.83),2(.07)
b) relaxed Mg_6^{++} + Mg			
(singlet)	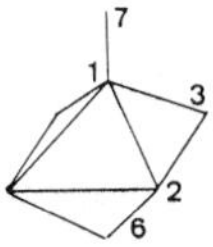 11.09	$\overline{12}$(3.15),$\overline{23}$(3.74) $\overline{17}$(3.05),$\overline{16}$(3.56)	1(.77),2(.15) 6(.58),7(.04)
(triplet)	11.21	$\overline{12}$(3.21),$\overline{23}$(3.92) $\overline{17}$(3.01),$\overline{16}$(3.24)	1(.82),2(.09) 6(.77),7(.06)
c) centred octahedron (singlet)	11.69	$\overline{12}$(2.93)	1(1.08),2(.15)
n=8			
a) relaxed Mg_6^{++} + 2Mg			
(triplet)	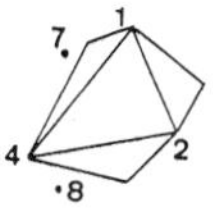 9.75	$\overline{16}$(3.14),$\overline{17}$(3.07) $\overline{12}$(3.26),$\overline{14}$(3.21)	1(.77),2(.04) 4(.13),7(.06)
(singlet)	9.76	$\overline{16}$(3.04),$\overline{17}$(3.01) $\overline{12}$(3.26),$\overline{14}$(3.27)	1(.80),2(.04) 4(.05),7(.11)
b) Mg_2^{++} + Mg_6 (chair conf.)			
(singlet)	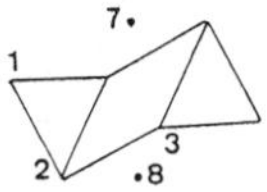 9.79	$\overline{13}$(5.19),$\overline{37}$(3.07) $\overline{78}$(3.01)	1(.09),7(.74)
(triplet)	9.94	$\overline{13}$(5.14),$\overline{37}$(3.07) $\overline{78}$(3.16)	1(.09),7(.72)
c) pentagonal bipyramid + Mg			
(triplet)	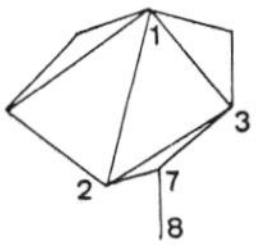 9.99	$\overline{17}$(3.09),$\overline{78}$(3.00) $\overline{12}$(3.28),$\overline{72}$(3.26)	1(.81),2(.06) 7(.84),8(.04)
(singlet)	10.02	$\overline{17}$(3.11),$\overline{78}$(3.01) $\overline{12}$(3.29),$\overline{72}$(3.26)	1(.79),2(.07) 7(.84),8(.04)

The stabilization of doubly charged clusters for $n > 7$ is evident from Figure 1 where the energies of the absolute minima (solid line) are reported together with

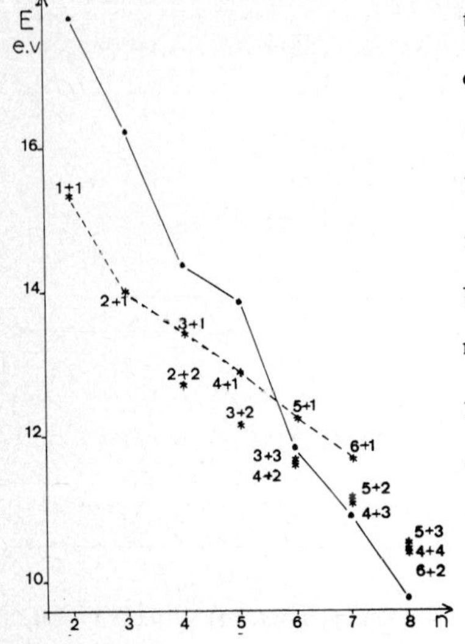

the energies of the different decomposition products of Mg_n^{++} as $Mg_{n-p}^{+} + Mg_p^{+}$. The broken lines ($Mg_{n-1}^{+} + Mg^{+}$) is simply a translation of the stability curve of the singly charged clusters. An interesting point for further studies of the possible decomposition of doubly charged clusters is the many possible routes. For instances, decomposition products for n=8 with p = 2,3,4 are in the same energy range.

Conclusions. A simple model Hamiltonian extracted from accurate diatomic calculations allows a complete and detailed study of small singly and doubly charged clusters of (s^2) metals. Even with such an easy to calculate energy functional, standard methods for studying potential surfaces seem to be limited to $n \cong 10-15$ atoms. Larger clusters studies will require statistical approaches. A similar model Hamiltonian is in progress for the study of charged rare gases clusters, including spin-orbit coupling.

References

(1) Bréchignac C., Broyer M., Cahuzac Ph., Delacretaz G., Labastie P. and Wöste L., Chem. Phys. Lett. 118 (1985) 174.

(2) For a recent and complete review of ab-initio calculations on small clusters see Koutecky J. and Fantucci P., Chem. Rev. 86 (1986) 539.

(3) Bauschlicher C.W. and Petterson L.M., J. Chem. Phys. 84 (1986) 2226.

(4) Durand G., Daudey J.P. and Malrieu J.P., J. Physique 47 (1986) 1335.

(5) Ellison F.O., J. Am. Chem. Soc. 85 (1963) 3540.

(6) Hesslich J. and Kuntz P.J., Z. Phys. D. 2 (1986) 251.

(7) Details of the ab-initio calculations, including diabatization transformations for Mg^{++} are reported in G. Durand and F. Spiegelmann, J. Phys. B, to appear. Fitting of the parameters can be found in ref. 4.

(8) G. Durand, J.P. Malrieu and J.P. Daudey, to be published.

Aknowledgment

The calculations reported have been made on a CRAY 1S computer through a grant from the Centre de Calcul Vectoriel pour la Recherche.

FROM MICRO-SYSTEMS TO MACRO-SYSTEMS: WHAT SIZE IS A METAL?

G. MAHLER

Institut für Theoretische Physik I
Universität Stuttgart
D-7000 Stuttgart 80
FRG

I. INTRODUCTION

Much of the progress in understanding condensed matter physics — despite its inherent complexity — may be traced back to the concept of model systems and model states, general enough to serve as a reference frame for more specific investigations. One of the most important examples is the macroscopic or thermodynamic limit[1]. The corresponding definition as a mathematical limiting process — though practically not executable — signals the fact that intensive physical properties should no longer depend on shape and volume; also any influence of the environment is excluded by definition. The resulting asymptotic "bulk-properties" are universal in the sense that least information is required for their specification. Homogeneous macroscopic systems may be defined as those which can within experimental accuracy be described by bulk properties. A metal is, strictly speaking, a macroscopic system.

Clearly, this concept is an idealization. Deviations occur away from thermal equilibrium and/or away from the macrolimit. Though non-equilibrium phenomena[2,3] also tend to be more dominant in bounded systems, we will presently restrict ourselves to thermal equilibrium. The violation of the macrolimit is of considerable current interest, e.g., in the context of cluster physics[4] and microstructured solid state devices[3].

II. STATEMENT OF THE PROBLEM

The influence of size can be systematically studied only for those physical properties which remain well-defined. One will expect that their variations cannot be predicted merely from general considerations but also require specific models. The latter include assumptions for size changes. The most transparent size changes are reduced to the change of a single control-parameter c (like radius of a sphere or slab thickness). The fractional shifts of scalar physical parameters α as a function of such control-parameters c are called scaling relations. Pertinent examples are the finite-size scaling[5] proposed for ordering transitions and the size-dependence of the classical work function[6]. The function $\delta\alpha(c)$ may, in particular describe invariance or a crossing-over between two asymtotic limits (see Fig. 1).

However, many properties are not just scalar parameters but rather fields $u(\underline{r}, t)$. Such fields can either be specified by their respective equation of motion (field equation) together with the appropriate boundary conditions or, directly, by the resulting temporal and spatial field patterns (or coherence functions). In this case the effect of size variation will be more complex. We will study examples related to a metal slab.

III. ELECTRONIC SUBSYSTEMS

While the infinite crystalline structure may readily be used as a given property, structural details of finite systems are still under consideration: This holds for the surface reconstruction of semi-infinite ideal crystals[7], but even more so for the equilibrium geometries of small particles[8].

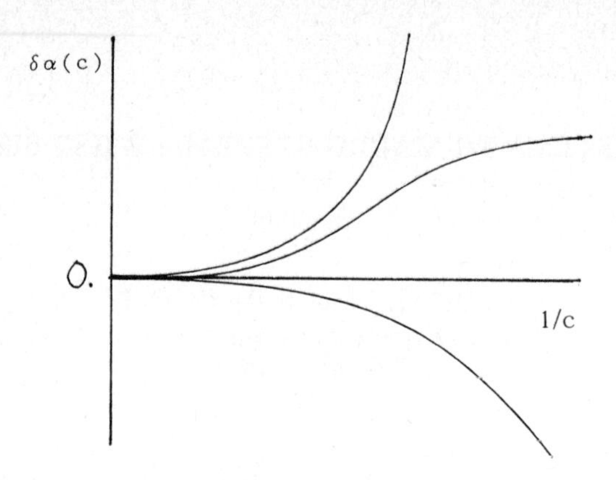

FIG. 1: Sketch of typical scaling relations $\delta\alpha(c)$, where $\delta\alpha$ is the fractional shift, c a size parameter

For the present study this issue will be set aside. The electronic subsystem is then defined by the Hamiltonian[9]

$$H_o = \sum_{j=1}^{N_o} \left\{ \frac{1}{2m} p_j^2 + e\phi(r_j) + \sum_l V(\underline{r}_j - \underline{R}_l) \right\} \quad , \tag{1}$$

$$\phi = \frac{1}{2}\phi_{\mathrm{int}} + \phi_{\mathrm{ext}} \quad . \tag{2}$$

Here $V(r) = \sum_l V_l$ is the single-particle potential created by the fixed array of ions at R_l, ϕ_{int} the potential due to the mutual Coulomb-interaction of the N_e electrons with charge density

$$\rho(\underline{r}) = e \sum_j \delta(\underline{r} - \underline{r}_j) = en(\underline{r}) \quad , \tag{3}$$

$$\Delta\phi_{\mathrm{int}} = -4\pi\rho/\epsilon_o \quad . \tag{4}$$

ϕ_{ext} is an external potential, and ϵ_o is the background dielectric constant. The Schrödinger-equation based on eq. (1) together with eq. (3) is the coupled matter-electric field representation of our charged electronic system. Putting (3) in (4) we can solve for ϕ_{int} so that the electronic field can be eliminated from (1): This representation is closed with respect to the matter field at the expense of non-linear interactions.

IV. SINGLE-PARTICLE-SCHRÖDINGER-FIELD

A simple model for the single-particle excitations is obtained for $\phi = 0$ and approximating V by a model potential. Such a procedure may even be thought to include the influence of ϕ_{int} on the level of a pseudo-potential. We remark in passing that these states by definition include the influence of the surface, and are therefore not subject to surface scattering.

In particular, for one-dimensional studies a finite Kronig-Penney-model with

$$V(x) = \begin{cases} \sum_n V_1 \delta(x - na_o), & \text{for } 0 \leq x \leq N_x a_o \quad , \\ V_o & \text{otherwise} \end{cases} \tag{5}$$

has been frequently applied recently. It can be shown to be equivalent to tight-binding models. Structural surface relaxation can be included; in the present case, a_o is taken to be a fixed lattice constant. With

$$\begin{aligned} \hbar K &:= (2mE)^{\frac{1}{2}} \\ Q &:= mV_1 K^{-1} \sin K a_o + \cos K a_o \end{aligned} \tag{6}$$

one finds, independent of size $L = N a_o$, two classes of states:

$$\begin{aligned} \mid Q \mid &\leq 1 : \text{extended states} \\ \mid Q \mid &> 1 : \text{surface states} \end{aligned} \quad . \tag{7}$$

The extended states can be classified by a band index $\lambda = 1, 2, \ldots$ and

$$k := \frac{1}{a_o} \arccos Q \quad ; \qquad 0 \leq k \leq \pi/a \quad . \tag{8}$$

In this sense the dispersion relation $E_\lambda(k)$ is size-invariant. The corresponding effective mass m_λ^*, defined by

$$\frac{1}{m_\lambda^*} = \frac{1}{\hbar^2} \frac{\partial^2 E_\lambda}{\partial k^2} \bigg|_{k=0} \tag{9}$$

can be used to approximate the dispersion relation by

$$E_\lambda(k) = \frac{\hbar^2 k^2}{2m_\lambda^*} \quad . \tag{10}$$

For $E < V_0$ and N finite, the allowed values of k are discrete (mode selection). For $V_0 \to \infty$ (infinite barrier) these allowed values become equidistant, $k = \nu\pi/L$, with ν integer. When this is put into eq. (10) the result is the same as obtained for a particle of mass m_λ^* in a box of length L, for which, however, ν was unrestricted. (In the limit $N = 1$ there is actually only one state left per band.) This one-dimensional box-model is a special case of Weyl's problem, reviewed in Ref. 11. The surface states, to be sure, depend on more details of size and surface region[12].

If

$$V(x, y, z) = V_x(x) + V_y(y) + V_z(z) \tag{11}$$

the above results would immediately carry over to three dimensions with

$$E = E_{\lambda_x}(k_x) + E_{\lambda_y}(k_y) + E_{\lambda_z}(k_z) \quad . \tag{12}$$

For isotropic bands, $\lambda = \lambda_x = \lambda_y = \lambda_z$ and $V_0 \to \infty$ we would thus obtain in effective-mass approximation

$$E_\lambda(\nu_x, \nu_y, \nu_z) = \frac{\hbar^2 \pi^2}{2m_\lambda^*} \left[\left(\frac{\nu_x}{L_x}\right)^2 + \left(\frac{\nu_y}{L_y}\right)^2 + \left(\frac{\nu_z}{L_z}\right)^2 \right] \tag{13}$$

where m_λ^* is a fixed bulk parameter. In particular, for $L_x, L_y \to \infty$ (slab geometry) and with m_λ^* taken from experiment this relation describes reasonably well the so-called electronic subband structure, being the subband index[13].

In general, however, the potential is not additive, and the band structure as a whole considerably deviates from eq. (13) and the implied simple scaling relations. Such deviations are enhanced if reduction in size is accompanied by a change of structural symmetry (compare, e.g., Ref. 4).

V. PARTICLE-DENSITY-FIELD

The ground state of spatially confined many-electron systems may be characterized by its particle-density field. In the simplest model one still neglects ϕ and just fills up the available single particle states according to the Pauli-principle. Typical examples are linear chain molecules or planar structures[14]. For later reference, we consider here a slab geometry (thickness L) with given electron bulk density n_b, based on the dispersion relation, eq. (13). The highest occupied subband is ν_F = integer$(k_F L/\pi\hbar)$, where $E_F = \hbar^2 k_F^2/2m^*$ is the Fermi-energy. The effective length is $L_{eff} = L - 3\pi/4k_F$. The resulting particle-density field is shown in Fig. 2.; the spatial pattern derives exclusively from the boudary conditions[15].

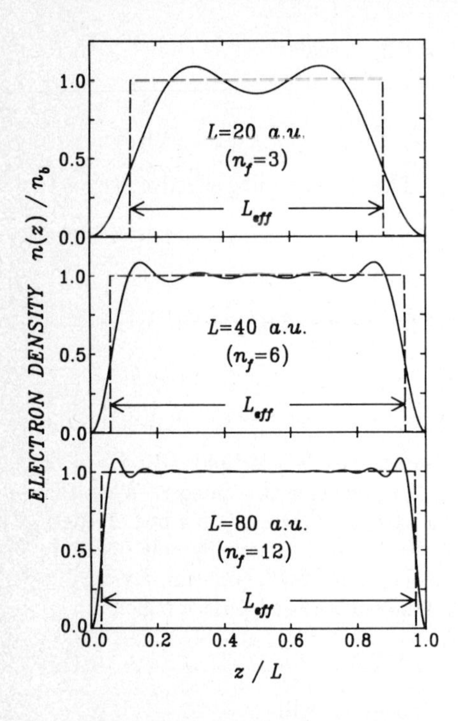

FIG. 2: Density-field $n(z)$ of a non-interacting electron system in its ground state (slab thickness L)

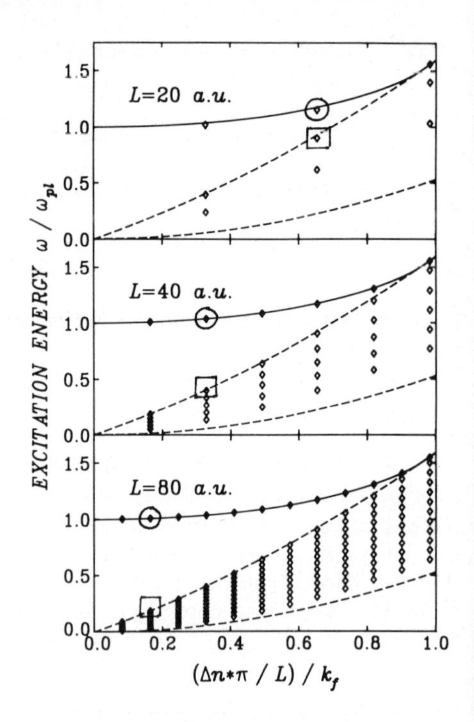

FIG. 3a: Intersubband exitations

Excited states of this model are pair excitations, obtained by replacing a single electron from an occupied to an unoccupied state. Such excitations are connected with a change of the local particle density, which, however, disappears with $L \to \infty$. Coherent superpositions thereof would correspond to density oscillations (compare the quantum beats known from atomic spectroscopy[16]).

VI. COUPLED CHARGE-DENSITY-AND ELECTRIC-FIELD

Classically, the plasma oscillation is a collective charge-density oscillation of an interacting electron gas. In an infinite homogeneous medium (or a medium with periodic boundary conditions) the excitation may be specified by its wave-vector $\underline{k}$, just like the single-particle excitations.

The long-wave-length bulk plasma frequency is

$$\omega_p = \left(\frac{4\pi n_b e^2}{\epsilon_o m^*} \right) \quad . \tag{14}$$

As for any hydrodynamic mode, damping increases with $\underline{k}$; these effects will not be considered here.

In quantum theory energy and time are complementary: Plasma-oscillations can thus be searched for as approximate eigenstates in terms of fixed energy quanta, leaving all observables time-independent, or as time-oscillations of the local density, for which the corresponding energy is not sharp. Both pictures refer to different experimental set-ups[17] as well as different theoretical approaches: The first or spectroscopy picture is usually developed from the closed matter-field representation (see Sect. III), the second one from the coupled field representation applying linear response, which we will use here.

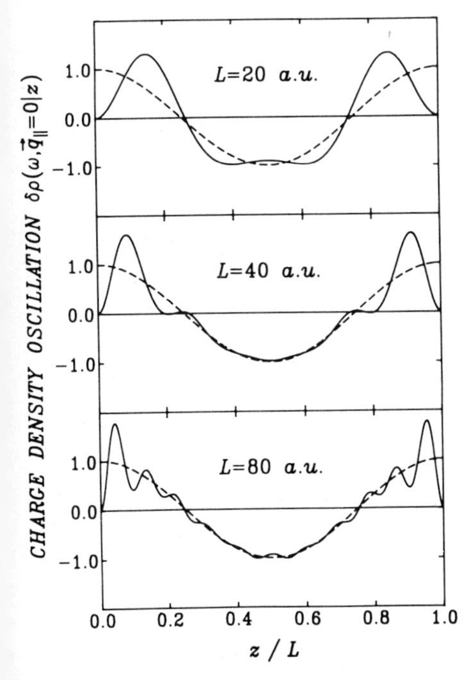

FIG. 3b Spatial pattern for the plasmon modes $\Delta n = 2$, (encircled in Fig. 3a)

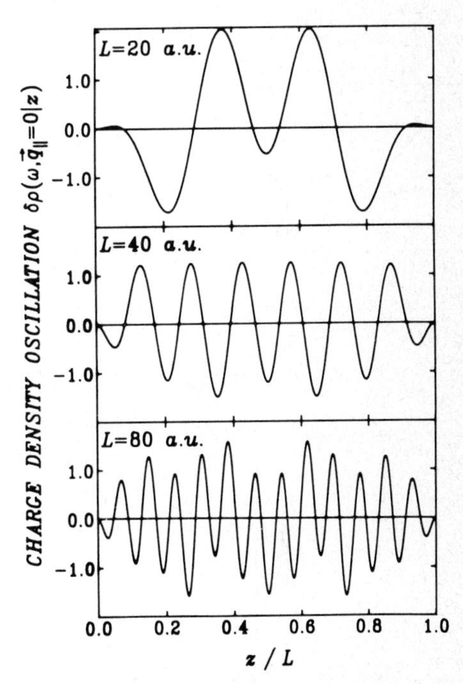

FIG. 3c Spatial pattern for pair excitation modes, $\Delta n = 2$, (squares in Fig. 3a)

In the inhomogeneous jellium model it is assumed that the positive ionic charge distribution exactly compensates the electronic charge of the ground state; microscopic structural details are thus neglected. Therefore, only the deviation $\delta\rho$ from the ground state charge density pattern acts as a source for the electric field.

For the slab geometry linear response is then defined by

$$\delta\rho(\omega, q_{\|}, z) = -\frac{1}{L} \int_o^L dz' \chi(\omega, q_{\|}, z, z') \phi(\omega, q_{\|}, z') \tag{15}$$

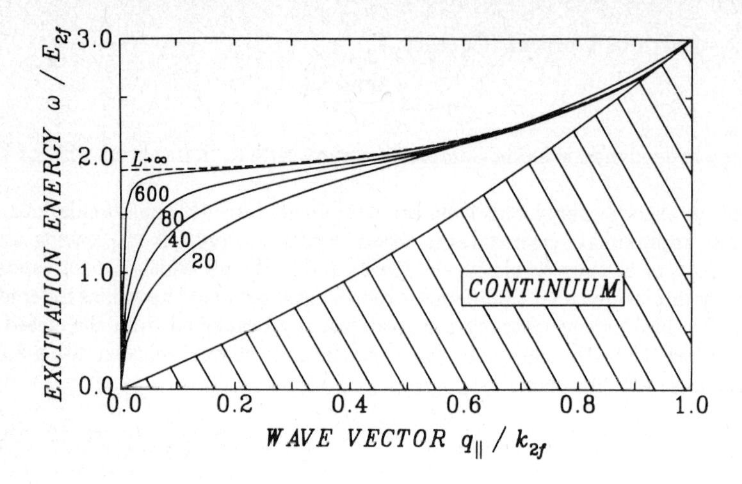

FIG. 4a Intrasubband excitations

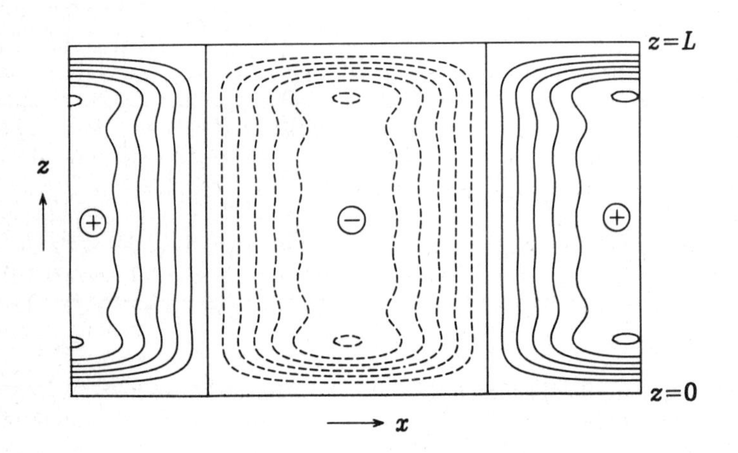

FIG: 4b Spatial pattern for the plasmon mode (encircled in Fig. 4a) Shown are lines of constant density

$$q_{\|} = (q_x, q_y, 0) \tag{16}$$

where

$$\left(q_{\|}^2 - \frac{d^2}{dz^2}\right) \phi_{\rm int}(\omega, q_{\|}, z) = 4\pi\delta\rho(\omega, q_{\|}, z) \tag{17}$$

and χ depends on the corresponding single-particle eigenvalues and eigenfunctions. In the absence of an external field, $\delta\rho \neq 0$ for self-sustained oscillation pattern. These excitations are characterized by a frequency and a spatial structure (charge amplitude field, electric field). A macroscopic spatial pattern is the generalization of the long wave-length limit in infinite systems.

Intersubband-excitations are obtained for $q_{\|} = 0$, i.e., there is no structure in the x- and y-direction. Excitations of the non-interacting system (compare Sect. V) are then characterized by the initial subband-index n_i and the subband-difference $\Delta n = n_f - n_i$. It is remarkable that this feature still holds for the interacting system, i.e., there is a definite relation between the spatial pattern (described by Δn) and the time-pattern (described by a frequency ω): The plasma-frequency, most strongly shifted by the inclusion of the Coulomb interaction, thus defines a dispersion-relation $\omega(\Delta n)$, see Fig. 3.

The major effect of the confinement is to allow only for discrete values of the wave-vector $q_{\perp} = \Delta n(\pi/L)$: Considering the complex nature of collective excitations this simple result could not have been expected. With $L \to \infty$ the usual bulk picture is recovered, provided the standing wave-representation used here is transformed into the travelling wave picture. In the small L-limit the approximate (13) as well as the Jellium-assumption tend to become unacceptable.

Fig. 3b (Fig. 3c) shows the spatial structure (even parity) of the charge density-field corresponding to a plasmon-mode (pair excitation mode). We see that with increasing size the plasmon pattern indeed approach that of the coherent bulk oscillation, while the pair excitation mode reduces to small scale fluctuations.

Intrasubband excitations correspond to a change of $q_{\|}$. The result is shown in Fig. 4. In this case the plasmon- dispersion remains continuous, but interpolates between 3-dimensional and 2-dimensional behaviour. As may have been expected, the limit of infinite wave-length always sensitively probes the finite size and therefore approaches the classical plasma frequency only asymptotically. For $10 \leq qL \leq 0.1 k_F L$ one finds for the fractional shift of ω^2 the scaling relation $\delta\omega^2 \sim 3\pi/4k_F L$.

In Fig. 5, finally, we demonstrate the influence of the environment (dielectric continuum ϵ_1). One should note that this influence derives only from the electric field, as the matter field, even for finite L, is decoupled from the surrouding by means of the infinite barrier $V_o \to \infty$. This point is clearly demonstrated in Fig. 5b: As the electric displacement field D within the slab approaches the constant value typical for an infinite system, also the dispersion becomes bulk-like.

Though truely low-dimensional model systems may also be addressed as "metals", actual slabs do depend on the environment, so that, e.g., the plasma-dispersion ceases to be an intrinsic property: Also finite "isolated" systems are an idealization of only limited validity.

VII. Summary and Conclusions

Having defined metal as a macroscopic (bulk) system, its minimum size is the size beyond which deviations form the bulk behaviour are no longer detectable. The size therefore depends on the physical property studied as well as on the sensitivity of the experimental apparatus.

We have presented some model treatments of a metallic slab. This system can be characterized by a number of scalar parameters as well as fields. As the slab thickness is varied the scalar parameters generally shift (scaling relations) while the field pattern change in space and time. Nevertheless, at least in the present model the time-pattern (frequency) and the space pattern are still found to be interrelated in terms of a dispersion relation.

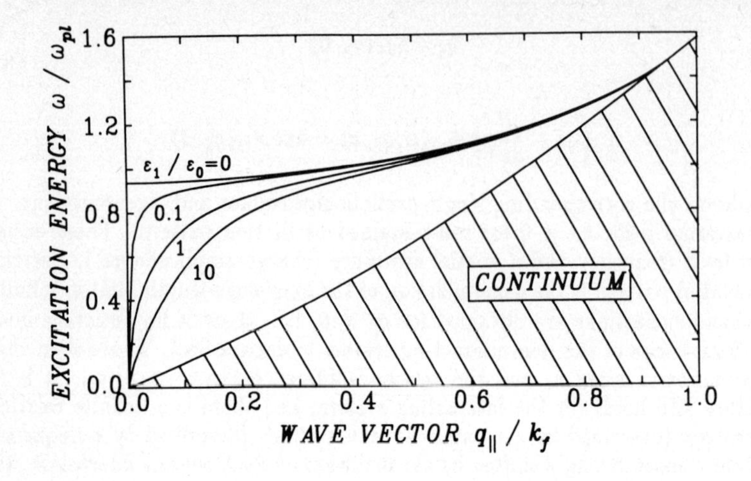

FIG. 5a Influence of the dielectric environment ϵ_1 on the intrasubbandexcitations ($L = 40$ u.a.)

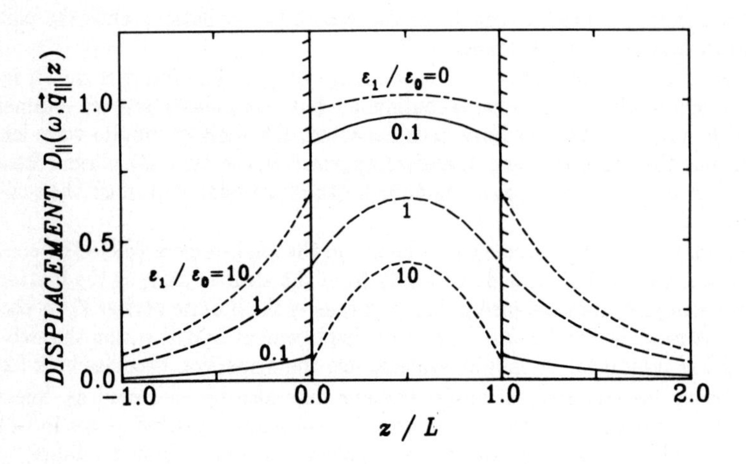

FIG. 5b Spatial pattern of the electric displacement field, $q = 0.1k_F$

REFERENCES

1. See, e.g., S. K. Ma, 'Statistical Mechanics', World Scientific 1985
2. G. Mahler and A. Fourikis, J. Lum. 30 (1985) 18
3. J. B. Barker and D. K. Ferry, Solid State Electron. 23 (1980) 519; 531; 545
4. T. Welker and T. P. Martin, J. Chem. Phys. 70 (1979) 5683
5. M. E. Fisher and M. N. Barker, Phys. Rev. Lett. 28 (1972) 1516
6. D. M. Wood, Phys. Rev. Lett. 46 (1981) 749
7. D. J. Chodi, Phys. Rev. Lett. 41 (1978) 1062
8. J. L. Martins, J. Buttet, and R. Car, Phys. Rev. Lett. 53 (1984) 655
9. H. Stolz, 'Einführung in die Vielelektronentheorie der Kristalle', Düsseldorf 1975
10. C. M. Soukulis, J. V. Jose, E. N. Economou, and Ping Shen, Phys. Rev. Lett. 50 (1983) 764
11. H. P. Balkes and E. R. Hilf, 'Spectra of finite systems', Mannheim 1976
12. H. M. Streib, Diplom Thesis, Stuttgart 1983
13. M. J. Kelly and R. J. Nicholas, Rep. Progr. Phys. 48 (1985) 1695
14. E. Heilbronner and H. Bock, 'Das HMO-Modell und seine Anwendungen', Verlag Chemie 1978
15. W. Teich and G. Mahler, Phys. Stat. Sol(b), to be publ.
16. K. Shimoda (ed.), 'High Resolution Laser Spectroscopy', Springer 1976
17. A similar situation as found for NMR is discussed by J. S. Ridgen, Rev. mod. Phys. 58 (1986) 433

ANALYTIC CLUSTER MODEL AS A BRIDGE BETWEEN MOLECULAR AND SOLID STATE PHYSICS

L. Skala

Faculty of Mathematics and Physics
Charles University
121 16 Prague 2, Czechoslovakia

H. Müller

Sektion Chemie der Friedrich-Schiller-
Universität Jena
DDR-6900 Jena

1. Introduction

Analytic cluster models (ACM) i.e. analytically solvable models of clusters are the only theoretical way which makes the investigation of the electronic properties of clusters of an arbitrary size and the convergence of their properties to those of the solids possible. The analytic solution of the Schrödinger equation is of course possible only for very simplified assumptions of the model. Two known approaches leading to analytic results are the HMO (or tight binding method) and FEMO (free electron molecular orbitals) methods [1-13]. More general arguments and experience gained till now indicate however that some conclusions following from the ACM have more general validity. This is the reason why we summarize the main results of the ACM together with some general results here and show in a few cases practical use of theoretical results.

2. Analytic Cluster Model

The larger part of the analytically solvable models has similar assumptions as the HMO: s-functions in the basis, the nearest neighbour interaction and the cubic form of the clusters (generalization to a rectangular parallelepiped is straightforward). Till now, the clusters with the SC [1-3], FCC [4-6], BCC [6-7] and diamond [10-11] lattice were investigated. The analytic one-electron energies and wave functions make possible to calculate the total energy, binding energy per atom, lower and upper "band" edge, Fermi level, bond orders and other quantities as a function of the number of atoms N. As an example we show in Fig. 1 the binding energy, Fermi level and lower and upper band edge of the FCC clusters for N up to 10^7.

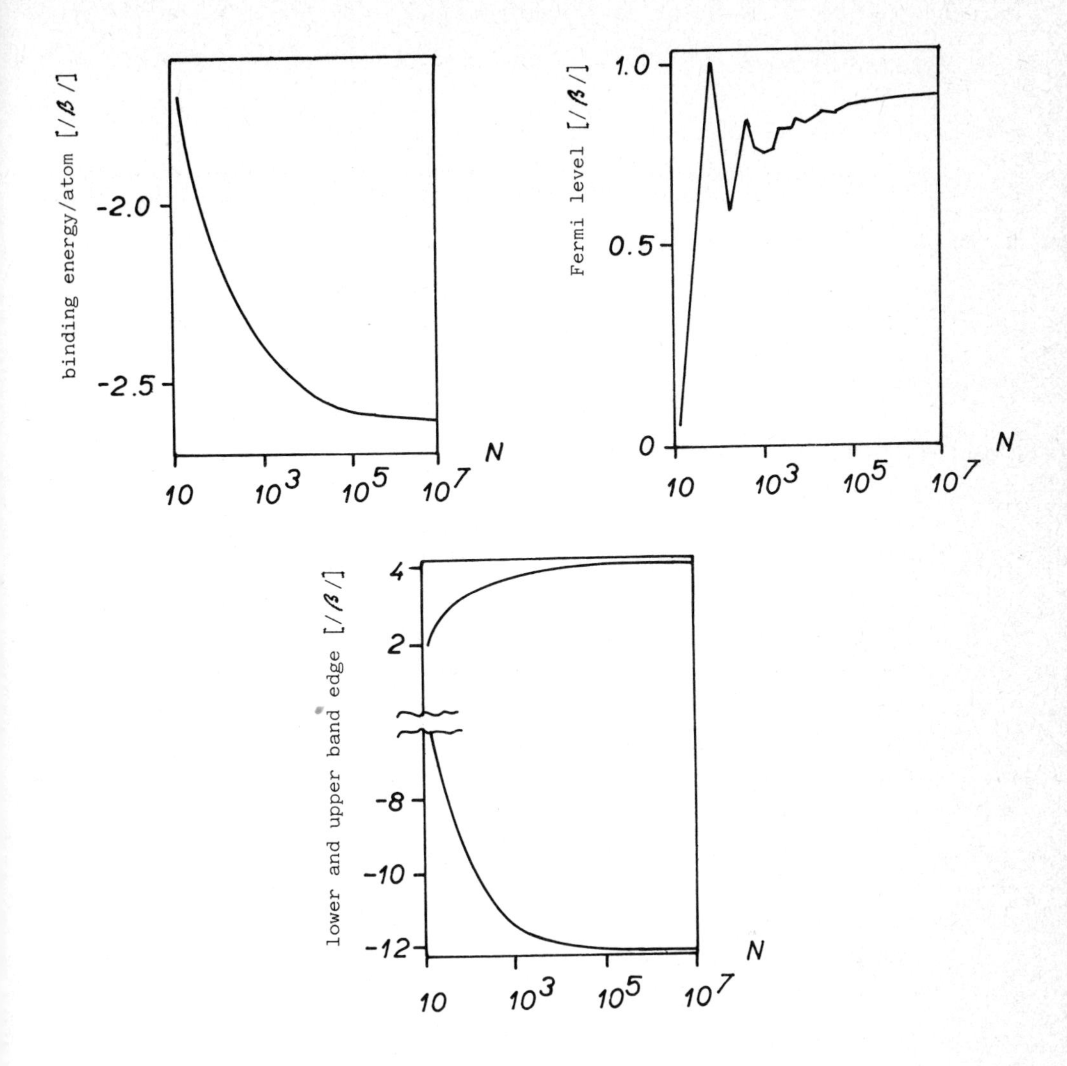

Fig. 1: The binding energy, Fermi level and lower and upper band edge of the cubic FCC clusters as a function of the number of atoms N. β is the nearest neighbour matrix element of the hamiltonian.

We see that the convergence of the quantities plotted in Fig. 1 is rather slow (in case of the Fermi level oscillatory) and that a large number of atoms (say 10^3 for the Fermi energy and lower/upper band edge and abou 10^5 for the binding energy) in the cluster has to be taken to achieve the convergence. Other examples of this kind are given in [3, 7, 11].

3. Interpolation Formulae

An analytic formula for the quantity G giving it as a function of N can be expanded in the powers of the parameter $1/N^{1/3}$ ($N^{1/3}$ is proportional to the number of atoms along the radius of a spherical particle) and truncated so that the asymptotically ($N \rightarrow \infty$) valid formulae

$$G(N)=G(\infty)(1-const/N^{1/3}) \quad (1a)$$

(for the binding energy) or

$$G(N)=G(\infty)(1-const/N^{2/3}) \quad (1b)$$

(for the Fermi level, bond orders, lower and upper band edge, ionization potential) can be obtained [3,7-8]. Equations (1) express the cluster property G(N) for arbitrary N by means of the same property for the solid $G(\infty)$ and one constant which can easily be calculated by fitting the experimental or calculated values of G for a few values of N. Formulae (1) do not of course give exact values of G(N). It is rather a simple way to interpolate G for large N. It is not surprising that eqs. (1) interpolate curves shown in Fig.1 well. Figures [2-5] show however that eqs. (1) can be applied more or less satisfactorily to calculated and experimental results lying out of the validity of the ACM.

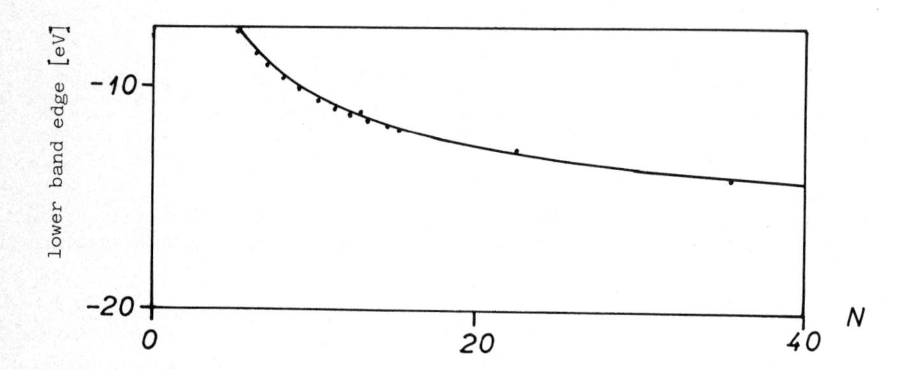

Fig. 2: The result of the interpolation (full line) of the calculated lower band edge (dots) of the BCC lithium clusters (CNDO/2, [8]).

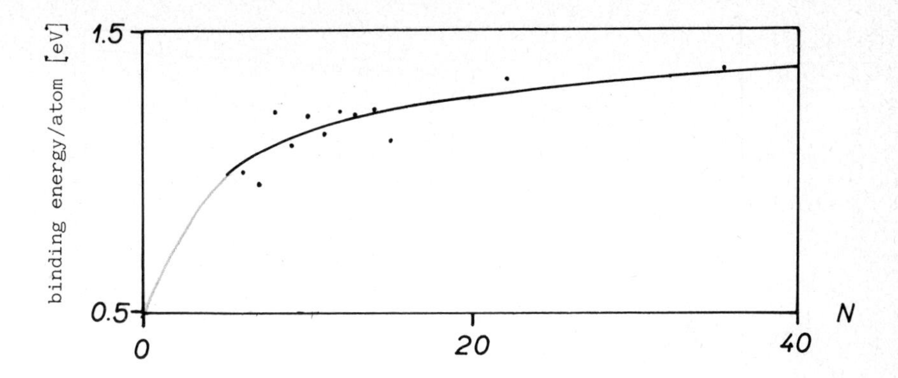

Fig. 3: The result of the interpolation (full line) of the calculated binding energy of the BCC lithium clusters (CNDO/2, [8]).

We see that the interpolation formula works for the lower band edge well. The agreement for the binding energy is less satisfactory because of the slower $1/N^{1/3}$ convergence and the oscillations of the Fermi level which are not in the interpolation formula taken into account. Figures 4 and 5 show the possibility of using eqs. (1) to interpolate experimental data [14]. In the first case the ionization potential of Na clusters is interpolated with a good result. The second figure shows the successful interpolation of the melting point of gold particles.

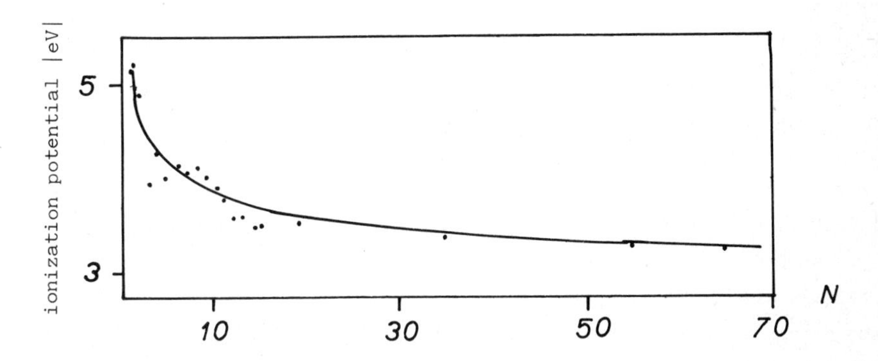

Fig. 4: The interpolation (full line) of the experimental ionization potential of Na clusters (dots) [15].

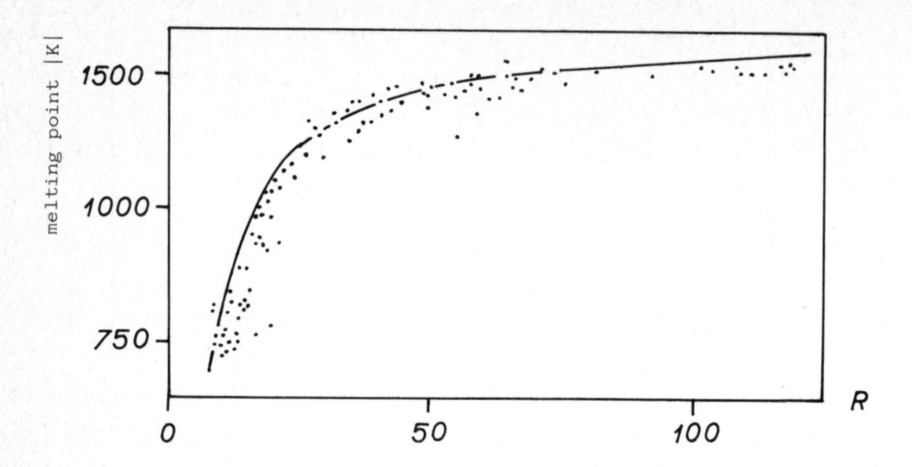

Fig. 5: The interpolation (full line) of the melting temperature of gold particles (dots) [16]. R is the radius of the particles.

We note that the number 3 appearing in eq. (1) has the meaning of the number of the dimensions d=3 in which the system increases. For instance, to get formulae valid for a thin film (N-the number of layers) one has to put d=1 [3, 12].

4. General Remarks

The analytic cluster model has a specific place between chemistry and solid state physics as it makes possible to introduce in full analogy with the solid state physics the k-vector which is, however, discrete (the surface boundary conditions lead to its quantization) and lies in 1/8 of the usual Brillouine zone (the boundary conditions lead to the standing wave functions so that the k and -k vectors give the equivalent states) [6]. This clarifies the character of the convergence of the cluster properties since, for example, the binding energy in the ACM leads to the summation over a regular mesh in the zone which becomes the integration in case of the solid. The convergence problem is therefore analogous to that met in the definition of the integral by the sum. These arguments have not unfortunately general validity since the dispersion energy relation E=E(k) for a cluster cutted out from the infinite crystal (with the so-called unrelaxed surface) is the same as for the crystal, however, has to be defined for the complex k-vector. The surface boundary conditions lead in general to a linear combination of the k-states so that the resulting sum is more complex. In case of the ACM, the k-vectors have particularly simple form so that the sum as well as other quantities can be evaluated analytically.

Despite of it, there are general arguments supporting the $1/N^{1/d}$ or $1/N^{2/d}$ convergence suggested by the ACM. It appears for example that the difference of the so-called integrated density of states I(E) of a cluster with the unrelaxed surface and the corresponding infinite crystal cannot be larger than a certain geometry-dependent value P [3, 12]

$$|I_{unrel}(E)-I_{bulk}(E)| \leq P. \tag{2}$$

The calculations show that $P \sim 1/N^{1/d}$ so that the largest possible difference of the integrated densities of states must go to zero similarly to the convergence of the binding energy in the ACM. Analogous inequalities can be derived for the clusters with a general (relaxed) surface as well as for the differential density of states and other quantities [3, 12]. The result is that the $1/N^{1/d}$ convergence is the slowest possible one. All above mentioned quantities must converge for $N \to \infty$ at least as quickly as $1/N^{1/d}$ to zero.

When using the interpolation formulae one has to bear in mind the nearest neighbour interaction assumed in their derivation. To get into the range of their validity one has to define the elementary cell (unit) large enough so that all other than the nearest neighbour interactions (in the sense of the units) can be with a reasonable accuracy neglected. The number N is the number of the units. Despite of this warning and in accordance with (2) Figs. 2-5 show that the interpolation formulaemay sometimes successfully be used even for a smaller number of atoms than required by this argument. The second comment regards the one-electron approximation supposed throughout the paper. If, for example, the electron correlation contributes significantly to the total energy the asymptotic interpolation formula for the binding energy has probably another form.

The parameter $1/N^{1/d}$ used in the derivation of the interpolation formulae can serve as a measure of the compactness used in the graph theory. From this point of view, its appearance in eqs. (1) is not accidental. We note also that a similar parameter (1/R, R-radius) appears in the theory of the surface tension of droplets.

References

1 T. A. Hoffmann: Acta Phys. Hungar. 1, 1(1951); 2, 97(1952)

2 R. P. Messmer: Phys. Rev. B15, 1811(1977)

3 L. Skala: Czech. J. Phys. B27, 171(1977)

4 O. Bilek, P. Kadura: phys. stat. sol.(b) 85, 225(1978)

5 P. Kadura, L. Künne: phys. stat. sol. (b) 88, 537(1978)

6 O. Bilek, L. Skala: Czech. J. Phys. B28, 1003(1978)

7 L. Künne, L. Skala, O. Bilek: Czech. J. Phys. B29, 1030(1979)

8 L. Skala: phys. stat. sol. (b) 109, 733(1982); 110, 299(1982)

9 G. B. Bachelet, F. Bassani, M. Bourg, A. Julg: J. Phys. C: Sol. St. Phys. 16, 4305(1983)

10 O. Bilek, L. Skala, L. Künne: phys. stat sol. (b) 117, 675(1983)

11 L. Künne, L. Skala, O. Bilek: phys. stat. sol. (b) 118, 173(1983)

12 L. Skala: phys. stat. sol. (b) 127, 567(1985)

13 H. Müller: Z. Chem. 12, 475(1972)

14 H. Müller, K. Strieckert: Wiss. Z. Friedrich-Schiller-Univ. Jena/Thür., Math.-Nat. R. (to be published)

15 E. Schumacher, M. Kappes, K. Marti, P. Radi, M. Schär, B. Schmidhalter: Ber. Bunsenges. Phys. Chem. 88, 220(1984)

16 Ph. Buffat, J. P. Borel: Phys. Rev. A13, 2287(1976)

MAGIC NUMBERS AND THEIR ORIGIN - ARE RIGID CORES REALISTIC?

L. Jansen and R. Block
Institute of Theoretical Chemistry
University of Amsterdam
Nieuwe Achtergracht 166
1018 WV Amsterdam
The Netherlands

TABLE OF CONTENTS.

I. INTRODUCTION AND BACKGROUND.

The concept of "magic numbers" in physics was not born in connection with clusters of atoms or molecules, the subject of this contribution. It originated in the late forties, when Maria Goeppert-Mayer[1] discovered on an empirical basis that nuclei with certain numbers of protons or neutrons (2, 8, 20, 50, 82, and 126; the last number refers to neutrons only) are particularly stable. The liquid-drop model and the uniform model proved to be inherently incapable of explaining such discontinuities. The same author approached this remarkable phenomenon herself theoretically[2] by assuming that strong spin-orbit forces exist, giving rise to a sequence of independent-particle states which match the experimentally observed irregularities. Practically simultaneously, Haxel, Jensen, and Süss[3] obs... the same phenomena and developed the same interpretation. The expression "magic numbers" came into use almost instantly[2,3]. These analyses, and those by several other nuclear physicists during the same period, led to the development of the nuclear shell model. Although the adjective "magic" suggests phenomena "seemingly requiring more than human power; startling in performance; producing effects which seem supernatural" (Webster's New College Dictionary), this non-scientific label has been adopted throughout as serving the purpose of identifying these numbers simply and conveniently. In the same vein, we will (ab)use "magic" in what follows.

Let us turn to atomic or molecular clusters. Experimentally, the determination of cluster-size distributions in a vapor is a very intricate task, and several different techniques are applied. For small metallic clusters, shock-tube and mass-spectroscopy methods have extensively been used. For non-metallic clusters, size distributions have been measured using nozzle-beam techniques, or electron diffraction. An intringuing observation has been that often such distributions are not continuous, but are marked by <u>local maxima</u>. Usually, such a maximum appears gradually with increasing cluster size, after which a steep decrease takes place. Following the nuclear physicists of the forties, it has become customary to call these numbers n* of maximal intensity "magic numbers".

Theoretically, the accurate evaluation of the concentration C_n of an n-particle cluster constitutes one of the principal, and most difficult, problems in nucleation physics. Especially in nozzle-beam applications the experimental situation is so complicated that it is practically impossible to retrace the thermal history of a given observed cluster. We briefly sketch two different avenues of approach:

A) At the time of measurement, the ensemble of clusters may, in good approximation, be treated as a multi-component system in thermodynamic equilibrium. The concentration C_n of a cluster of $\underline{n}$ particles is then given by (Boltzmann distribution)

$$C_n = C_1 \exp[-\Delta G(n,p,T)/k_B T] \quad , \tag{1}$$

where C_1 is independent of $\underline{n}$, at given pressure $\underline{p}$ and temperature $\underline{T}$. The quantity ΔG is the <u>free enthalpy</u> (<u>Gibbs free energy</u>) <u>of formation</u> of the cluster from $\underline{n}$ isolated particles. Plotting $\Delta G(n,p,T)$ against $\underline{n}$, maxima in C_n ("magic numbers" n*) correspond to "<u>dips</u>" in the free enthalphy of formation.

B) At the time of measurement, the ensemble of clusters is neither in thermodynamic equilibrium nor in a steady state, i.e. C_n depends on the time $\underline{t}$. The time evolution of the size distribution is, in the absence of condensation, and assuming monomer evaporation to be dominant,

$$dC_n/dt = R_{n+1}C_{n+1} - R_nC_n \quad , \tag{2}$$

where R_n is the evaporation rate of a cluster of $\underline{n}$ particles. The quantity R_n is simply taken proportional to $\exp[-(\Delta G(n-1)-\Delta G(n))/k_BT]$, where we have omitted the variables p,T attached to ΔG. The difference $\Delta G(n-1)-\Delta G(n)$ will be called "sublimation[4] free enthalphy" of the cluster of $\underline{n}$ monomers. To calculate the concentration C_n at time $\underline{t}$, we have to solve eq. (2) and then follow approach (1) to evaluate the free enthalpies. Maxima in the cluster-size distribution will now correspond to maxima in the <u>sublimation free enthalpy</u> as a function of cluster size $\underline{n}$, at given $\underline{p}$ and $\underline{T}$. We remark here that minima in the free enthalpy of formation necessarily imply maxima in the sublimation free enthalpy, but that the converse of this statement <u>is not true</u>: maxima in sublimation free enthalpy at given p,T can occur without "dips" in ΔG, namely, if ΔG is a "<u>quasi-monotonic</u>" function (i.e. ΔG is continuous with increasing $\underline{n}$, but the gradient function exhibits discontinuities) of cluster size $\underline{n}$. Examples will be given later on. The general theoretical problem of finding maxima in the cluster-size distribution (i.e. the magic numbers) of "either type", i.e. either those associated with dips in the free enthalpy of formation $\Delta G(n,p,T)$, or those corresponding to maxima in the sublimation free enthalpy, is much too complicated to be solved along either one of the above-sketched routes. To render our task as simple as possible, we limit ourselves to clusters of rare-gas atoms containing fewer than, say, 100 atoms. The first magic numbers in this range were observed in 1981, in an already classical nozzle-beam experiment on xenon gas by Echt, Sattler, and Recknagel[5]. Pronounced maxima where found for the values n* = 13, 19, 25, 55, 71, 87, and 147, with less-marked effects at n* = 23, 81, 101, and 135. These maxima are observed after ionization of neutral clusters, and the question whether they are characteristic for ionized aggregates only and bear little relation to neutral-cluster properties has been, and still is, frequently discussed. Harris <u>et al.</u>[6] found maxima for argon at n* = 13, 19, 23, 26, 29, 32, and 34 ionizing the gas before it enters the nozzle. This range of sizes is thus much more densely populated than for xenon clusters. Ding and Hesslich[7] observed peaks for argon at n* = 14, 16, 19, 21, 23, 27, ..., i.e. in part different from those established by Harris <u>et al.</u>[6]. Sáenz <u>et al.</u>[8] mention as a possible explanation of these discrepancies incomplete thermalization of argon clusters during their time of flight. Kreisle *et al.*[9] using various retarding potentials, found little evolution with time of the intensity maxima for e.g. water clusters and those of CO_2 , but for argon no such experiments have as yet been undertaken. The above facts are stated principally to illustrate some of the experimental intricacies which beset reliable measurements of cluster-size distributions.

We must here disregard such confusing complications, restricting ourselves to thermally stabilized, neutral, rare-gas clusters. To simplify the problem even further, we will in first instance assume the clusters to be near zero K and neglect zero-point energies. Under those assumptions, the *potential* energy determines relative stability of clusters and their most favourable configurations. For rare-gas atoms, we have a reasonably accurate knowledge concerning their interactions in terms of a Lennard-Jones (6,12) potential $4\varepsilon[(\sigma/R)^{12}-(\sigma/R)^{6}]$, where $\varepsilon>0$ is the depth of the potential well, reached at distance $R_0 = (\sqrt[6]{2})\sigma$, with σ the interatomic separation at which the potential is zero. The first term is a rather steep repulsion, the second term the well-known long-range Van der Waals attraction. We should warn against interpreting this potential at any R as the sum of two *physical* effects: it is just a mathematical representation of an interaction which is attractive at large distances and repulsive at short distances, and which appears to fit gas-, liquid-, and (some) solid-state properties rather well. For argon and xenon, to give two examples, ε/k_B is 120 and 220 K, respectively; σ = 3.4 Å for argon and 4.1 Å for xenon. The total cohesive energies for the solids are only of the order of a few kcal/mole, i.e. about 0.1 eV, and thus very weak on the chemical scale.

In nozzle-beam experiments the clusters are in a region of very low temperatures, a few times 10 K. Then the probability for two atoms to find themselves in the repulsive part of their potential is very small. It is thus tempting to replace this part by an infinitely steep wall, i.e. the atoms are rigid spheres interacting through an attractive R^{-6} potential. To find structures of especially low potential energy we must look for close packings of the rigid spheres, as each atom attempts to be in contact with a maximum number of its neighbours. We will follow this track for a while, retaining the idea that interactions between rare-gas atoms are strictly of two-body type, i.e. strictly "additive.

II. CLOSE PACKINGS OF RIGID SPHERES, AND THEIR DEFICIENCIES.

The start of the magic-number series for both argon and xenon is n* = 13. This particular value can be readily explained in several different ways. Bulk crystals of the rare gases, except He, are definitely face-centered cubic (fcc), in which a central atom is surrounded by twelve nearest neighbours. Also a hexagonal close packing (hcp) will do, since again the number of nearest neighbours is 12. An icosahedron, with 20 equilateral triangles as faces, is an even better candidate, being more spherical. These three main competitors for explaining n* = 13 are sketched in the following Figure 1.

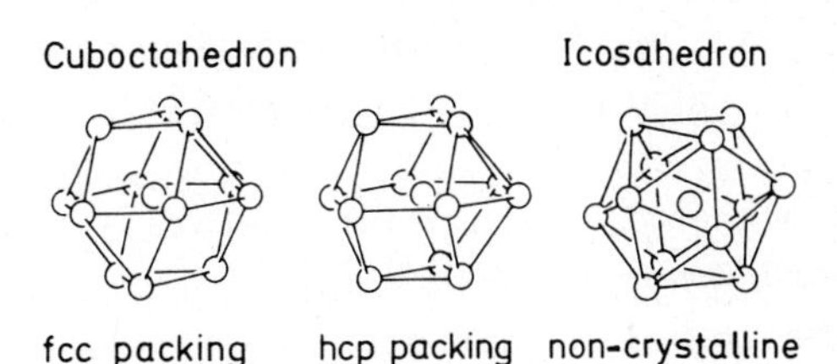

Figure 1. Face-centered cubic (fcc), hexagonal close packing (hcp) and the n = 13 icosahedron.

In 1962, Mackay[10] developed icosahedral packings of spheres by consecutively adding shells of icosahedra around a central one. Such structures, albeit non-crystallographic (note the five-fold symmetry axis in Figure 1), certainly come closest to a sphere and, moreover, achieve this with virtually close-packed surfaces. A closely spherical packing is intuitively synonymous with strongest binding per atom in a rigid-sphere-plus-R^{-6}-potential, since the number of point contacts is a maximum. The Mackay packings count 13, 55, 147, 309, 561, ... atoms, of which the first three are indeed magic numbers for xenon. However, magic numbers in-between cannot be explained on this simple basis. Farges _et al_.[11] proposed to fill these gaps by developing interpenetrating icosahedral structures. For example, n* = 19 is interpreted in terms of a double icosahedron sharing seven atoms. From their electron diffraction experimental studies on argon, these authors had earlier concluded[12] that argon clusters grow in terms of polyicosahedra (PIC's), until in the region of a few thousand atoms the "signal" is given to transform to the fcc bulk structure. If one assumes that the magic numbers observed are associated with positively charged clusters, then $n^* = 19$ for Xe or Ar can be readily constructed by placing a Xe_2^+ or Ar_2^+ dimer at the center of the two interpenetrating icosahedra. The following magic number, $n^* = 25$ for Xe, deviates definitely from the fcc or hcp series, since their third shell counts 24 atoms, which would take us to 43 atoms. Neither is 25 a Mackay number, nor can it easily be fitted into the Farges *et al.*[12] series of interpenetrating icosahedra. The conclusion thus far must be that close packing of rigid spheres, plus R^{-6} attraction, does offer *some* insight into the observed magic numbers, but this insight is much too crude to be called sufficient. Especially the "fine structure" found with argon clusters, if real, remains a complete mystery.

III. MAGIC NUMBERS OBTAINED WITH REALISTIC PAIR POTENTIALS.

The next step in our analysis consists naturally in replacing the rigid-sphere by a Lennard-Jones R^{-12}-repulsion, in the hope of obtaining more "structure" in the stability series. Very extensive work on this basis was performed especially by Hoare and Pal[13]; for an outstanding account of their work and that by many other researchers we refer to the penetrating survey article written by Hoare[14]. Here, we can mention only a few results. A systematic attempt to enumerate all existing potential-energy minima under a Lennard-Jones (6,12) potential (and under a Morse potential) was undertaken by McInnes[15] for $n>6$. Between $n=6$ and $n=13$ this number (not counting enantiomorphs) grew as 2 ($n=6$), 4, 8, 18, 57, 145, 366, and 988 ($n=13$), a staggering sequence. Of the 988 minima for $n=13$, the icosahedron was indeed found to constitute the structure with the lowest potential energy. More striking is undoubtedly the sensitivity of the results to the form of pair potential: with a Morse potential and a "reasonable" value for the parameter in the exponentially decreasing repulsion, 952 of these minima *collapse*; the remaining ones correspond to 36 structures also obtained under the Lennard-Jones pair interaction. In view of this complexity, the accurate evaluation of *thermodynamic* cluster functions (internal energy, free enthalphy, entropy) becomes at best excessively complicated and otherwise practically impossible. To obtain, nevertheless, results of hopefully more-than-accidental validity, drastic simplifications are unavoidable. The quickest route to cluster thermodynamics is through the so-called harmonic oscillator/ rigid rotor (HO/RR) approximation applied to the supposedly (at zero K) most stable, isolated, configuration for given number *n* of atoms (i.e. describing the whole ensemble of cluster types by a single "representative" structure). In this model, the translational and rotational modes of the clusters are treated classically, whereas the vibrational modes are calculated quantum mechanically. Full details can be found in refs 14 and 16. Of special interest to us are the results as far as these pertain to the magic-number phenomenon. Here, we quote Hoare[14] (p. 113): "... Apart from a small inflection at n=13 for the icosahedron, the dependence of zero-point energy, internal energy, and Helmholtz free energy are all virtually monotonic in the cluster size when reduced per atom. Thus earlier suggestions that "magic numbers" may exist ... cannot be substantiated. Certainly the numbers n = 13, 55, 115, and 147 are "magic" in the sense that they mark crucial geometric features, but these do not cause sudden shifts in any of the thermodynamic functions, because they emerge and assemble gradually with increasing *n*". Hoare continues to conclude, albeit somewhat *ad hoc*, that "... although thorough studies of [the above properties] under a variety of interatomic [pair] potentials have yet to be published, there is no reason ... to doubt that the same trends will prevail when alternative force laws are introduced. However, this is a point on which further confirmation would be desirable, particularly in relation to aspects of crystal growth morphology".

To the same category of analyses belongs the investigation by Farges *et al.*[18] regarding stability of cluster models made of double-icosahedron units, among which the n* = 19 structure, already mentioned, is an example. These authors presented a cluster sequence in which a primitive 13-atom icosahedron is grown in such a way that the maximum number of double icosahedra is formed, leading to 33 constructed models containing 13 to 45 atoms. Each of these structures was then relaxed under a Lennard-Jones (6,12) potential, until the forces on the atoms (argon) were negligibly small. The calculated sublimation energies (T = 0 K) exhibited sharp peaks at n* = 13, 19, 23, 26, 29, 32, 34, 37, ... Most surprisingly, these magic numbers match those observed by Harris *et al.*[5], although in their experiment the gas was *ionized prior to* entering the nozzle. Each peak appears just when a new icosahedron is formed interpenetrating the primitive one, thus providing a perfect illustration of Hoare's "crucial geometric features". Their calculations are of interest also in a different sense. In Figure 2 we present the (binding) energy per atom as calculated from Farges *et al.*[12] sublimation energies.

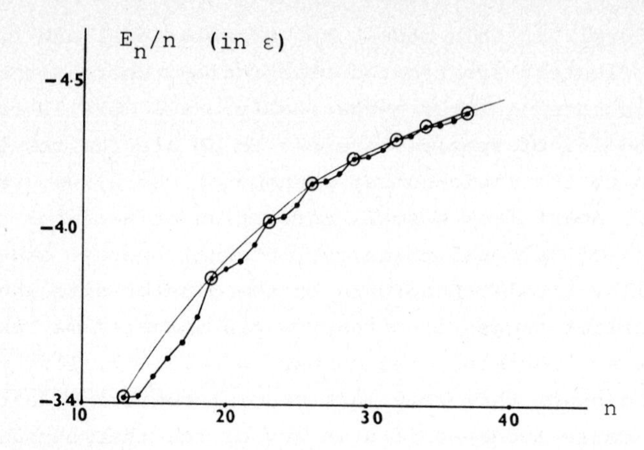

Figure 2. Binding energies per atom, as a function of number of cluster atoms n, for the growth sequence of Farges et al. (ref. 12).

We see that the energy per atom increases quasi-monotonically with n, exhibiting no dips, although it does lead to maxima in the sublimation energy and, thus, in the cluster-size distribution. This is an example[12] of "magic numbers" arising from a virtually (quasi-)monotonic dependence of energy on n. We note that, on the basis of approach A (equilibrium clusters) the ratio C_n/C_{n-1} is exponentially proportional to the sublimation free enthalpy and, thus, this ratio is a maximum for clusters of highest sublimation enthalpy. However, whenever $\Delta G(n) < \Delta G(n-1)$, the ratio $C_n/C_{n-1} > 1$, i.e. no maximum C_{n^*} occurs.
The magic numbers delivered by the Farges et al. sequence for argon also coincide with those calculated by Sáenz, Soler, and Garcia[8] for positively charged argon clusters $(Ar_n)^+$, taking into account the formation of a molecular dimer ion within the cluster. Their interaction function consisted of three components: a Lennard-Jones (6,12) potential between neutral atoms, the interaction between the charge on the dimer ion and the induced dipoles, and the intramolecular potential for the two atoms forming the ion. On the other hand, no explanation could be offered for some of the peaks observed by Ding and Hesslich[7] (n* = 14, 16, 21, 27, ...). As we mentioned earlier, Sáenz et al.[8] ascribe these discrepancies to possibly incomplete thermal stabilization of the Ar clusters.

All of the calculated magic numbers discussed heretofore are of the "sublimation-energy-induced" type, and the question arises whether those of the "first type", associated with dips in the free enthalpy of formation $\Delta G(n,p,T)$, occur at all. We mentioned the negative outcome of analyses based on the HO/RR approximation[14,16]. Recently, Freeman and Doll[17] undertook a very detailed study of the thermodynamic properties of argon clusters by a combination of classical and quantum Monte Carlo methods. The argon atoms interact through a Lennard-Jones potential; internal energies, free energies, and entropies were calculated as a function of pressure and of cluster size. To isolate a given cluster, a continuous, strongly repulsive, potential at a certain distance from the center of mass is added, avoiding vaporization.
For each cluster of size n an initial configuration was chosen corresponding to the minimum potential-energy structure of Hoare and Pal[18]. Full details are found in ref. 17. The results for the free enthalpy of formation at T = 10 K and three different pressures are presented in Figures 3a, 3b, and 3c (see next page).
The three pressures were chosen such that a maximum in ΔG occurs at $n \approx 8$-10. The results are most illuminating: whereas a dip in Figure 3a is found in the quantum results at $n^* = 7$ (the Hoare-Pal structure is a pentagonal bipyramid), it is absent from the classical curve. In Figure 3b, $n^* = 7$ has disappeared, and $n^* = 13$ emerges as magic in both the quantum and classical calculations. At still lower pressures, Figure 3c, $n^* = 13$ remains, and $n^* = 19$ tends to become magic on the quantum curve. At very much lower pressures still, ΔG increases rapidly with n, and no dips are discernible. On the other extreme, at very high supersaturation, ΔG decreases steeply, (quasi-)monotonically, with increasing n. Local maxima in the sublimation free enthalpy however are, also in the last case, present at $n^* = 7$, 13, and 19. Such magic numbers are not observable at high supersaturation, since the sublimation enthalpy is strongly positive for any n, implying excessively slow evaporation. We note that $n^* = 7$, 13, and 19 also mark special points on the free enthalpy versus n curve in the HO/RR approximation[16], without the free enthalpy exhibiting a dip. This "coincidence" will be discussed further in the final section, where it will be shown that both "types" of magic number have the same origin. We will, consequently, be in a position to drop the distinction between the two types.
Freeman and Doll[17] dissected, at $p = 3.34 \times 10^{-16}$ atm., their results in order to analyze which one of the thermodynamic functions: internal energy, enthalpy, or entropy, causes the dip at $n^* = 13$. It was established that only the internal energy exhibits this irregular behaviour,

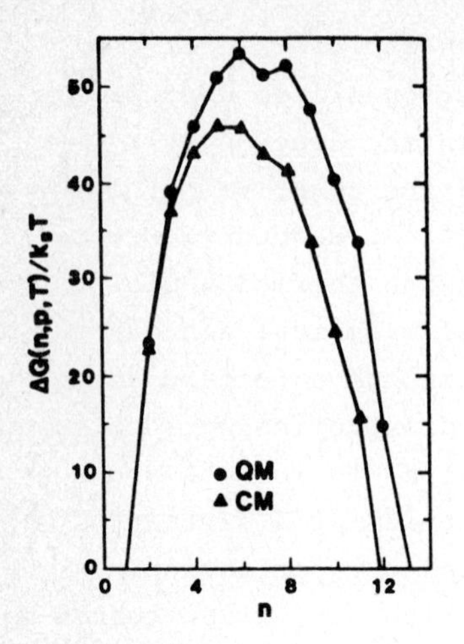

Figure 3a

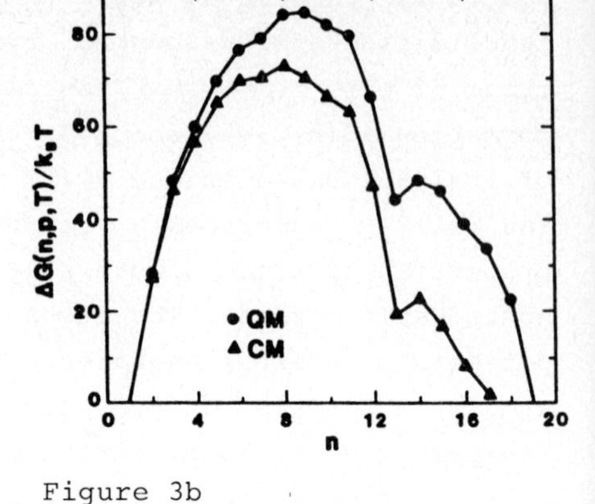

Figure 3b

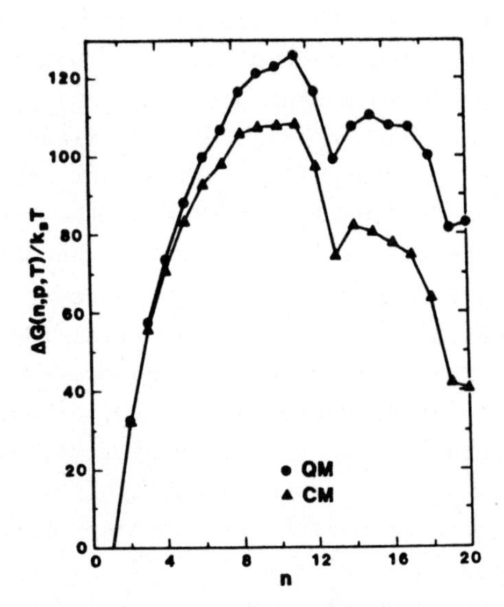

Figure 3c

Figures 3a, 3b, 3c. The Gibbs free energy of formation as a function of argon cluster size at T = 10 K and p = 33.4 fatm (3a), 0.334 fatm (3b), and 3.34 aatm (3c), respectively.
The circles are the quantum results and the triangles are the classical results. The points are connected by straight lines. In no case is the error bar larger than the circle or triangle. (From: D.L. Freeman and J.D. Doll, J. Chem. Phys. 82, 462 (1985).)

the entropy changing much more regularly[19] as a function of *n*. These results are shown in the following Figures 4a and 4b (see next page) for the enhalpy change $\overline{\Delta H}$ of the process $Ar_{n-1}+Ar \rightarrow (Ar)_n$ and the corres ponding entropy change $\overline{\Delta S}$. The authors find that the occurrence of magic numbers is very sensitive also with respect to temperature, but they provide no further details on this point.

All-in-all, we may conclude that magic numbers associated, at pressure p and fixed temperature T, with a dip in the free enthalpy of formation $\Delta G(n,p,T)$, emerge or disappear with pressure in a rather erratic way; the dips which do occur correspond to maxima in the sublimation energy of the clusters. The agreement between observed (Harris *et al.*[6]), constructed (Farges *et al.*[12]) and calculated (Sáenz *et al.*[8]) magic numbers for neutral[12] and ionic[6,8] argon clusters remains a mystery, as is the occurrence of some additional n* found by Ding and Hesslich[7]. It is also to be noted that the theoretical magic number n* = 7, corresponding to a pentagonal bipyramid on a Lennard-Jones (6,12) basis, has not been found experimentally.

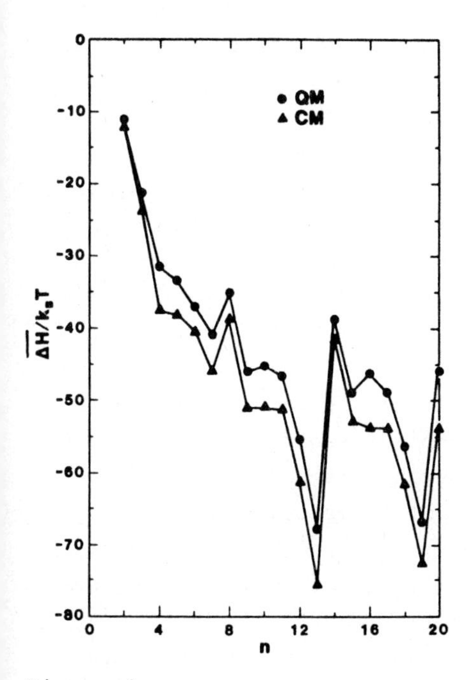

Figure 4a

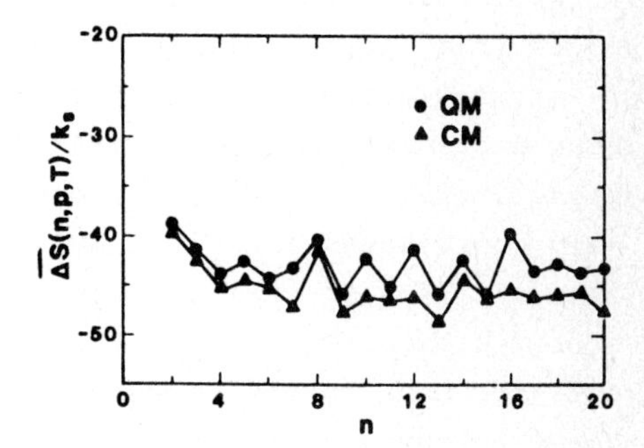

Figure 4b

Figures 4a, 4b. The enthalphy change (4a) and entropy change (4b) for the process $Ar_{n-1(g)}+Ar \rightarrow Ar_{n(g)}$ at T = 10 K and p = 0.334 fatm. The circles are the quantum results and the triangles are the classical results. The points are connected by straight lines. In no case is the error bar larger than the circle or triangle. (From: D.L. Freeman and J.D. Doll, J. Chem. Phys. 82, 462 (1985).)

IV. ARE PAIR POTENTIALS SUFFICIENT TO EXPLAIN MAGIC NUMBERS AND MICROSTRUCTURES?

In the previous Section we had ample opportunity to extol the Lennard-Jones (6,12) potential in providing a relatively solid basis for explaining the occurrence of maxima in cluster densities and of microcluster static morphology at low T. Given also its simple form, it is not surprising that practically all microcluster calculations involving rare-gas atoms are based on that type of interaction.

There is, however, one fundamental flaw with the Lennard-Jones potential, a defect it shares with all other physically acceptable pair interactions: the stability of the face-centered cubic bulk-crystal structure cannot be explained on this basis. To show this we can, of course, exclude the icosahedra, as these are non-crystallographic. The problem is thus to decide between a fcc and a hcp crystalline configuration. In both structures, a central atom is surrounded by 12 nearest neighbours in a first shell and 6 in a second shell. The third shell counts 24 atoms; in the fcc structure they are at $3^{\frac{1}{2}}$ times the nearest-neighbour distance from the central atom, in the hcp configuration somewhat closer, viz. at $(2+2/3)^{\frac{1}{2}}$ times that distance. The rest of the crystal only contributes a negligibly small amount to the cohesive energy per atom. It is seen that the hcp structure will have a small advantage over fcc because of its third shell. Numerical calculations, carried out more than thirty years ago[20-24], have shown that for a Lennard-Jones potential, or any physically acceptable pair interaction, the hcp lattice is favoured over fcc to the order of 0.01% of the cohesive energy. This difference has the wrong sign, and is much too small to be of any significance. Inclusion of zero-point energy does not change the outcome. It is clear that the only possible remedy is to look for short-range, many-atom (at least three-atom) interactions, i.e. those depending simultaneously on the coordinates of more than two atoms.

Before presenting a more quantitative analysis, let us return to the rare-gas and other clusters. Against a quite different background, Hoare[14] arrives at much the same conclusions, in conjunction with crystal growth mechanisms. We can do no better than to quote the relevant passages from his survey article (pp. 127-129): "... we should point to the fundamental difference between the genesis of a crystalline or crystalloid microcluster from a liquid drop cooled by evaporation or in a carrier gas and the corresponding process under conditions of atom-by-atom deposition at a seed structure. In the former case there is little difficulty in imagining a substructure such as the 55-icosahedron forming first by solidification of its 13-icosahedron core, then by cooperative rearrangement of the second and third shells about it. For the same structure to form from the vapor atom by atom, a definite reconstructive rearrangement would be required to put the second shell in place. The first icosahedron could certainly form [by the Werfelmaier sequence], but the second shell could not, for the simple reason that atoms would have to sit waiting balanced on a tetrahedral edge for their supporting partners to arrive. ... We have serious doubts whether on very cold substrates such structures would assemble by vapor deposition if two-body forces alone would be active". "A similar mystery attends the growth of the less stable multiply twinned pentagonal bipyramidal crystallites which ... tend to form with some readiness in deposits of metal smokes. ... A number of authors ... seem[s] to have assumed that the pentagonal nuclei for these structures simply assemble themselves, each atom moving unerringly to its required place. It would seem to us that the only situation in crystal growth where atoms might appear to "know where they are supposed to go" is when the final state is of appreciably lower free energy than all others and where sufficient thermal fluctuations are present for each atom to wander between alternative positions before it finally settles. Even then the surfeit of kinetic energy must be removed before the structure will become permanent. The data ... show beyond doubt that the free energies for small multiply twinned structures of $n = 10$, 14, 18, and 22 are quite unable to compete with those of tetrahedral/icosahedral type in the same size range. ... Thus there can be no free-energy driving force to favour such a growth route...". "The easy way out of this conflict between theory and observation is to find fault with the Lennard-Jones potential. Once the necessity of two-body central forces is abandoned, it is not difficult to imagine the occurrence of several-atom forces which, through some peculiarity of surface densities of states, actually favour the formation of, say, a protruding tetrahedral vertex...".

That many-atom interactions, if they exist, generally cause differences between structures which are identical on a pair-potential basis, is evident. As a simple example, consider a central atom and its 12 nearest neighbours in a fcc and hcp crystal, respectively. We suppose that also <u>three</u>-atom interactions act between the central atom and two nearest neighbours, forming a triplet of atoms. There are $12 \times 11/2 = 66$ of such triplets in either structure. Further analysis shows that 57 of these triplets are shared by the two structures, <u>but 9 are different</u>.

These 9 different (isosceles) triangles, with opening Θ, are listed in the following Table I.

	a^2	b^2	c^2	θ	No.	a^2	b^2	c^2	θ	No.
hcp	1	1	8/3	110°	3	1	1	11/3	146°	6
fcc	1	1	3	120°	6	1	1	4	180°	3

Table I. The nine different triangles a, b, c with a = b = 1, between the hcp and fcc structures.

The fcc structure has 3 triangles with $\theta = 180°$ because it is centro-symmetric. It depends on the angular part of the three-atom interaction in the region $110° \leqq \theta \leqq 180°$ which one of the two configurations is more favourable. In the next Section we will examine the existence of such three-atom interactions.

V. THREE-ATOM INTERACTIONS, CRYSTAL STABILITY, AND MAGIC CLUSTERS.

The three-atom potential which, because of its simplicity, readily presents itself is the Axilrod-Teller, or triple-dipole, interaction(25), hereafter denoted as AT. The AT potential is straightforwardly obtained as the first term, triple-dipole, of a multipole series for the interaction between three spherically symmetric atoms in third order of conventional Rayleigh-Schroedinger perturbation theory. This means that it is still of relatively long range, and calculated without exchange, i.e. without taking the Pauli principle into account. Its validity is, therefore, limited to that region of interatomic distances where the overlap for each of the three pairs of atoms forming the triplet, may be neglected. Even then, higher-multipole terms should be taken into account for a realistic estimate since multipole series are, at any distance between two atoms, asymptotically divergent(26). The AT potential has the simple form (we divide by the Lennard-Jones well depth ε to obtain dimensionless quantities)

$$V_{AT}/\varepsilon = Z^*(1+3\cos\theta_i\cos\theta_j\cos\theta_k)/R_{ij}^3R_{ik}^3R_{jk}^3 \quad , \tag{3}$$

with
$$Z^* = (9/16)V\alpha^3/\varepsilon R_o^9 \quad , \tag{4}$$

with V the first ionization potential of the atom (estimate of the Unsöld average excitation energy), and α the atomic polarizability. The interatomic distances R_{ij} , R_{ik} , and R_{jk} for the triplet (i,j,k) are expressed in units of $R_o = (\sqrt[6]{2})\sigma$. Finally, θ_i , θ_j , θ_k are the internal angles of the triangle formed by the triplet. Values for the different parameters and for Z* are given in the following Table II (see below).

The value of Z* can be taken as a measure of the relative strength of the AT potential with respect to the Lennard-Jones pair interaction. Z* is very small for neon, increases steeply to Ar and to Kr, and then flattens off. We note that V_{AT} can have either sign, depending on the geometry of the triangle: for θ_i , θ_j , $\theta_k < 90°$ it is positive (repulsive), whereas for a linear array $1+3\cos\theta_i\cos\theta_j\cos\theta_k = -2$, and thus V_{AT} is attractive. For an isosceles triangle, $V_{AT} = 0$ at opening angle $\theta = 117°$. This potential was designed by Axilrod and Teller explicitly in trying to stabilize fcc over hcp for the bulk crystal. Axilrod(27) found that, although the AT potential does favour the fcc structure, the difference with the hcp configuration is again of the order of 0.01%, much too small to tip the scales unequivocally.

	V(eV)	$\alpha(10^{-24}$ cm$^3)$	ε/k_B(K)	R_o(Å)	$V\alpha^3/R_o^9(10^{-16}$ erg)	Z* (average)
Ne	21.56	0.391/0.377	34.9	3.08	0.839/0.752	0.009
Ar	15.76	1.635/1.614	119.8	3.82	6.28 /6.04	0.021
Kr	14.00	2.48	171	4.10	10.76	0.027
Xe	12.13	4.03	221	4.47	16.86	0.029

Table II. Values for the different parameters occurring in the expression for the Axilrod-Teller (triple-dipole) potential between three rare-gas atoms. The values for V (first ionization potential), α (polarizability), ε (depth of the Lennard-Jones (6,12) potential), and R_o ($=\sqrt[6]{2})\sigma$ in a Lennard-Jones (6,12) potential) are taken from standard literature. The dimensionless parameter Z* is equal to $(9/16)V\alpha^3/\varepsilon R_o^9$.

Applications to properties of dilute gases, e.g. third virial coefficients, may be carried out on the basis of the AT potential, but solid-state or microcluster phenomena, involving nearest-neighbour atoms, may well lie outside the range of validity of V_{AT}. We must here express the same warning as that stated before with respect to a "physical" interpretation of the Lennard-Jones (6,12) potential: V_{AT} is not just a correction to the (sum of) (6,12) potential(s), valid at any R. With the above reservations in mind, we note that also for microclusters there exist several papers using the AT potential, by Halicioglu and White[28,29] and by Oksuz[30]. Most recently, Garzon and Blaisten-Barojas[31] added another three-atom term to the AT potential, varying exponentially with the sum-of-distances between the three atoms, and of negative sign, all this in combination with a (6,12) potential. This gives sufficient parametric freedom for obtaining a wide range of results concerning stability of microclusters, whatever these results may signify.

We here consider, in particular, the ranges of values for the parameter Z^* explored by the authors mentioned above. Halicioglu and White[28,29] varied Z^* between 0 and 1.6; Oksuz[30] between 0 and 0.7, whereas Garzon and Blaisten-Barojas[31] study the range $Z^* = 0 - 1.2$. For cluster sizes n between n = 3 and 8, and the highest values of Z^*, a linear configuration of atoms is energetically favoured, in view of the form of the AT potential. At about $Z^* = 0.3$, interesting changes happen: the close-packed 13-icosahedron is no longer the structure of lowest potential energy: it first gives way to a close-packed two-dimensional structure, and at $Z^* \approx 0.9$ to a regular polygon. This is illustrated in the following Figure 5, taken from reference 29.

The authors quoted aimed their calculations mainly at a possible explanation of observed geometries of metal-atom clusters. A look at Table II shows at once that the "interesting Z^*-region" is totally outside the range for rare-gas atoms. For those clusters V_{AT} is, in view of its weakness and long-range character, inherently incapable of "guiding" the growth of microcrystals.

A three-atom exchange potential was proposed by one of us[32], again in connection with rare-gas crystal stability. This potential is obtained from model calculations, applying perturbation theory in first and second orders, including exchange. In the model the electrons on each of the three atoms are replaced by one electron in an "effective orbital"; the spins of the three electrons are taken parallel, to avoid chemical bonding. The orbitals of these electrons are chosen of 1s-Gaussian form with a parameter β such as to reproduce the R^{-6} part of

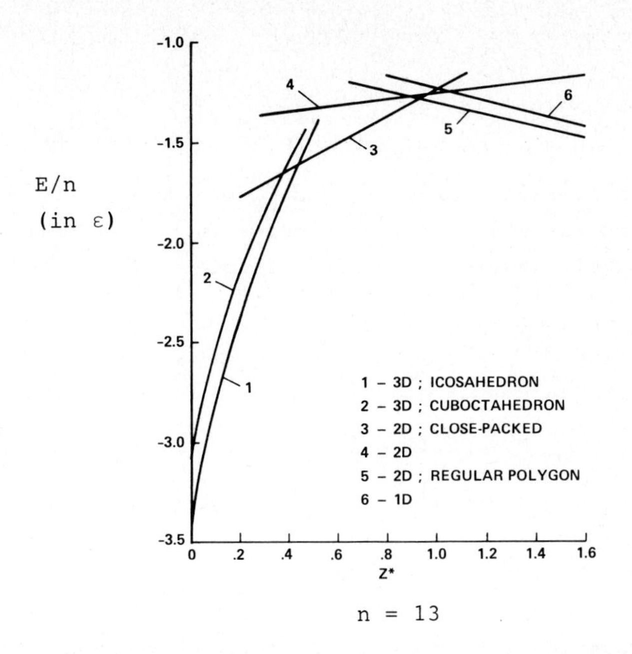

Figure 5. Cluster (potential) energy per atom, in units of the depth ε of a Lennard-Jones (6,12) potential, as a function of the AT parameter Z*, for arrangements of 13 atoms. (Taken from T. Halicioglu and P.J White, Surf. Sci. 106, 45 (1981).)

the Lennard-Jones (6,12) potential, or from a fit to the diamagnetic susceptibilities of the atoms [23,24]. A perturbation procedure is followed in which the different orders are defined in terms of powers of a smallness parameter λ as in the usual Schrödinger perturbation theory without exchange. This necessitates a redefinition of the unperturbed Hamiltonian H_o and the perturbation H' which, for the special case of two hydrogen atoms a and b, are defined as follows:

$$H_o = H_o(1,2)\Lambda_{12} + H_o(2,1)\Lambda_{21} ;$$

$$H' = H'(1,2)\Lambda_{12} + H'(2,1)\Lambda_{21} ,$$

where the notation (1,2) means that electron 1 is associated with nucleus a, electron 2 with nucleus b; (2,1) implies the reverse assignment. Λ_{12} and Λ_{21} are linear operators which project from the permutation-symmetrized wavefunctions to the associated simple-product functions. In this formalism, the first-(E_1) and second-(E_2) order perturbation energies are given by:

$$E_1 = \langle\psi^{(0)}|H'\psi^{(0)}\rangle + C_1 \;;$$

$$E_2 = -\varepsilon^{-1}[\langle\psi^{(0)}|H'^2\psi^{(0)}\rangle - \langle\psi^{(0)}|H'\psi^{(0)}\rangle^2] + C_2 \;,$$

with ε the Unsöld average energy. C_1 and C_2 are correction terms which are neglected with respect to the main terms in the equations. The unperturbed (orbital) wavefunction for a triplet (abc) of atoms is written as

$$\psi^{(0)}(abc) = N \det[\phi_a(1)\phi_b(2)\phi_c(3)] \;,$$

where

$$N = [3!(1 - \Delta^2_{abc})]^{-\frac{1}{2}} \;;$$

$$\phi = (\beta/\pi^{\frac{1}{2}})^{3/2}\exp(-\beta^2 r^2/2) \;,$$

and

$$\Delta^2_{abc} = \Delta^2_{ab} + \Delta^2_{ac} + \Delta^2_{bc} - 2\Delta_{ab}\Delta_{ac}\Delta_{bc} \;,$$

in terms of the overlap integrals Δ_{ab} , etc., between the different pairs of atoms.

We may call H_o and H' "label-free" operators in (anti-)symmetrized space. The above procedure was earlier implicitly adopted by one of us and Lombardi[23,24,34] in the analysis of stability for rare-gas and ionic crystals.

Let E_1 , E_2 , denote the first- and second-order interaction energies for a given triplet, and $E_1^{(0)}$, $E_2^{(0)}$, the sum of pair interactions in these orders. Then $\Delta E_1 \equiv E_1 - E_1^{(0)}$ and $\Delta E_2 \equiv E_2 - E_2^{(0)}$ are the first- and second-order three-atom interactions, and $\Delta E_1/E_1^{(0)}$, $\Delta E_2/E_2^{(0)}$ their values relative to pair interactions. For isosceles triangles of atoms, opening angle θ, two pairs at nearest-neighbour distance in the solids, from neon to xenon, it is found that (details are given in Ref. 24)

$$\Delta E_1/E_1^{(0)} \simeq \Delta E_2/E_2^{(0)} \;.$$

These relative three-atom energies are negative (10 to 20 percent) at $\theta = 60°$ (equilateral triangle), positive (4 to 7 percent) at $\theta = 180°$ (linear array of atoms) from neon to xenon. We need to know the total three-atom exchange energy ΔE per triplet, with $\Delta E \equiv \Delta E_1 + \Delta E_2$. Denoting the total pair interaction for a triplet by $E^{(0)} \equiv E_1^{(0)} + E_2^{(0)}$ we have, from the equality mentioned above,

$$\frac{\Delta E}{E^{(0)}} = \frac{\Delta E_1 + \Delta E_2}{E_1^{(0)} + E_2^{(0)}} \simeq \frac{\Delta E_1}{E_1^{(0)}} \simeq \frac{\Delta E_2}{E_2^{(0)}}$$

and thus

$$\Delta E \simeq \frac{\Delta E_1}{E_1^{(0)}} E^{(0)} \simeq \frac{\Delta E_2}{E_2^{(0)}} E^{(0)} \quad . \tag{5}$$

For the total pair interaction $E^{(0)}$ of a given triplet it is most accurate to take the sum $\Sigma(LJ)$ of Lennard-Jones (6,12) potentials for the three pairs of atoms; the relative three-atom exchange interaction is obtained from the model calculations. In the rare-gas solids, non-isosceles triangles as well as those involving atoms from further shells around a central atom play a minor role for stability. On the basis of the above results it is easily established that the face-centered cubic lattice has the higher stability, by a few percent of the cohesive energy, compared with the hexagonal close-packed structure Details are given in Ref. 24.

It is instructive to compare $\Delta E_1/E_1^{(0)}$ (or $\Delta E_2/E_2^{(0)}$) with the Axilrod-Teller potential relative to the sum (all distances in units R_o)

$$- \sum_{i<j} \tfrac{3}{4} V(\alpha/R_o{}^3)^2/R_{ij}^6$$

of the London-Van der Waals induced-dipole interactions for the three pairs. Abbreviating this sum by $\Sigma(VdW)$ we obtain, with (3) and (4),

$$V_{AT}/\Sigma(VdW) = - \frac{3(\alpha/R_o^3)(1+3\cos\theta_i\cos\theta_j\cos\theta_k)/R_{ij}^3R_{ik}^3R_{jk}^3}{4(1/R_{ij}^6+1/R_{ik}^6+1/R_{jk}^6)} \quad . \tag{6}$$

In case of an isosceles triangle with opening angle θ, the above ratio is negative for $\theta < 117°$, positive for larger θ. In the following Figure 6 this comparison is made, multiplying eq. (6) by a factor 20.

We see that, apart from the factor 20 (and reminding the reader that validity of the AT potential for triplets of atoms involving nearest neighbours in a crystal is doubtful), the two curves look much alike: the relative three - atom interactions are negative for small-opening triangles, positive for those with large opening. The exchange curve rises somewhat more steeply at small θ, and flattens off more markedly for large θ. Curve (a) also demonstrates, in conjunction with eq. (5), that three-atom exchange interactions favour large-opening (isosceles) triangles (as does the AT potential, albeit much more weakly): as $\Sigma(LJ) < 0$ for all triplets, the three-atom energy $\Delta E = \Delta E_1 + \Delta E_2$ is

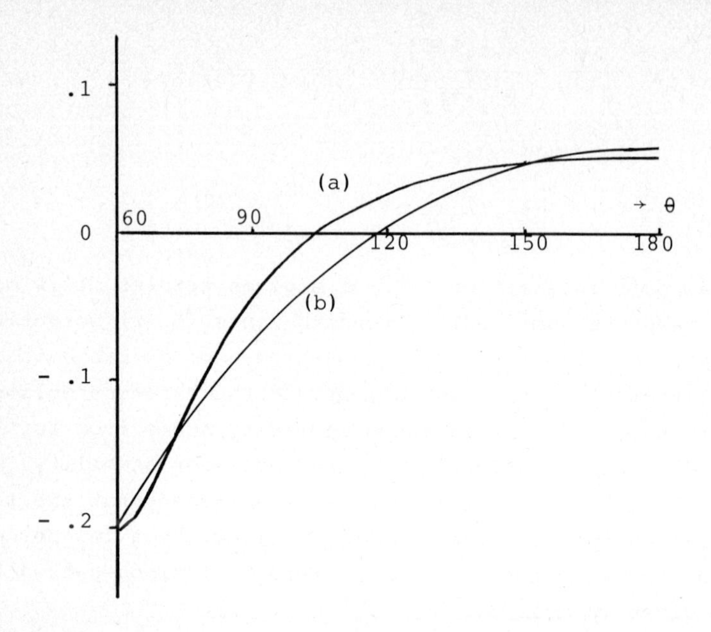

Figure 6. Comparison between Axilrod-Teller and exchange three-atom interactions for an isosceles triangle, with opening angle θ, of argon atoms; two of the atoms are nearest neighbours in the solid. The three-atom exchange interactions, curve (a), are given relative to the sum of pair interactions. The AT results, curve (b), are plotted relative to the sum of Van der Waals induced-dipole interactions Σ(VdW) for the three pairs, i.e. eq. (6), and multiplied by 20.

< 0 for $\Delta E_1/E_1^{(0)}$ (or $\Delta E_2/E_2^{(0)}$) > 0, i.e. for large opening angles θ. In connection with the magic-number issue, we have calculated the effect of three-atom exchange interactions on the relative stability of different structures for a few very small clusters, with n = 4, 7, and 13, at 0 K, of argon atoms. We compare the Hoare-Pal structures of minimal Lennard-Jones (6,12) potential energy, viz. the tetrahedron for n = 4, the pentagonal bipyramid for n = 7, and the icosahedron for n = 13, with a square, a hexagon-plus-center, and a cuboctahedron, respectively. In the following Figure 7, the potential energy per atom, E/n, is given as a function of three-atom exchange multiplied by a factor varying between 0 and 1. In each structure the nearest-neighbour distance is always that for the argon crystal (3.83 Å) except for the icosahedron, where nearest surface atoms are at that distance. The central atom is then at 0.951 times the nearest-neighbour separation in the solid. A "breathing" procedure yields a minimal potential per atom (Lennard-Jones + three-atom exchange) at a = b = 0.948, c = 0.998, of - 1.3685 ε, whereas a = b = 0.951, c = 1, yields -1.3679 ε.

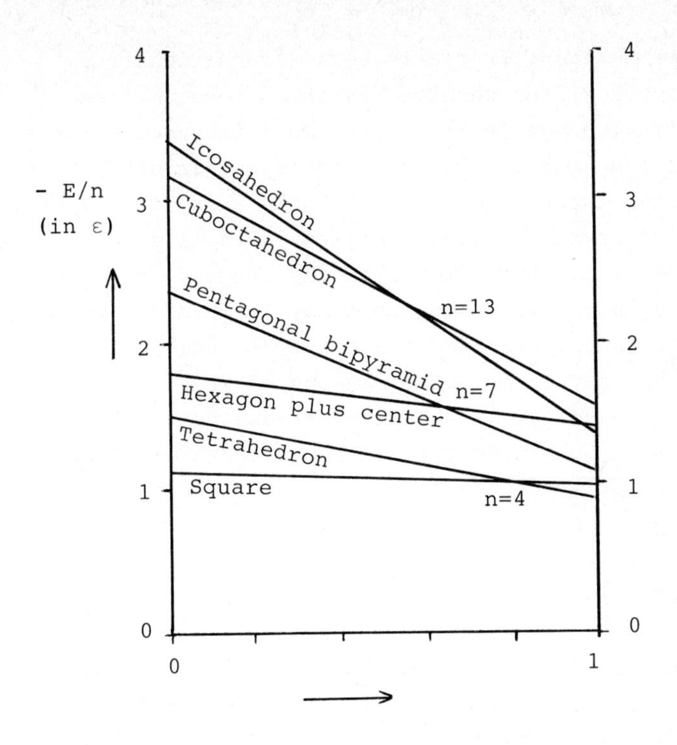

Figure 7. Potential energy per atom, E/n, in units of ε, as a function of the three-atom exchange interaction multiplied by a factor between 0 (only Lennard-Jones (6,12) pair potentials) and 1 (full three-atom exchange energy) for clusters of n = 4, 7, and 13 atoms. Compared are (n = 4) square and tetrahedron, (n = 7) pentagonal bipyramid and hexagon-plus-center, and (n = 13) icosahedron and cuboctahedron. Closest atoms are at nearest-neighbour distance of solid argon except with the icosahedron, where nearest surface atoms are at that distance.

Let us ponder the results presented in Figure 7. With growing three-atom exchange, closest packing looses the advantage of minimal potential energy, giving way to two-dimensional structures for n = 4, and 7 (square, and hexagon-plus-center, respectively), and to the "less-spherical" cuboctahedron for n = 13. Further, the decrease in -E/n becomes the more dramatic, the higher n: it amounts to about 50 percent of the pair energy for the icosahedron and cuboctahedron! We have verified that for n = 8, the additional atom takes its place above the center of one of the isosceles triangles formed by the central atom and two nearest neighbours on the hexagon. In one respect (transitions from 3- to 2-dimensional structures) the results resemble those obtained(28-31) with the AT potential, although in the latter case the necessary values of the parameter Z* are unphysically high for rare-gas atoms. Three-atom exchange interactions are probably sufficiently strong to serve in "guiding" crystal growth. In view of their strength, it is very important to investigate the effects on the cohesive energy of the bulk crystals, to ascertain that no unacceptably large differences are obtained compared with the results of a Lennard-Jones (6,12) potential. This question was analyzed in detail by one of us(24). If we limit ourselves to a central atom and its twelve nearest neighbours, then the cohesive energy, E_{coh}, of the fcc and hcp structures is given by

$$E^{fcc}_{coh}/N = 8.4\ \varepsilon + 8\Delta_{60} + 12\Delta_{90} + 24\Delta_{120} + 6\Delta_{180}\ ;$$

$$E^{hcp}_{coh}/N = 8.4\ \varepsilon + 8\Delta_{60} + 12\Delta_{90} + 3\Delta_{110} + 18\Delta_{120} + 6\Delta_{146} + 3\Delta_{180}\ ,$$

where N is the total number of atoms in the crystal; Δ_θ denotes the three-atom energy ΔE for a triangle with opening angle θ. In the summation, we have taken into account that each equilateral triangle is counted three times, the others only once. Substituting the numerical values(24) for Δ_θ, one obtains for the argon crystal:

$$E^{fcc}_{coh} = 8.4\ N\varepsilon(1 - 0.206)\ ;$$

$$E^{hcp}_{coh} = 8.4\ N\varepsilon(1 - 0.251)\ ,$$

implying that the cohesive energies for the cubic and hexagonal structures are decreased in absolute value by 21 and 25%, respectively, because of three-atom interactions. The magnitude of this effect is surprising, and not acceptable, since it is generally acknowledged that the values of intermolecular potential parameters (ε and σ, in our case) are in fair agreement with a pair-potential interpretation of the cohesive energies of rare-gas solids. It appears, however, that the three-atom component of the cohesive energy is extremely sensitive to the precise values of $\Delta E/E^{(0)}$ for small values of θ, whereas the relative stability of the fcc and hcp structures is totally inert to this region. For example, if we decrease for argon the second-order relative three-atom energy $\Delta E_2/E_2^{(0)}$ at $\theta = 60°$ by 10% from 0.18 to 0.16, and at $\theta = 90°$ from 0.06 to 0.05, then we obtain for the cohesive energies:

$$E_{coh}^{fcc} = 8.4\ N\varepsilon(1 - 0.028)\ ;$$

$$E_{coh}^{hcp} = 8.4\ N\varepsilon(1 - 0.073)\ ,$$

i.e. with a relatively small change in the three-atom part for such triangles the cohesive energy decreases by only 3 and 7%, respectively. This decrease is of the same order of magnitude as that obtained with the Axilrod-Teller potential[27] (from 2 to 9% for solid neon to xenon). The reason is that, although the AT interaction is very much weaker than the three-atom exchange for the 66 triplets considered, the AT potential is of much longer range, leading to a non-negligible total decrease in the cohesive energy.

Following the above changes in relative three-atom interactions we have, for the icosahedron-cuboctahedron pair, decreased $\Delta E/E^{(0)}$ by 10% for isosceles triangles with opening angle $\theta < 90°$. The result is an increase in binding energy of about 18% for the icosahedron and 10% for the cuboctahedron, i.e. the difference between the two structures becomes smaller, although the cuboctahedron remains the somewhat more stable one.

As a final remark, we note that the energy differences for the three pairs of structures tend to become smaller, comparing the case of pair potentials only and full three-atom interactions. In addition, for $n = 13$ the difference is small at both extremes. This might well imply that in a rigorous calculation the "best single cluster" approximation is of doubtful validity. It was mentioned earlier that in the model three-atom interactions are very small when one goes beyond the first shell around a central atom. Also, strongly non-isosceles triangles contribute very little to the three-atom energy. This is illustrated in detail, for each type of triangle, in the following Table III for the isosahedron and the cuboctahedron, which consist of 8 and 10 different types of triangle, respectively. Listed are the sides a, b, c of each triangle, the number of each type, and the relative non-additive contribution $\Delta E/E^{(0)}$, taken equal to $\Delta E_1/E_1^{(0)}$.

The drastic weakening of the binding energy E/n per atom in both structures is mainly due to the large number of nearest-neighbour equilateral (or practically equilateral) and isosceles triangles in the two configurations. This implies that in a bulk crystal, three-atom interactions can be evaluated "shell-by-shell" around a central atom. In ionic solids this is a particularly simple and transparent prescription for the interpretation of relative stability of crystal structures since, due to electrostatic compression, further shells do contribute to the crystal

energy. For a review, see the last article quoted in Ref. 34.
Together with Prof. Brickmann and collaborators at the TH Darmstadt, molecular dynamics (MD) calculations including the three-atom exchange part of the interatomic potential, were recently carried through for

Icosahedron

total pair energy	-44.020	(in ε)
three-atom contribution	26.238	
E/n	- 1.368	

	a	b	c	No.	Rel. Non-Add.
1	.951	.951	1	30	- .202
2	.951	.951	1.618	30	.026
3	.951	.951	1.902	6	.061
4	1	1	1	20	- .204
5	1	1	1.618	60	.008
6	1	1.618	1.618	60	- .005
7	1	1.618	1.902	60	- .001
8	1.618	1.618	1.618	20	0

Cuboctahedron

total pair energy	-40.744	(in ε)
three-atom contribution	20.607	
E/n	- 1.549	

	a	b	c	No.	Rel. Non-Add.
1	1	1	1	32	- .204
2	1	1	1.414	36	- .047
3	1	1	1.732	48	.025
4	1	1.414	1.732	48	- .006
5	1	1.732	1.732	24	- .002
6	1	1	2	6	.047
7	1	1.732	2	48	0
8	1.414	1.414	2	12	0
9	1.414	1.732	1.732	24	0
10	1.732	1.732	1.732	8	0

Table III. Effect of three-atom exchange interactions, listed as $\Delta E/E^{(0)}$ and denoted by "Rel. Non-Add." in the last column, for the 8 and 10 different types of triangle in the icosahedral and cuboctahedral configurations, respectively, of argon atoms. All distances are in units of nearest-neighbour distance in solid fcc argon.

argon and xenon[35,36]. Prior to this work, Polymeropoulos and Brickmann[37,38] had undertaken a pioneering MD analysis, comparing results obtained with Lennard-Jones potentials alone and those resulting from adding Axilrod-Teller three-atom interactions. For details of these analyses, we refer to the corresponding papers. As an illustration of the results obtained, we present in the following Figure 8 (see next page) cluster densities versus cluster-size and the average lifetimes of clusters at 120 K and at reduced densities of 0.0553 and 0.0088 for argon and xenon, respectively. Clusters made up of the same atoms at subsequent sampling steps of the simulation are counted only once, thus giving a distribution of unique cluster sizes. Results obtained for the Lennard-Jones plus Axilrod-Teller ("LJ-plus-AT") potential, and those for the Lennard-Jones plus exchange ("LJ-plus-EX") potential are plotted side by side for each cluster size. Black columns refer to the LJ-plus-AT, white columns to the LJ-plus-EX potential.

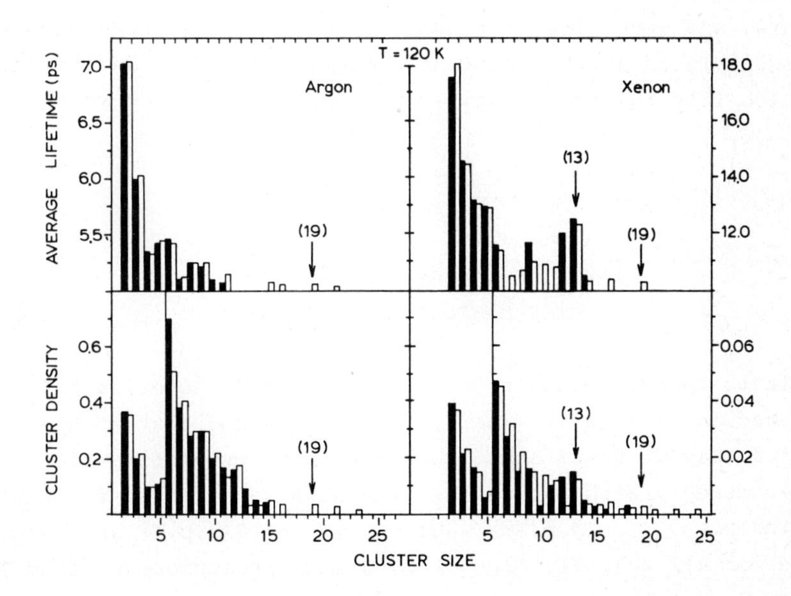

Figure 8. Cluster density vs cluster size distribution (bottom) and average lifetimes (top) at a temperature of T = 120 K and reduced densities of 0.0553 and 0.0088 for Ar and Xe, respectively. Density of clusters 2-5 and the average lifetimes for Ar clusters are given on the left-hand side while the density of clusters larger than 5 and average lifetimes for Xe clusters are given on the right-hand side of the figure. Dark columns represent clusters being built under the influence of a LJ-plus-AT potential; light columns represent clusters being built under the influence of a LJ-plus-EX potential.

The general observation from the figure is that the LJ-plus-EX potential leads (in most cases) to more stable clusters than the LJ-plus-AT potential, even though the temperature chosen (120 K) is high. This is, in particular, reflected in the relative densities of clusters corresponding to "magic numbers" compared to the densities of their immediate neighbours. For argon, no cluster corresponding to Ar_{19} was found in previous work with the LJ-plus-AT potential[38] and it was concluded that, even for xenon, no reasonable-size calculations with this potential would provide information on that particular cluster. The LJ-plus-EX potential, on the other hand, does yield a (small) peak at $n = 19$ as well as some peaks in its neighbourhood. Calculations at much lower temperatures could not be performed in view of the excessively high computing time necessary. For xenon the same general observations can be made. In particular, one sees that the average lifetime and the density of Xe_{13} are much higher relative to its neighbours with the LJ-plus-EX than with the LJ-plus-AT interaction. At the beginning of each calculation, 108 atoms were placed on the lattice points of a fcc structure, and were given a random velocity distribution. All further details are found in the references quoted above. An analytical expression, which fits the three-atom exchange potential very well, is given in Ref. 36.

VI. ON CAUSES AND EFFECTS: THE MAGIC KNOT UNTIED.

We return to Lennard-Jones potentials and the (theoretical) "coincidence" established in Section IV. There, it was remarked that at high supersaturation, say at p = 1 atm., the free enthalpy of formation $\Delta G(n,p,T)$ decreases steeply with increasing cluster size $\underline{n}$. Thus, the sublimation free enthalpies $G_{subl}(n,p,T) \equiv \Delta G(n-1,p,T) - \Delta G(n,p,T)$ are large and positive for all $\underline{n}$ (≥ 2). Also at such high pressures, we find local maxima in G_{subl} at $n^* = 7$, 13, and 19 in the range 2-20, although there occur no discontinuities in $\Delta G(n,p,T)$. We noted that Freeman and Doll[17], in their classical and quantum Monte Carlo calculations, find dips at certain values of the pressure $\underline{p}$ in $\Delta G(n,p,T)$ at 10 K. Such minima emerge and disappear somewhat "mysteriously" when the pressure is changed. The cluster sizes in question are again 7, 13, and 19. The question asked was: is this a "coincidence"?

It is not difficult to unravel the phenomenon described. Suppose a dip in $\Delta G(n,p,T)$ occurs at a value $n = n^*$. This implies that

$$\Delta G(n^*-1,p,T) > \Delta G(n^*,p,T) < \Delta G(n^*+1,p,T) \ . \qquad (7)$$

Given the ideal-gas approximation, ΔG at any pressure p_o can be obtained from $\Delta G(n,p,T)$ by[16]

$$G(n,p,T) = G(n,p_o,T) - k_B T(1-n)\ln(p/p_o) \ . \qquad (8)$$

We will choose p_o = 1 atm., since Hoare *et al.*[16] and Freeman and Doll[17] tabulate argon values at that pressure. Substituting (8) into (7), one obtains:

$$\Delta G(n^*-1,1,T) + k_B T\ell np > \Delta G(n^*,1,T) < \Delta G(n^*+1,1,T) - k_B T\ell np \ , \qquad (9)$$

from which it follows that

$$\Delta G(n^*-1,1,T) - \Delta G(n^*,1,T) > -k_B T\ell np > \Delta G(n^*,1,T) - \Delta G(n^*+1,1,T) \ ,$$

i.e.

$$G_{subl}(n^*,1,T) > -k_B T\ell np > G_{subl}(n^*+1,1,T) \ . \qquad (10)$$

It follows that the sublimation energy at 1 atm. of the cluster of size n* must be larger than that of size n*+1. In addition, since all sublimation energies at 1 atm. are strongly positive[16,17], these inequalities can only be satisfied at very low pressures, where $-k_B T\ell np \gg 0$. The curves for $\Delta G(n,1,T)$ obtained from the tables of refs 16 and 17 are precisely of the form given in Figure 2, taken from the work by Farges *et al.*[11], i.e. at n = n* we have a local *maximum* in the sequence of positive sublimation energies at 1 atm.

Which one (or ones) of the n* in the local maximum in $E_{subl}(n^*,1,T)$ will cause a *dip* in the free enthalpy $\Delta G(n,p,T)$ depends strongly, *but trivially*, on the pressure *p*, at a given T. Let us give a few numerical examples. Freeman and Doll[17] tabulate $\Delta G(n,p = 1 \text{ atm.},T)/k_B T$, at T = 10 K, for cluster sizes 2 to 20, calculated from their classical and quantum Monte Carlo analyses, and also the results obtained by Hoare *et al.*[16] at 10 K. Using eq. (8), one can find ranges of $-\ell np$ for which a local minimum in $\Delta G(n,p,10 \text{ K})$ *will* occur. In the following Table IV we list values of the sublimation enthalpies $G_{subl}(n^*, 1\ atm., T)/k_BT$, at T = 10 K, for n = 7,8; n = 12,13; and n = 18,19 on the basis of the classical and quantum Monte Carlo (ref. 17), and the HO/RR (ref. 16) methods.

		G_{subl} (n,p = 1 atm.,T)/k_BT ; T = 10 K.		
		classical MC	quantum MC	HO/RR
n = 7		34.8	33.4	35.6
	Figure 3a		(31.0)	(31.0)
8		32.4	30.3	29.2
n = 13		62.2	57.6	59.2
	Figure 3b	(35.6)	(35.6)	(35.6)
	Figure 3c	(40.2)	(40.2)	(40.2)
14		32.9	31.7	30.9
n = 19		62.1	59.0	59.4
	Figure 3c		(40.2)	
20		41.3	38.5	42.9

Table IV. Values of the quantity

$$G_{subl}(n,p = 1\ atm.,T)/k_BT, \quad at\ T = 10\ K,$$

on the basis of the classical MC, quantum MC, and HO/RR methods, for cluster sizes n = 7, 8; 13, 14; 19, 20. The $-\ell np$ values chosen by Freeman and Doll are indicated in parentheses if the dip-conditions (10) are fulfilled.

Freeman and Doll selected pressures 3.34×10^{-14} atm., 3.34×10^{-16} atm., and 3.34×10^{-18} atm.; their $-\ell np$ values are 31.0, 35.6, and 40.2, respectively. By comparing with the tabulated figures, we see that at n=7 the classical curve will show no dip at $-\ell np$=31.0, whereas on the quantum MC curve a minimum is found (it occurs as well in the HO/RR approximation), cf. Figure 3a. For $-\ell np$=35.6 (Figure 3b) and 40.2 (Figure 3c), all three approximations cause a local minimum in $\Delta G(n,p,T)$ at n*=13. The value $-\ell np$=40.2 lies outside the range at n=19 required for a classical minimum, inside in the quantum MC method, and again outside in the HO/RR approximation. Classically, minima at n=7 and 13 can occur together only in the small range $32.9 < -\ell np < 34.8$, and pairs at n=13 and 19 will be present for $41.3 < -\ell np < 62.2$. In the quantum MC case, dips at n=13 as well as 19 are calculated for $38.5 < -\ell np < 57.6$, whereas in the

HO/RR approximation, local minima at n=13 and 19 go together for 42.9 $< -\ell np < 59.2$. Note the very narrow range of $-\ell np$ values for n=(7,8) and the broad ranges for the pairs n=(13,14) and (19,20). No pressure p exists for which local minima at n*=7, 13, and 19 will manifest themselves together, according to any of the three methods.

In summarizing the results of this Section, it has been shown that local maxima in the sublimation energy of clusters at high (e.g. 1 atm.) pressure are inseparably connected with local minima in the free enthalpy of cluster formation in certain (different) regions of pressure p, and conversely: they are different effects with the same origin. The link between the two phenomena is provided through the ideal-gas term $(1-n)k_BT\ell np$ in the free enthalpy change, different for different cluster sizes n. If we replace "free enthalpy" by "energy at 0 K", apparently a fair approximation at low T, then, except for zero-point vibrations, we fall back on Hoare's[14] "crucial geometrical factors" as the unique source. Then the potential energy, i.e. the interaction between atoms, plays the dominant role. Undoubtedly, the Lennard-Jones (6,12) potential provides a good basis, but fundamental difficulties remain: stability of the cubic bulk structure (Section IV), crystal growth mechanisms[14], the missing n*=7 in nozzle-beam experiments with rare gases, and the experimental occurrence of additional cluster-size maxima. Three-atom exchange interactions were shown to stabilize the fcc bulk configuration over hcp to the order of one percent of the cohesive energy. They may be expected to yield positive results also with respect to the remaining problems. This should apply, in the first place, to the theoretical magic number n*=7 for argon, which is experimentally not observed (neither for xenon). The narrow range of $-\ell np$ for the pair n=(7,8) relative to (13,14) or (19,20) in Table IV leads us to expect that n*=7 (at low T) may indeed disappear under three-atom interactions.

ACKNOWLEDGEMENTS.

One of us (L.J.) wishes to thank the Organizers of the "1st INTERNATIONAL WORKSHOP ON PHYSICS OF SMALL SYSTEMS" for their kind invitation to him to speak on this subject.
The authors express their gratitude to Mrs Désirée Jansen-Huijsmans for her truly magic performance in typing the text and solving the many intricate composition problems.

REFERENCES.

1. M.G. Mayer, Phys. Rev. 74, 235 (1948).
2. M.G. Mayer, Phys. Rev. 75, 1969 L (1949) (jj-coupling).
3. O. Aaxel, J.H.D. Jensen, and H.S. Süss, Phys. Rev. 75, 1766 L (1949); also Naturw. 35, 376 (1948), 36, 153, 155 (1949), Phys. Rev. 75, 1766 (1949), Z. Physik 128, 295 (1950).
4. The name "sublimation free enthalpy" is not quite accurate, since the temperature of a cluster falls as it evaporates, i.e. the free-enthalpy distribution changes in a sublimation process.
5. O. Echt, K. Sattler, and E. Recknagel, Phys. Rev. Lett. 47, 1121 (1981).
6. I.A. Harris, R.S. Kidwell, and J.A. Northby, 17th International Conference on Low Temperature Physics, Karlsruhe 1984 (North-Holland, Amsterdam, 1984).
7. A. Ding and J. Hesslich, Chem. Phys. Lett. 94, 54 (1983).
8. J.A. Sâenz, J.M. Soler, and N. Garcia, Chem. Phys. Lett. 114, 15 (1985).
9. D. Kreisle, O. Echt, M. Knapp, and E. Recknagel, Surf. Sci. 156, 321 (1985).
10. A.L. Mackay, Acta Cryst. 15, 916 (1962).
11. J. Farges, M.F. de Feraudy, B. Raoult, and G. Torchet, Surf. Sci. 156, 170 (1985).
12. J. Farges, M.F. de Feraudy, B. Raoult, and G. Torchet, J. Chem. Phys. 78, 5067 (1983).
13. M.R. Hoare and P. Pal, Nature 230, 5 (1972); 236, 35 (1972); Adv. Phys. 20, 161 (1971); J. Cryst. Growth 17, 77 (1972); Adv. Phys. 24, 645 (1975).
14. M.R. Hoare, Adv. Chem. Phys. 40, 49 (1979).
15. Refs 94-96 in Hoare, ref. 12.
16. M.R. Hoare, P. Pal, and P.P. Wegener, J. Colloid Interface Sci. 75, 126 (1980).
17. D.L. Freeman and J.D. Doll, J. Chem. Phys. 82, 462 (1985).
18. M.R. Hoare and P. Pal, Adv. Phys. 20, 161 (1971).
19. Reliability of the values for ΔS may be in doubt, however, on the basis of one "representative" cluster at every n replacing an ensemble of structures. Especially for larger clusters, this approximation may be of questionable validity.
20. T. Kihara and S. Koba, J. Phys. Soc. Japan 7, 348 (1952).
21. T. Kihara, Revs. Mod. Phys. 25, 831 (1953).
22. T.H.K. Barron and S. Domb, Proc. Roy. Soc. (London) A227, 447 (1955).
23. L. Jansen, Phys. Rev. 125, 1798 (1962).
24. L. Jansen, Phys. Rev. 135, A1292 (1964).
25. B.M. Axilrod and E. Teller, J. Chem. Phys. 11, 299 (1943).
26. See, e.g. L. Jansen, Phys. Rev. 110, 661 (1958).
27. B.M. Axilrod, J. Chem. Phys. 19, 719, 724 (1949).
28. T. Halicioglu and P.J. White, J. Vacuum Sci. Technol. 17, 1213 (1980).
29. T. Halicioglu and P.J. White, Surf. Sci. 106, 45 (1981).
30. I. Oksuz, Surf. Sci. 122, L585 (1982).
31. I.L. Garzon and E. Blaisten-Barojas, Chem. Phys. Lett. 124, 84 (1986).
32. L. Jansen, Phys. Rev. 162, 63 (1967).
33. Since solid neon, argon, krypton, and xenon have the same (face-centered cubic) structure, only the range of β values is of primary importance for relative stability. In addition (see text), ratios of the three-atom interactions, relative to the respective pair contributions in first and second order, are calculated in the model.
34. E. Lombardi and L. Jansen, Phys. Rev. 136, A1011 (1964); 140, A275 (1965); 151, 694 (1966); for a review, see L. Jansen and E. Lombardi, Disc. Faraday Soc. 40, 78 (1965).

35. E.E. Polymeropoulos, J. Brickmann, L. Jansen, and R. Block, Phys. Rev. A 30, 1593 (1984).
36. E.E. Polymeropoulos, P. Bopp, J. Brickmann, L. Jansen, and R. Block, Phys. Rev. A 31, 3565 (1985).
37. E.E. Polymeropoulos and J. Brickmann, Chem. Phys. Lett. 92, 59 (1982); 96, 273 (1983).
38. E.E. Polymeropoulos and J. Brickmann, Ber. Bunsenges. Phys. Chem. 87, 1190 (1983).

GROUND STATE PROPERTIES OF SMALL MATRIX-ISOLATED MOLECULES: FeH, FeN, $FeCO$ AND $Fe(CO)_2$

V.R. MARATHE, A. SAWARYN, A.X. TRAUTWEIN

Institut für Physik
Medizinische Universität Lübeck
2400 Lübeck
FRG

1. INTRODUCTION

Small molecules or fragments of molecules, which are generally unstable under normal conditions, can be created using special laboratory techniques such as (a) cocondensation with solid noble gas (1) or (b) thermal or photochemical decomposition of larger molecules trapped in matrices (2.3). Several of these molecules or fragments have also been studied using spectroscopic techniques such as Mössbauer-, I.R.-, ESR spectroscopy etc.

These systems are ideal for rigorous quantum chemical calculations; because of their small sizes information about their structure and bonding can be derived without too serious approximations in the theory. Particularly in cases where these molecules contain transition metal ions, they can also be considered as models for studying chemisorption and catalytic properties (4,5). Hence there is a wide interest of the quantum chemistry of such systems which may be studied at various levels of sophistication.

Most of the theoretical efforts are concentrated on calculating total energy and optimized geometry for various electronic configurations of free molecules ignoring the effect of matrices. There are no special efforts to check whether ground state configurations obtained from these calculations really exist in the matrices. In this respect, consideration of spectroscopic properties, which are particularly sensitive to the choice of electronic configuration should prove important.

Two of the experimentally observed quantities in Mössbauer spectroscopy, i.e. isomer shift (I.S.) and quadrupole splitting (ΔE_Q) are known to be strongly dependent on the electronic configuration of Mössbauer atoms (6). We have, therefore, used results of Mössbauer spectroscopy on some iron containing molecules along with ab initio molecular orbital calculations as tools for understanding the electronic configurations of molecules trapped in matrices.

At this initial stage of our calculations, we pose ourselves with four important questions:

(i) Which electronic configuration of the molecule exhibits the observed spectroscopic properties?

(ii) Is this configuration also lowest in energy or is the matrix responsible for stabilizing it?

(iii) Is the conventional Hartree-Fock SCF procedure good enough for obtaining proper energy sequence of various configurations as well as in the calculations of one-electron properties or are configuration interaction calculations necessary?

(iv) Can the differences, if any, between the calculated and observed spectroscopic properties of a molecule be, at least qualitatively, be understood in terms of interaction of that molecule with the surrounding matrix or is explicit consideration of the matrix in the ab initio calculations a necessity?

With these questions in mind we have performed ab initio MO calculations on a number of iron containing molecules, and we present here some of the results for those molecules which exhibit the influence of the matrix on their spectroscopic properties.

2. COMPUTATIONAL METHODS

The all-electron calculations, presented here, were performed using the GAUSSIAN 82 ab initio program (7), modified suitably to generate initial guesses for molecules containing iron. In order to assure convergence of the SCF procedure the method of direct minimization (SCFDM) was used throughout (8). The correlation energy was evaluated using Möller Plesset perturbation theory up to fourth order including all substitutions (MP4SDTQ) (9).

The quadrupole splittings observed in Mössbauer spectra are directly proportional to the electric field gradient (EFG) at the site of Mössbauer nuclei. The EFG is calculated after SCF as well as MP4SDTQ procedure using one-electron properties package of GAUSSIAN 82. The value of the nuclear quadrupole moment for ^{57}Fe was taken as 0.15 barns (10).

The isomer shifts observed in Mössbauer spectra are linearly related to the electron density, $\rho(0)$, at the Mössbauer nuclei. Exact values of $\rho^{MO}(0)$ were calculated using the bond order matrix available after SCF as well as MP4SDTQ procedures. The nonrelativistic $\rho^{MO}(0)$ values, thus obtained, were scaled by a relativistic correction factor S' to obtain relativistic electron densities $\rho^{MO}(0)$. This factor was derived by comparing atomic $\rho^{MO}(0)$ values for various electronic configurations of iron (such as $3d^6 4s^2$, $3d^6 4s^1$, $3d^6$, $3d^5 4s^1$ etc.) with corresponding relativistic values, $\rho(0)$ (11). The resulting interpolation formula for S' depends on the choice of basis set and on the number of s and d electrons of iron:

$$S' = a + b(n_s - 6) + c(n_d - 6) \quad .$$

The coefficients a, b and c are basis set dependent. For the basis we have used in the present calculation, these coefficients are $a = 1.3888898$; $b = -0.0000123$ and $c = 0.0001636$.

3. CHOICE OF BASIS SETS

In order to check the quality of basis sets needed to obtain satisfactory results for total energy calculations as well as for one-electron properties, we performed ab initio *MO* calculations using various different basis sets with *FeCO* as a test molecule. The results of these calculations will be published elsewhere (12). We have observed that basis sets with at least double zeta representation for valence orbitals are required in order to obtain a clear decision about the electronic ground state and about one- electron properties of *FeCO*. A similar observation has recently been made by Huzinaga et al (13,14) after investigating several molecules containing transition metal atoms (excluding iron). Basis sets used in the present calculations have 6s, 4p and 3d functions for *Fe*, 3s and 2p functions for *C*, *N* and *O*, and 2s and 2p functions for *H*, respectively.

4. ELECTRONIC CONFIGURATONS OF *FeH*

Dendramis et al (15) have studied *FeH* molecules trapped in solid argon using UV-, VIS-, I.R.- and ESR spectroscopy. They find from their I.R. studies that binding energy of the molecule is 1.7 eV, however, they fail to observe an ESR spectrum indicating that the ground state is probably orbitally degenerate. Pasternak et al (16) have observed from their studies that for the *FeH* molecule trapped in solid hydrogen, the isomer shift is $+0.59 mm/s$ relat. to iron metal while the quadrupole splitting is $\pm 2.4 mm/s$. Scott and Richards (17,18) have performed ab initio molecular orbital calculations for various configurations of *FeH* to predict $^6\Delta$ as the electronic ground state with $^6\Sigma^+$ and 6π lying within about $1620 cm^{-1}$. This ground state is also in agreement with the absence of an ESR spectrum for *FeH* and *FeD* molecules. $^6\Delta$ as well as other sextets, however, will have more than one electron in the 4s orbital, and the calculated values of $\rho(0)$ would be higher than that predicted from the observed isomer shift (from I.S. we expect the 4s population of the order of 0.5 electrons). We have therefore, performed ab initio *MO* calculations for various configurations of *FeH* in order to study the variation of Mössbauer parameters with configuration.

The schematic energy level diagram for FeH is shown in Fig. 1. The three electronic configurations which merit considerations as a ground state are $^6\pi(1\sigma^2 1\pi^3 1\delta^2 2\sigma^1 3\sigma^1)$, $^6\Delta(1\sigma^2 1\pi^2 1\delta^3 2\sigma^1 3\sigma^1)$ and $^4\phi(1\sigma^2 1\pi^3 1\delta^3 2\sigma^1)$.

In Table 1 we give the SCF and MP4SDTQ results for the sextet and quartet configurations. For all these configurations geometry is optimized to obtain minimum energy at SCF level. It is seen from this table that the sextets are lower in energy compared to quartet; $^6\Delta$ has the lowest energy and 6π is lying close by. These results are similar to the calculations by Scott and Richards (17,18). For both sextets the $4s$ population is greater than one electron, while for 4ϕ it is 0.57 which is closer to that expected from experimental I.S.

In Table 1 we also give the calculated values of the Mössbauer parameters for various configurations of FeH. As expected the s electron densities for both sextets are much higher than those observed experimentally, and the calculated values of quadrupole splitting also differ significantly from the experiment. The calculated values of quadrupole splitting (under low-symmetry condition) and of electron density for $^4\phi$ match closely with experiment.

We have also performed MO calculations for other possible quartet states of FeH. For all these states, $4s$ population lies between 0.50 and 0.66 electrons. We can, therefore, conclude from our calculations that the sextet configurations are affected significantly by the interaction with the solid H_2 matrix because of the direct occupation of the Fe $4s$ orbital thus stabilizing a quartet ground state.

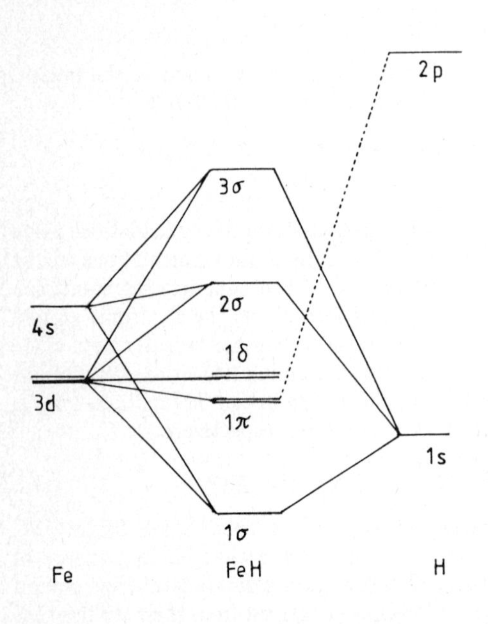

FIG. 1: Schematic energy level diagram for FeH showing valence orbitals

5. ELECTRONIC CONFIGURATIONS OF FeN

Heyden et al (19) have studied Mössbauer spectra of FeN using ^{57}Co embedded in frozen N_2 as a source. The experimental values of isomer shift and quadrupole splitting are $-0.46mm/s$ and $3.90mm/s$, respectively. The observed value of the isomer shift indicates a high value of $\rho(0)$ at Fe site which may be either due to the direct occupation of Fe $4s$ orbital and/or significant deshielding of Fe $3s$ orbitals due to rel. low Fe $3d$ occupancy, i.e. due to a strong $Fe - N$ bond. We have considered various probable configurations of FeN in order to understand the relatively high value of $\rho(0)$.

Table 1: SCF and MP4SDTQ calculations for various configurations of FeH

Configuration	bond distance in Å[a]	total energy in hartree SCF -1262.0	total energy in hartree MP4SDTQ -1262.0	Mulliken population of Fe 3d and 4s orbitals 3d	Mulliken population of Fe 3d and 4s orbitals 4s	ΔE_Q in mm/s[b] SCF	ΔE_Q in mm/s[b] MP4SDTQ	η[b] SCF/MP4SDTQ	$\rho(0)$ in a.u. SCF 15000	$\rho(0)$ in a.u. MP4SDTQ 15000
$^6\pi$	1.7392	-0.64065	-0.74230	6.04	1.10	-4.48 (+5.35)	-4.42 (+5.25)	0 (0.89)	69.5	69.8
$^6\Delta$	1.7403	-0.65142	-0.75222	6.05	1.06	+0.31	+0.22	0	69.2	69.6
$^4\phi$	1.7011	-0.55352	-0.71341	7.00	0.57	+0.75 (-2.60)	+0.74 (-2.47)	0 (0.40)	66.0	66.4
Experimental						(±)2.4[c]			66.4 for I.S. = +0.59 mm/s	

a) Optimized at SCF level

b) Values in bracket correspond to the low-symmetry situation with liftet orbital symmetry, which likely occurs for FeH in solid H_2

c) Sign of EFG experimentally undetermined

Fig. 2 gives a schematic energy level diagram for *FeN*. There are two electronic configurations for *FeN* which merit consideration as ground state, $^6\Sigma^+(1\sigma^2, 2\sigma^2, 1\pi^4 1\delta^2 2\pi^2 3\sigma^1)$ and $^4\Delta(1\sigma^2, 2\sigma^2, 1\pi^4 1\delta^3 2\pi^2)$. The calculated values of the total energy for these two configurations, given in Table 2, show that $^6\Sigma^+$ is 74 mhartrées lower than $^4\Delta$. In both these configurations the *Fe* 4*s* orbital is not directly occupied, however, $^6\Sigma^+$ shows a large contribution of 4*s* through a strong σ bond between *Fe* and *N*. The calculated $\rho(0)$ value (Table 2) for $^6\Sigma^+$ is 15071.7 a.u. which is close to the value of 15070.4 a.u. derived from experimental isomer shift of $-0.46mm/s$. The calculated value of quadrupole splitting, $-2.80mm/s$, is small compared to the experimentally observed value of $(\pm)3.9mm/s$.

We have also examined the possibility of direct occupation of a 4*s* orbital to either $^8\pi(1\sigma^2 2\sigma^2 1\pi^3 1\delta^2 2\pi^2 3\sigma^1 4\sigma^1)$ or $^8\Sigma^+(1\sigma^2 2\sigma^1 1\pi^4 1\delta^2 2\pi^2 3\sigma^1 4\sigma^1)$ states. However, geometry optimization for both these configurations diverged to very large distances indicating that the *FeN* molecule is not stable in these two configurations.

We can,therefore, conclude that the ground state for *FeN* is $^6\Sigma^+$. We also observe that since the *Fe* 4*s* orbital is not directly occupied, the matrix has very little influence on the electron density, and the value of $\rho(0)$ derived from experimental I.S. matches well with the calculated one. The matrix, however, seems to affect the EFG at the *Fe* site to a small extent.

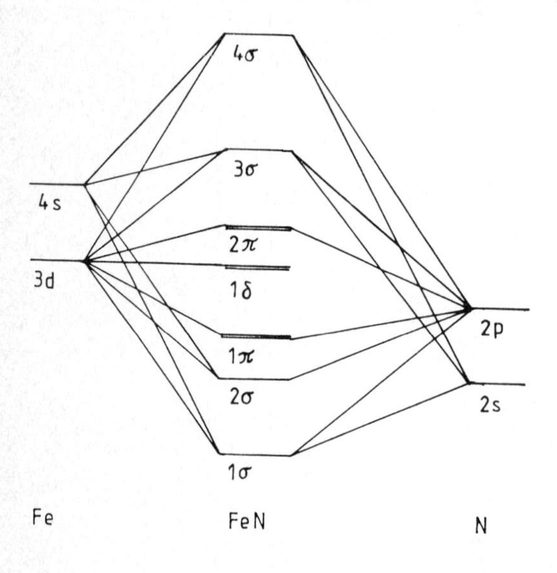

FIG. 2: Schematic energy level diagram for *FeN* showing valence orbitals

6. ELECTRONIC CONFIGURATIONS OF *FeCO*

Peden et al (20) have studied *FeCO* molecules trapped in solid *Ar* and *Kr* matrices using cocondensation technique. Their I.R. studies show a reduction in *CO* stretching frequency for *FeCO* by $240cm^{-1}$ with respect to that observed for free *CO* molecules (which correspond to a reduction in force constant by $4.2mdynes/\text{Å}$) indicating a strong $Fe - CO$ bond while the observed Mössbauer spectra show an isomer shift of $-0.60mm/s$.

These two observations seem to be contradictory, because

(i) with a bond between *Fe* and *CO* one does not expect more than one electron in the *Fe* 4*s* orbital, however

(ii) from the observed large and negative isomer shift one expects a *Fe* 4*s* population close to $4s^2$ (21).

Table 2: SCF and MP4SDTQ calculations for two configurations of FeN .

Configuration	bond distance in Å[a]	total energy in hartree SCF	total energy in hartree MP4SDTQ	Mulliken population of Fe 3d and 4s orbitals 3d	Mulliken population of Fe 3d and 4s orbitals 4s	ΔE_Q in mm/s SCF	ΔE_Q in mm/s MP4SDTQ	$\rho(0)$ in a.u. SCF	$\rho(0)$ in a.u. MP4SDTQ
		-1316.0	-1316.0					15000	15000
$^6\Sigma^+$	1.7010	-0.36528	-0.62400	6.08	0.86	-2.80	-2.84	71.3	71.7
$^4\Delta$	1.7099	-0.32151	-0.54977	6.30	0.17	+3.32	+3.29	65.7	66.0
Experimental						(±)3.9[b]		70.4 for I.S. = -0.46 mm/s	

a) Optimized at SCF level

b) Sign of EFG experimentally undetermined

It is therefore interesting to check whether ab initio CI calculations are able to solve this discrepancy or whether again the matrix may play an important role.

Fig. 3 shows a schematic energy level diagram for $FeCO$ while Fig. 4 depicts various electronic configurations with different spin values, which merit consideration as a ground state. The spin degeneracy can vary from 0 to 4 depending upon the energy separation between 4π orbitals, which have mainly $Fe(3d\pi)$ character, and 5π orbitals, which have mainly $CO(\pi^*)$ character. Table 3 gives the results of our calculation for five spin states defined in Fig. 4. It is interesting to note that the SCF procedure predicts $^7\Sigma^-$ as a ground state while MP4SDTQ calculations after SCF predicts $^5\Sigma^-$ as a ground state. These results are similar to those obtained from MCSCF calculation by Bagus et al (22,23). Expect for $^9\Sigma^-$ state none of the configurations show more than one electron in the $4s$ orbital; also the $3d$ orbital population is always larger than 6 electrons which indicates that the deshielding effect due to the backdonation would be insufficient. The calculated values of $\rho(0)$, except for $^9\Sigma^+$, are much smaller than the experimentally expected value of 15071.7 a.u. The $^9\Sigma^-$ state which is about 125 mhartrée above the ground state is not expected to stabilize as a ground state because of the influence of the rare gas matrix. It is, therefore, not possible to explain, even qualitatively, a large value of $\rho(0)$ at Fe site. If we rule out any ambiguity in the analysis of the experimental Mössbauer spectra by Peden et al (20), then we are forced to conclude that the solid noble gas matrix affects the molecular properties of $FeCO$ to a higher extent than believed so far, because neither the $\rho(0)$-value nor the calculated quadrupole splitting of the ground state of the free $FeCO$ molecule corresponds to observed Mössbauer parameters of matrix-isolated $FeCO$.

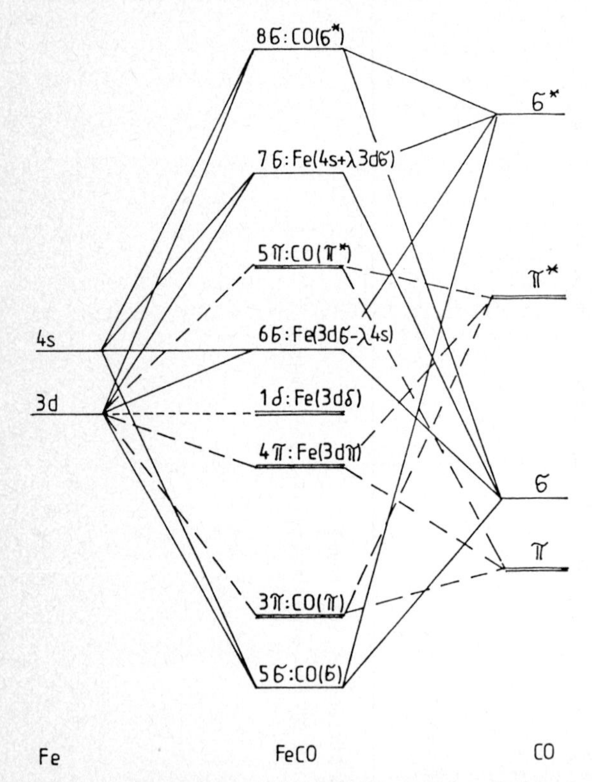

FIG. 3: Schematic energy level diagram for $FeCO$ showing only the valence orbitals of Fe and relevant orbitals of the CO molecule. (For simplificity, the notation used for MOs represents predominantly metal or ligand contributions).

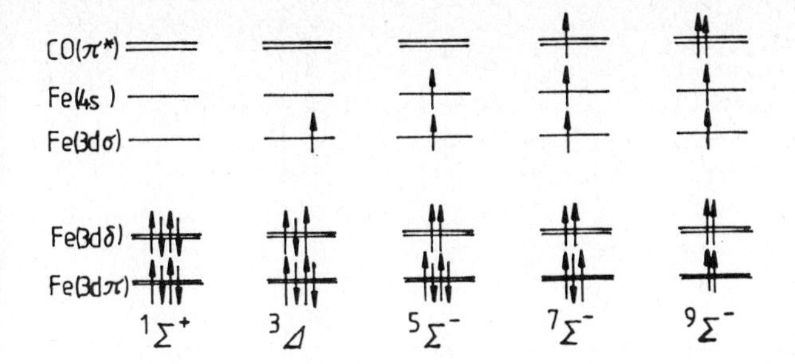

FIG. 4: Schematic energy level diagram showing MOs with predominantly iron and $CO(\pi^*)$ character, which the eight valence electrons should occupy. The five electronic configurations, which merit consideration in understanding the properties of $FeCO$ are also shown.

7. ELECTRONIC CONFIGURATION FOR $Fe(CO)_2$

Peden et al (20) have also observed the existence of $Fe(CO)_2$ species in solid Ar and Kr matrices which they have studied using I.R. and Mössbauer spectroscopy. The experimental values of isomer shift and quadrupole splitting are $-0.28mm/s$ and $3.63mm/s$, respectively. Since only a single infrared absorption peak at $1860cm^{-1}$ can be attributed to $Fe(CO)_2$, Peden et al. have concluded that $Fe(CO)_2$ is a linear molecule with $D_{\infty h}$ symmetry. The calculated value of $C-O$ stretching force constant is $14.8mdynes/\AA$ which is considerably lower than that for free CO ($18.5mdynes/\AA$). The problem is, therefore, very similar to that observed in $FeCO$. The expected $\rho(0)$ value from the measured I.S. is about 15069.7 a.u. which can arise either through high $4s$ population or significant $d\pi$ backdonation.

For a linear geometry of $Fe(CO)_2$ the electronic energy level diagram is similar to the one obtained for $Fe(CO)$. Four different electronic configurations which merit consideration as ground state are shown in Fig. 5. Geometry optimization for $^5\Sigma^-$ state diverged to large $Fe-C$ distance showing that the molecule is unstable in this configuration. The MO results for $^1\Sigma^+$, $^3\Delta$, and $^5\pi$ are shown in Table 4: We observe that for all the three configurations the $4s$ population is less than 0.5 electrons while the $3d$ population is larger than 6.7 electrons, and the calculated $\rho(0)$ values therefore lie between 15066.8 a.u. to 15067.2 a.u. These values are much lower than the expected value of 15069.7 a.u. The calculated value of quadrupole splitting for $^5\pi$ is also higher than the experimental value of $(\pm)3.63mm/s$.

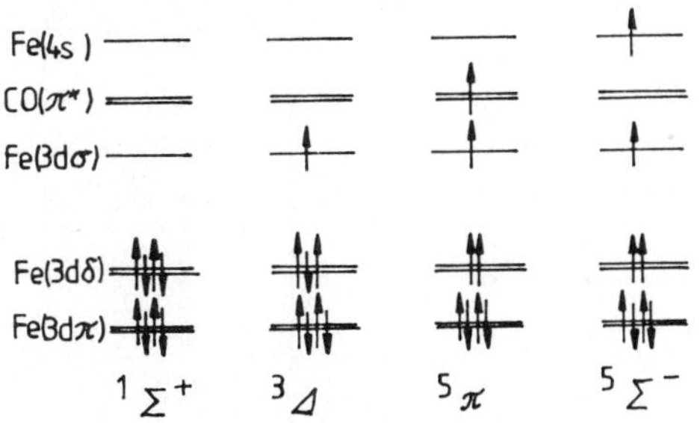

Fe(4s)
CO(π*)
Fe(3dσ)
Fe(3dδ)
Fe(3dπ)
$^1\Sigma^+$ $^3\Delta$ $^5\pi$ $^5\Sigma^-$

FIG. 5: Schematic energy level diagram showing relevant MOs with predominantly iron or $CO(\pi^*)$ character for $Fe(CO)_2$ in $D_{\infty h}$ symmetry. The four electronic configurations which merit consideration in understanding the properties of $Fe(CO)_2$ are also shown.

Table 3: SCF and MP4SDTQ calculations for various configurations of FeCO.

Configuration	bond distances in Å		total energies in hartree		Mulliken population of Fe 3d and 4s orbitals		ΔE_Q in mm/s		$\rho(0)$ in a.u.	
	Fe-C[a]	C-O[a]	SCF -1374.0	MP4SDTQ -1374.0	3d	4s	SCF	MP4SDTQ	SCF 15000	MP4SDTQ 15000
$^1\Sigma^+$	1.8105	1.1573	-0.43646	-1.04577	7.69	0.10	+5.15	+4.99	64.2	64.6
$^3\Delta$	1.8202	1.2166	-0.62632	-0.98224	6.59	0.37	+1.78	+1.45	66.9	67.3
$^5\Sigma^-$	1.8923	1.1611	-0.68437	-1.06102	6,42	0.82	-4.25	-4.42	68.8	69.2
$^7\Sigma^-$	2.0740	1.1767	-0.72535	-1.02183	6.05	0.92	-3.35	-3.21	68.6	68.9
$^9\Sigma^-$	1.9923	1.1617	-0.62295	-0.90187	5.10	1.02	+0.05	+0.03	72.6	72.8
Experimental							(±)3.39[b]		71.7 for I.S. = -0.60 mm/s	

a) Optimized at SCF level

b) Sign of EFG experimentally undetermined

Table 4 : SCF and MP3 [a] calculations for various configurations of $Fe(CO)_2$

Configuration	bond distances in Å		total energy in hartree		Mulliken population of Fe 3d and 4s orbitals		ΔE_Q in mm/s [c]	η [c]	$\rho(0)$ in a.u.
			SCF	MP3			SCF	SCF	SCF
	Fe-C [b]	C-O [b]	-1487.0	-1487.0	3d	4s			15000
$^1\Sigma^+$	1.8920	1.1464	-0.13920	-0.69363	7.62	0.21	+4.15	0	66.8
$^3\Delta$	1.9383	1.1666	-0.30165	-0.85055	6.96	0.46	-0.94	0	67.1
$^5\pi$	1.9161	1.1526	-0.33793	-0.87057	6.76	0.45	-4.54 (-4.74)	0 (0.50)	67.2
Experimental							(±)3.63 [d]		69.7 for I.S. = -0.28 mm/s

a) MP3 = configuration interaction calculation using Møller-Plesset perturbation procedure up to third order

b) Optimized at SCF level

c) Values in bracket correspond to the low-symmetry situation with lifted orbital symmetry, which likely occurs for $Fe(CO)_2$ in solid Ar

d) Sign of EFG experimentally undetermined

We have also performed geometry optimization for quintet states with angle $C - Fe - C$ as a parameter. However, minimum energy is obtained for an angle of 180° which agrees well with the results of I.R. spectroscopy.

Thus, as in $FeCO$ we have experimental Mössbauer results which can not be explained unless the noble gas matrix takes significant influence on the molecular properties on $Fe(CO)_2$.

GENERAL CONCLUSIONS

Through the four examples for small molecules trapped in matrices we observe that the matrices may significantly influence the ground state properties, hence the observed spectral parameters. This effect arises mainly due to unsaturated coordination of the transition metal ion and is expected to be small for bigger molecules (or fragments of molecules) for which coordination conditions are improved, for example, $Fe(CO)_5$.

Correlation corrections are certainly important in obtaining appropriate total energies. In the case of $FeCO$, the MP4SDTQ calculations even altered the energy sequence of various configurations. However, Mössbauer parameters calculated before and after CI corrections differ only slightly. Though calculated Mössbauer parameters are strongly sensitive to the choice of an electronic configuration they are less sensitive to more rigorous CI calculations.

REFERENCES

1. M. Pasternak; Hyperfine Interactions 27 (1986) 173
2. F. Seel, B. Wolf, U. Gonser, R. Klein, G. Doppler, E. Bill, A. X. Trautwein; Z. anorg. allg. Chem. 534 (1986) 159
3. M. A. de Paoli, S. M. de Oliverira, E. B. Saitovitch, D. Guenzburger; J. Chem. Phys. 80 (1984) 730
4. M. Wrighton; Chem. Rev. 74 (1974) 401
5. E. W. Plummer, W. R. Salaneek and J. S. Miller; Phys. Rev. B18 (1978) 1673
6. P. Gütlich, R. Link, A. Trautwein; "Mössbauer spectroscopy and transition metal chemistry", Springer- Verlag Berlin (1978)
7. J. S. Binkley, M. Frisch, K. Raghavachari, D. De Frees, M. B. Schlegel, R. Whiteside, E. Fluder, R. Seeger, J. A. Pople; GAUSSIAN 82, Release H, Department of Chemistry , Carnegie-Mellon University Pittsburg, P.A.
8. R. Seeger, J. A. Pople; J. Chem. Phys. 65 (1976) 265
9. a) C. Möller, M. S. Plesset; Phys. Rev. 46 (1934) 618
   b) J. S. Binkley, J. A. Pople; Int. J. Quantum Chem. 9 (1975) 229
   c) R. Krishnan, J. A. Pople; Int. J. Quantum Chem. 14 (1978) 91
10. S. Lauer, V. R. Marathe, A. X. Trautwein; Phys. Rev. A19 (1979) 1852
11. R. Reschke, A. X. Trautwein, J. P. Desclaux; J. Phys. Chem. Solids 38 (1977) 837
12. V. R. Marathe, A. Sawaryn, A. X. Trautwein, M. Dolg, G. Igel-Mann, H. Stoll; Hyperfine Interactions (submitted for publication)
13. S. Huzinaga, M. Klobukowski, Z. Barandiaran, L. Seija; J. Chem. Phys. 84 (1986) 6315
14. Z. Barandiaran, L. Seijo, S. Huzinaga, M. Klobukowski; Int. J. Quantum Chem. 29 (1986) 1047
15. A. Dendramis, R. J. Van Zee, W. Weltner Jr.; Astrophys. J. 231 (1979) 632
16. M. Pasternak, M. Van der Heyden, G. Langouche; Chem. Phys. Letters 104 (1984) 398
17. P. R. Scott, W. G. Richards; J. Chem. Phys. 63 (1975) 1690
18. P. R. Scott, W. G. Richards, Molecular Spectroscopy 4 (1976) 70
19. M. Van der Heyden, M. Pasternak, G. Langouche; Hyperfine Interactions 29 (1986) 1315
20. C. H. F. Peden, S. F. Parker, P. H. Barrett, R. G. Pearson; J. Phys. Chem. 87 (1983) 2329
21. T. K. McNab, H. Micklitz, P. H. Barrett; Phys. Rev. B4 (1971) 3787
22. P. S. Bagus, C. J. Nelin, C. W. Bauschlicher Jr.; Vacuum Science and Technology A2 (1984) 905
23. C. W. Bauschlicher Jr., P. S. Bagus, C. J. Nelin, B. O. Roos; J. Chem. Phys. 85 (1986) 354

Computer Simulation of Cluster Thermodynamics

G. FRANKE

Fachbereich Physik
Universität Oldenburg
D-2900 Oldenburg
FRG

ABSTRACT

For the computer-simulation of the quantum-statistical mechanics of rare-gas-clusters, a Path Integral Monte-Carlo method is developed and thermodynamical and geometric properties of the clusters are investigated. First results show the important influence of zero-point-energy on the structure for low temperature and lead to a discussion of "structural transitions" in clusters.

INTRODUCTION

In the physics of clusters, the thermodynamics of small systems is an important field, because all clusters that can be produced experimentally are more or less warm. This is true especially for weakly bound systems — such as Van-der-Waals clusters — where thermal effects are not negligible even at low temperature. As an example, for the discussion of cluster-size-distributions in molecular-beam experiments, the size- and temperature-dependent free energy is necessary to determine the equilibrium-distribution and to derive from it nucleation- (and aggregation-) rates. Another important question is the kind of "phase transition" that finite systems can undergo. How broad is the transition region, what shift of the transition temperature one gets, what does "liquid" mean for small clusters?

Up to now, the full explicit consideration of electron-exchange-energies (as necessary for metallic clusters) is only possible by quantum-chemical methods [1], which do not allow the introduction of non-zero temperature and neglect the kinetic energy of the nuclei. For Van-der-Waals systems, where the interaction may be approximated by some effective two (or more-)body-potential, two general computational methods for the calculation of thermodynamic properties are known: Molecular Dynamics (MD) and Monte-Carlo (MC) methods. Classical MC-calculations — where the thermodynamic phase space is sampled by a stochastic process for fixed temperature — have been recently applied to clusters for lattice and continuum formulations [2] as well as MD-methods [3], where the classical dynamics of the system is simulated and thermodynamic results can be obtained by reason of ergodicity. The main disadvantage of these methods is to be not extendable to a quantum mechanical formulation.

For rare gases and other Van-der-Waals systems, the classical treatment is not sufficient because of the weakness of the interaction: For example, a one-dimensional two-body Lenard-Jones system with realistic masses and potentials has only 2-4 bound states and the zero-point-energy is a large fraction of the whole binding energy.

For the full consideration of quantum-mechanics, Path Integral Monte-Carlo techniques (PIMC) [4] can be used. Especially from Lattice-gauge-theoretical calculations they are known as a powerful tool for the simulation of bound quantum systems.

THE PATH INTEGRAL METHOD

For the evaluation of thermodynamical averages of an Observable O

$$<O>=\frac{1}{Z}Tr(\rho O)=\frac{1}{Z}\iint dx_o dx_1 <x_o \mid O \mid x_1><x_1 \mid \rho \mid x_0> \tag{1}$$

one needs to determine the elements of the density matrix $< x_1 \mid \rho \mid x_0 >$ of the statistical operator $\rho = e^{-\beta H}$ with hamiltonian

$$H=\sum_{i=1}^{N}\frac{1}{2m}p_i^2+V(x) \quad , \quad \begin{array}{r} i = \textit{particle index} \\ x = \textit{N-particle-coordinate vector} \end{array} \quad . \tag{2}$$

These matrix elements are related to the transition matrix-elements of the wellknown quantum mechanical time propagation operator e^{iHt} by Wick-rotation. One can exploit this analogy by following Feynman's argumentation [5] and describing the density matrix by a path integral

$$< x_1 \mid \rho \mid x_0 >= const. \cdot \int Dy(\tau)\ e^{-\frac{1}{\hbar}S[y(\tau)]} \quad . \tag{3}$$

$Dy(\tau)$ runs over all paths in real space from x_1 to x_0. In constrast to usual quantum mechanics, here the exponent is real and the "Euclidian action" has to be used:

$$S[y(\tau)]=\int_0^{\hbar\beta} d\tau \left(V(y(\tau))+\frac{1}{2}m\dot{y}(\tau)^2\right) \quad , \quad \beta=\frac{1}{kT} \quad . \tag{4}$$

The path integral formalism is known to be adequate for systems with a large number of degrees of freedom. One of its main advantages is the formulation in a quasiclassical configuration space with the additional "time"-direction, which makes the comparison to classical results easy.

To calculate average values of x- or p- dependent operators, one has to close the paths. For example,

$$O=O(x) \quad \Longrightarrow \quad <O> \ = \frac{1}{Z}\int Dy(\tau)\ O(y)\ e^{-S} \tag{5}$$

where $Dy(\tau)$ now runs over all closed paths. For p-dependent Operators, the continuum description is not unique [5] but nevertheless the practical calculation of averages is not problematic.

EVALUATION OF THE PATH INTEGRAL

The computation of the functional integrals (5) can be performed after splitting the paths into m discrete steps at "times" τ_k, so that a finite- (i.e. $3 \cdot N \cdot m$) dimensional integral remains. For the calculation of the "moments" $O(y)$ resp. $O(\dot{y})$ of the weight function e^{-S}, the Metropolis-algorithm [6] is a convenient method because no explicit normalization by the partition function Z is required. In realization of this algorithm, two problems arise:

1. A starting-configuration for the MC-process with low "action" S is required in order to avoid long "warming-up" times and evaporation of particles during it.
2. The random step from one configuration to the next one has to be optimized in order to get fast convergence.

The first problem we solve by using a classical "annealing" program before starting the main MC-procedure. Here, by slowly lowering the temperature, the "classical groundstate" can be found as a good starting point for the quantum mechanical calculation. The second — and more serious — problem is due to the "kinetic" part of the action $S_k = \int \frac{m}{2}\dot{y}^2 d\tau$: The direct method of varying the particle-coordinates $y_i(\tau_k)$ of each time step was found to cause extremely poor convergence. The reason is, that the changes of the configurational and the kinetic terms due to the statistical motion in one time-step are not in the same order of magnitude. Especially at high temperature, the dynamics of the stochastic MC-process is determined only by the second one and the change of configurations becomes very slow.

To avoid this problem, one has to "decouple" potential and kinetic energies by fourier-transformation of the coordinate-step in time-direction. Then the change in configurational energy is mainly determined by the zero-component that does not contribute to the kinetic part and one can choose different step sizes for the fourier-components. In contrast to the method of Doll and Freemann [7,8], the transformation is applied only to the production of new configurations.

RESULTS

Because we want to obtain information on the geometric structure of the cluster and single configurations used for the integration process are physically meaningless, one has to extract geometrical data from the system that are invariant under transformations like rotations and particle exchange that don't influence the physical properties of the system. Therefore we investigate the pair correlation function:

$$\Gamma(r) = \langle \frac{2}{N(N-1)} \sum_{i<j}^{N} \delta(\mid x_i - x_j \mid - r) \rangle \tag{6}$$

where the x_i are particle-positions. For strongly ordered systems, i.e. in the classical limit $T \to 0$, this function is very characteristic for the geometric shape. Fig. 1 shows this plot for a configuration produced by the classical "annealing program" at $T \approx 0$ for a 13-particles LJ-system with potential minimum at $3\mathring{A}$. Positions and height of the peaks correspond to the famous icosahedral structure.

For the same LJ-model system with high mass but weak potential that we have investigated up to now, Figs. 2-5 show the correlation function for several temperatures while Figs. 6-8 compare it with classical curves that can be obtained by the same calculation scheme by suppressing all Fourier-coefficients except $k = 0$. Some qualitative properties are obvious:

For low temperature, the curves are very similar while they differ strongly from the classical ones. The structure is dominated by the zero-point energy and is not very sharply determined. Even near the ground state, the system can not make full use of the existence of highly symmetric structures like the icosahedron.

In the entire temperature region a slow "melting process" takes place: Additional to the broadening of the peaks due to vibrational motion, the gaps between them are filled smoothly. This means that the number of configurations contributing to the partition function is growing continuously and no transition point can be recognized. This may be named as quantum-amorphous in the sense that sharply defined configurations only play a minor role as in amorphous substances, where different configurations are found at different sites.

PERSPECTIVES

Besides technical improvements of the method — especially the introduction of "partial averaging" [9] and explicit consideration of collective motions — we will extend this calculations to larger clusters and various interatomic potentials. The method should be useful to test the reliability of different discussed two- or more-body atomic interactions [10]. It will be interesting to compare the correlation functions to those obtained by other methods [11] and to try to distinguish between configurations mainly contributing to the statistical ensemble.

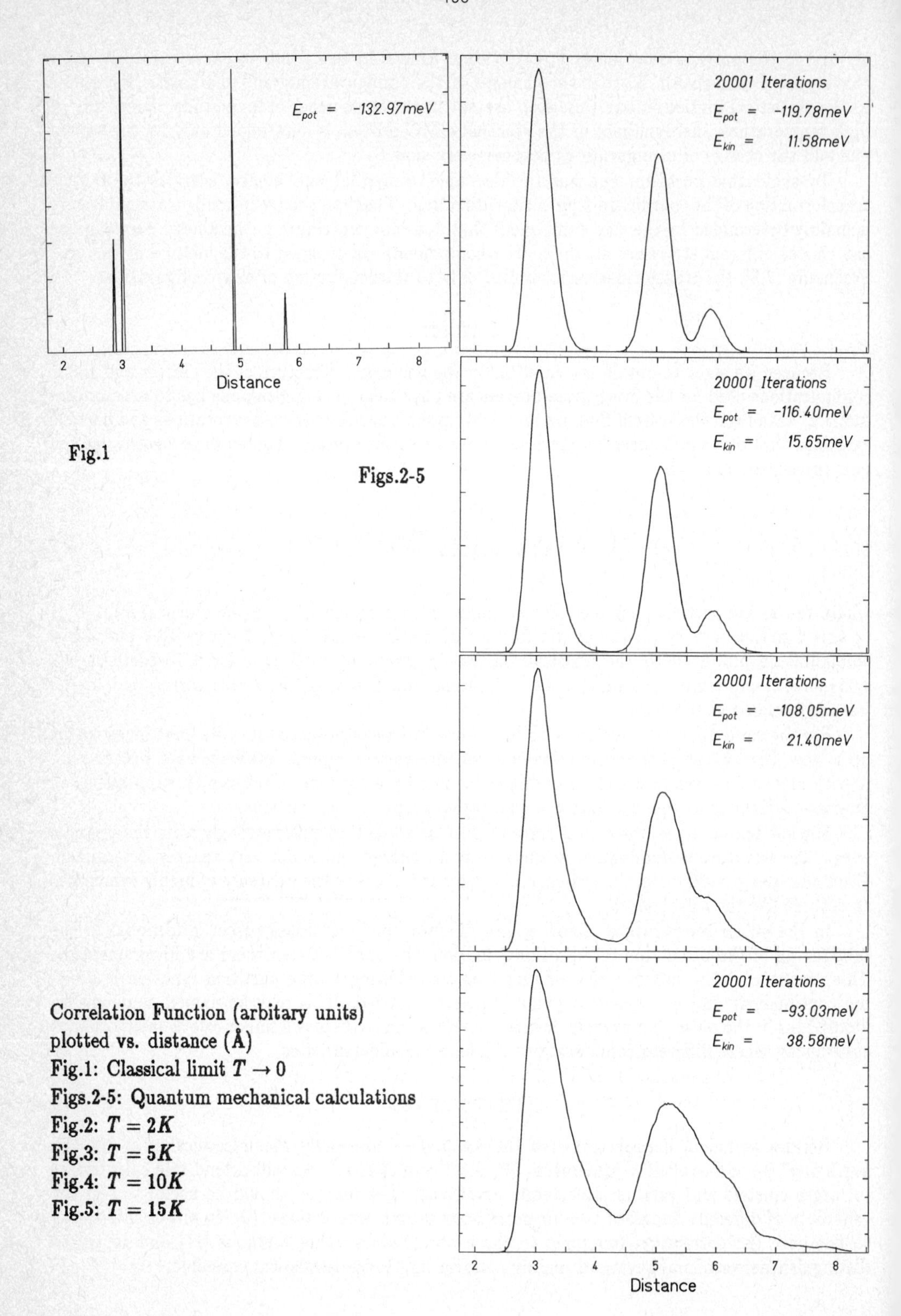

Fig.1

Figs.2-5

Correlation Function (arbitary units) plotted vs. distance (Å)
Fig.1: Classical limit $T \to 0$
Figs.2-5: Quantum mechanical calculations
Fig.2: $T = 2K$
Fig.3: $T = 5K$
Fig.4: $T = 10K$
Fig.5: $T = 15K$

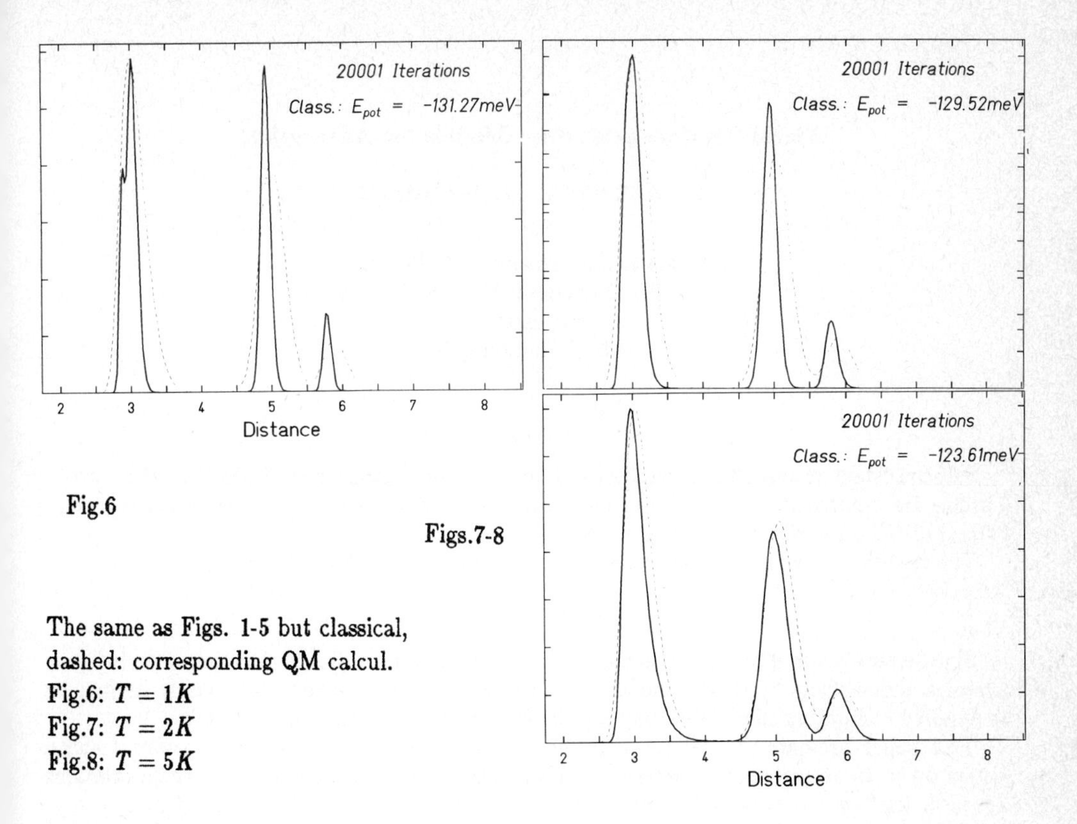

Fig.6

Figs.7-8

The same as Figs. 1-5 but classical,
dashed: corresponding QM calcul.
Fig.6: $T = 1K$
Fig.7: $T = 2K$
Fig.8: $T = 5K$

Acknowledgement

This work was made possible by collaboration with L. Polley and E. Hilf which is appreciated hereby. Also some very interesting and stimulating suggestions by K. Sattler are acknowledged. Numerical computations have been done on the excellent computer facilities of the Gesellschaft für Schwerionenforschung (GSI) in Darmstadt.

References

[1] J. Koutecký, G. Pacchioni: J. Chem. Phys. 81 (1984) 3588
[2] H. Müller-Krumbhaar in: K. Binder (Ed.): MC Methods in statistical Physics, Springer 1979
[3] E.E. Polymeropulos, J. Brickmann: Chem. Phys. Lett. 92 (1982)
[4] G. Jacucci in: M.H. Kalos (Ed.): MC-Methods in Quantum Problems, D. Reidel 1984
[5] R.P. Feynman, A.R. Hibbs: Quantum Mechanics and Path Integrals, McGraw Hill 1965
[6] N. Metropolis, A.W. + M.N. Rosenbluth, E. Teller: J. Chem. Phys. 21 (1953) 1087
[7] J.D. Doll, D.L. Freeman: J. Chem. Phys. 80 (1984) 2239
[8] J.D. Doll, D.L. Freeman: J. Chem. Phys. 82 (1985) 465
[9] J.D. Doll, D.L. Freeman, R.D. Coalson: Phys. Rev. Lett. 55 (1985) 1
[10] E.E. Polymeropulos, J. Brickmann, L. Jansen, R. Block: Phys. Rev. A 30 (1984) 1593
[11] C.L. Briant, J.J. Burton: J. Chem. Phys. 63 (1975) 2045

Nickel Clusters as Surface Models for Adsorption

O. KÜHNHOLZ, M. GRODZICKI

I. Institut für Theoretische Physik
der Universität Hamburg
Jungiusstr. 9
D-2000 Hamburg 36
FRG

ABSTRACT

Self-consistent charge $X\alpha$ calculations of the electronic properties of nickel clusters up to 19 atoms are reported. The geometries were chosen according to adsorption sites and represent (100), (110) nickel surfaces, and bulk nickel.

The results are discussed with respect to bulk properties of these clusters.

INTRODUCTION

Since several years efforts have been made to understand theoretically the interaction between molecules and surfaces by cluster model calculations (1,2). Such studies are mainly limited by the required computing time depending on the number of atoms and basis orbitals.

This paper presents the results of molecular orbital (MO) calculations on several nickel clusters up to 19 atoms which can serve as substrates for adsorbed molecules. The calculations were performed on the basis of the self-consistent charge $X\alpha$ ($SCC - X\alpha$) method described in chap. 1. Comments on computational details are given in chap. 2 and the results of the different cluster calculations and their discussion are presented in chap. 3.

1. THEORY

The self-consistent charge $X\alpha$ method (1,2) can be characterized as a semi-empirical valence-electron-only MO method. It has been developed in particular for calculating the electronic structure of large molecules or clusters. The accurary of the results is comparable to more sophisticated $X\alpha$ methods, e.g. $SW - X\alpha$ (3) and $DV - X\alpha$ (4), while, at the same time, the computation time exceeds that of an extended Hückel method by 3 - 4 times only.

On the basis of Born-Oppenheimer, Hartree-Fock and $X\alpha$ approximations the following Schrödinger equation for the k th one-particle state is obtained (in Rydberg-a.u.)

$$[-\Delta + V(\underline{r})]\,\psi_k(\underline{r}) = \varepsilon_k \psi_k(\underline{r}) \quad .$$

The molecular orbital $\psi_k(\underline{r})$ is represented as a linear combination of atomic orbitals (LCAO).

$$\psi_k(\underline{r}) = \sum_{\mu i} \phi_i^{(\mu)}(\underline{r} - \underline{R}_\mu) c_{ik}^\mu \quad .$$

$\phi_i^{(\mu)}(\underline{r} - \underline{R}_\mu)$ denotes an atomic wavefunction characterizing the (n_i, l_i, m_i)th atomic orbital of the μ th atom at the position $\underline{R}_\mu$. The unknown potential $V(\underline{r})$ is constructed by a superposition of atomic potentials

$$V(\underline{r}) = \sum_\kappa V_{\text{at}}^{(\kappa)}(|\,\underline{r} - \underline{R}_\kappa\,|) \quad .$$

These additional approximations lead to the secular equation (1)

$$\sum_{\nu j}\left(H_{ij}^{\mu\nu}-\varepsilon_k S_{ij}^{\mu\nu}\right)c_{jk}^{\nu}=O$$

with the Hamilton operator

$$H_{ij}^{\mu\nu}=\frac{1}{2}\left(\varepsilon_i^{\mu}+\varepsilon_j^{\nu}\right)S_{ij}^{\mu\nu}+\frac{1}{2}\left(V_{ij}^{\mu\nu}+V_{ji}^{\nu\mu}\right) \quad ,$$

the overlap matrix

$$S_{ij}^{\mu\nu}=\int \phi_i^{(\mu)*}(\underline{r}-\underline{R}_{\mu})\phi_j^{(\nu)}(\underline{r}-\underline{R}_{\nu})d^3r \quad ,$$

and the potential matrix

$$V_{ij}^{\mu\nu}=\sum_{\kappa\neq\nu}\int \phi_i^{(\mu)*}(\underline{r}-\underline{R}_{\mu\nu})V_{\mathrm{at}}^{(\kappa)}(|\,\underline{r}-\underline{R}_{\kappa\nu}\,|)\phi_j^{(\nu)}(\underline{r})d^3r \quad ; \quad \underline{R}_{\mu\nu}=\underline{R}_{\mu}-\underline{R}_{\nu} \quad .$$

The valence-state ionization potential of the μ th atom and the i th atomic orbital ε_i^{μ} is assumed to be a quadratic function of the effective charge Q_{μ} (5)

$$\varepsilon_i^{\mu}=\varepsilon_{io}^{\mu}+\varepsilon_{i1}^{\mu}Q_{\mu}+\varepsilon_{i2}^{\mu}Q_{\mu}{}^2 \quad .$$

For facilitating the evaluation of the respective matrix elements further approximations are introduced. The total electron density is separated in a core and a valence part

$$\rho^{(\kappa)}(\underline{r}-\underline{R}_{\kappa})=\rho_{\mathrm{core}}^{(\kappa)}(\underline{r}-\underline{R}_{\kappa})+\rho_{\mathrm{val}}^{(\kappa)}(\underline{r}-\underline{R}_{\kappa}) \quad .$$

Using the point charge approximation, the expression for the core density is simplified to

$$\rho_{\mathrm{core}}^{(\kappa)}(\underline{r}-\underline{R}_{\kappa})=N_{\mathrm{core}}^{(\kappa)}\delta(\underline{r}-\underline{R}_{\kappa})$$

where $N_{\mathrm{core}}^{(\kappa)}$ denotes the number of core electrons of the κ th atom. The valence density is approximated by a single exponential

$$\rho_{\mathrm{val}}^{(\kappa)}(r)=\frac{N_{\kappa}\eta_{\kappa}^3}{8\pi}\exp(-\eta_{\kappa}\cdot r) \quad ; \quad r=|\,\underline{r}-\underline{R}_{\kappa}\,| .$$

N_{κ} represents the effective number of valence electrons, and the exponential parameter η_{κ} is assumed to be a linear function of Q_{κ}

$$\eta_{\kappa}=\eta_{\kappa o}+\eta_{\kappa 1}Q_{\kappa} \quad .$$

Such an approximation for the valence density turns out to be sufficiently good in the valence region even of larger atoms (6). As the final result a simple atomic potential (1) is obtained

$$V_{\mathrm{at}}^{(\kappa)}(r)=-\frac{2Q_{\kappa}}{r}-\frac{2N_{\kappa}\exp(-\eta_{\kappa}r)}{r}-\eta_{\kappa}\exp(-\eta_{\kappa}r)\cdot\left(N_{\kappa}+\alpha'\cdot N_{\kappa}^{\frac{1}{3}}\right) \quad ;$$

$$\alpha'=1.5\alpha\,(3/\pi)$$

allowing the rapid computation of the potential matrix. The potential depends on the effective charge $Q_{\kappa}=Z_{\kappa}-N_{\kappa}$ where Z_{κ} is the atomic number of the κ th atom. The α coefficient of the

$X\alpha$ term is taken to equal 0,7 for all atoms. Finally, the radial part of the atomic wavefunction is approximated by a single Slater-type function

$$R_i^{(\mu)} = a_i r^{(n_i-1)} \exp(-\varsigma_i^\mu \cdot r) \quad .$$

The parameter ς is also assumed to be charge dependent.

$$\varsigma_i^\mu = \varsigma_{io}^\mu + \varsigma_{i1}^\mu \cdot Q_\mu \quad .$$

The last simplification lies in using a self-consistent-charge iteration scheme instead of a full self-consistent field procedure. The basis of the approximation is the assumption that constructing an 'atom in a molecule' is a reliable concept. The required effective atomic charges are determined according to a weighted population analysis. Namely, the overlap charge is divided proportionally to the population of the i th atomic orbital of the μ th atom and the respective orbital radius $n_i^\mu / \varsigma_i^\mu$. The expression for the effective occupation number x_i^μ is (7)

$$x_i^\mu = 2a_i^\mu \sum_{\nu j} S_{ij}^{\mu\nu} P_{ij}^{\mu\nu} / (a_i^\mu + a_j^\nu)$$

with the bond order matrix

$$P_{ij}^{\mu\nu} = \sum_k n_k c_{ik}^\mu c_{jk}^\nu \quad ,$$

(Here: n_k denotes the occupation number of the k th molecular orbital $n_k \in [0,2]$) and the weighting factors

$$a_i^\mu = \frac{n_i^\mu}{\varsigma_i^\mu} \cdot \frac{x_{n_i l_i}^\mu}{2l_i + 1} \quad .$$

The iteration procedure uses the atomic occupation numbers as convergence criterion, instead of the effective charges, in order to gain more stable results. The form of an iteration cycle is as follows

$$\{Q_\mu\}_{old} \longrightarrow H_{ij}^{\mu\nu}(\{Q_\mu\}) \longrightarrow \varepsilon_k \quad ; \quad \psi_k(\underline{r}) \longrightarrow P_{ij}^{\mu\nu} \longrightarrow \{x_i^\mu\} \longrightarrow \{Q_\mu\}_{new} \; .$$

This process terminates if the condition

$$\max_{\mu,i} | x_i^\mu(\text{in}) - x_i^\mu(\text{out}) | < \varepsilon \quad ; \quad \varepsilon \geq 10^{-4}$$

is fulfilled.

The critical point of the $SCC - X\alpha$ method consists in the determination of the various parameters, i.e. the atomic orbital energies ε_{in}^μ the atomic orbital exponents ς_{in}^μ and the atomic potential parameters $\eta_{\mu n}$ (8). They have to be fitted by test calculations on small molecules starting from experimental or theoretical atomic data. For ε_{in}^μ such data can be obtained from optical and photoelectron spectra or Hartree-Fock calculations.

The wave function parameter ς is obtained by calculating the $< r >$ expectation values and the potential parameter η either also by calculating $< r >$ and $< r^2 >$ expectation values or by the slope of the function $\ln \rho(r)$ (6). For details see Ref. 9 and a description of the computer program is given in Ref. 10 .

2. Computational Details

For each nickel atom, a valence basis consisting of the 3d, 4s and 4p orbitals has been chosen, yielding nine MOs per atom. Test calculations showed that the 4p atomic orbital cannot be neglected. The required parameters have been optimized by comparison with experimental data of several nickel molecules like NiH, $NiCl_2$, $NiCO$, and $Ni(CO)_4$. The parameter set used in this calculation is shown in Table 2.1 where respective atomic data are also listed (1).

Table 2.1 Nickel parameter set

a) Atomic orbital parameters

Orbital	ε_0(Ry)	BE(Ry)	ε_1(Ry)	ε_2(Ry)	$\zeta_0(a_0^{-1})$	$\zeta_1(a_0^{-1})$
3d	0.63	0.74	0.60	0.13	2.70	0.15
4s	0.54	0.56	0.21	0.07	1.36	0.55
4p	0.10	(<0.44)	0.66	0.07	1.95	0.71

BE: free atom subshell binding energy

b) Potential parameters

$\eta_0(a_0^{-1})$	$\eta_1(a_0^{-1})$
3.4	0.7

All nickel clusters are artificial fragments of the bulk fcc crystal structure with a nearest neighbour distance of $R = 2.49\text{Å}$ (2). To get an idea of how much time is needed for the calculation of a cluster with the $SCC - X$ method, Table 2.2 lists the average CPU time per iteration vs. cluster size. The energy levels of d character are narrowly spaced causing interchanges during the iterations.

Table 2.2 Ni cluster size, number of molecular orbitals, average CPU time per iteration (Siemens 7.882/Fujitsu M 200)

Number of atoms	Number of MOs	CPU time (sec)
2	18	0.5
4	36	7
5	45	8
8	72	23
9	81	35
13	117	74
14	126	86
19	171	157

3.1 Ni_2

The results of the calculation on Ni_2 are shown in Table 3.1 for two different nickel-nickel distances where the larger one corresponds to the nearest neighbour distance in bulk nickel.

As expected, the shorter distance yields a lower lying ground state; however, a determination of an equilibrium distance failed. The reason is the change of the closed shell configuration to an open shell for shorter distances arising from the occupation of the antibonding σ_u orbital. Regarding multiplet states, six lowest lying closely spaced Ni_2 states, i.e. ${}^1\Sigma_g^+, {}^1\Gamma_g, {}^1\Sigma_u^-, {}^3\Sigma_g^-, {}^3\Sigma_u^+$, and ${}^3\Gamma_u$ $(R = 4.35a_o)$ (1), are reported (2,3,4). The resulting ground state strongly depends on the chosen internuclear distance e.g. ${}^1\Sigma_g^+$ $(R = 4.35a_o)$ and ${}^3\Sigma_u^+$ $(R = 5a_o)$ (1).

Table 3.1 Ionization energies and atomic orbital contributions of Ni_2

Ni-Ni distance(Å)				2.17	2.49
Ground state				$^1\Sigma_g^+$	$^1\Sigma_g^+$
Total energy(Ry)				-14.7021	-14.0005
R=2.17Å Contribution(%)			Orbital	Energy(eV)	
s	p	d	(Occupation)		
0	99	1	π_u(0)	1.547	1.493
9	88	3	σ_g(0)	2.545	1.909
79	13	8	σ_u(0)	6.777	5.391
40	13	47	σ_u(0)	8.081	8.345
0	0	100	π_g(4)	8.232	8.458
0	0	100	δ_u(4)	8.520	8.555
0	0	100	δ_g(4)	8.611	8.580
0	0	100	π_u(4)	8.925	8.682
35	0	65	σ_g(2)	8.996	8.583
43	2	55	σ_g(2)	9.964	9.125

With regard to the ground state of Ni_2, the present calculation agrees with the result of the CI calculation by V. Shim et. al. (1), i.e. $^1\Sigma_g^+$, and deviates, for example, from the LSDF result $^3\Sigma_g^-$ $(R = 4.12a_o)$ by J. Harris et. al. (3).

N. Rösch et. al. (7) reported a $^1\Sigma_g^+$ $(R = 2.49\text{Å})$ ground state for a $SCF-SW-X\alpha$ treatment yielding a similar order of energy levels as our calculation. The $d\sigma_g$ level lies, in deviation to our result, nearly at the top of the d manifold.

Ab initio calculations show a different order of energy levels due to relaxation effects (5). Regarding the characters of the MOs in our calculation the order of the orbitals still resembles those of the Ni atom with strong interactions between the s and d_{z^2} atomic orbital.

Table 3.2 Optical transitions(eV) of Ni_2

Transition	Calc.(R=2.17Å) 4pε_0=0.1	(4pε_0=0.4)	Exp. (Ref.8)
B $s\sigma_g \rightarrow s\sigma_u$	3.19	(5.22)	(3.02)
A $d\pi_u \rightarrow p\pi_u$	6.69	(2.39)	2.35
C $d\sigma_g \rightarrow p\pi_u$	7.45	(3.14)	3.29
Δ_{CA}	0.76	(0.75)	0.94

For comparison with the calculated results experimental data of optical transitions (8) are available shown in Table 3.2. The calculated transitions A and C disagree with the experimental ones using the 4p energy parameter ε_o = 0.1 Ry. By a larger value of ε_o the $p\pi_u$ is shifted downwards resulting in a coincidence between calculated and experimental transitions. The transition B, $s\sigma_g \longrightarrow s\sigma_u$ causes great changes in the electronic structure which go beyond the one-particle approximation. However, the choice of the smaller $4p\varepsilon_o$ parameter is preferred according to results on other nickel test molecules.

3.2 $Ni_4(T_d)$

The geometry of the tetrahedral Ni_4 cluster is shown in Figure 1, and the main results in Table 3.3.

The lowest orbital is of s character followed by a narrow spaced sequence of d orbitals. The next unoccupied orbitals are of s and p character. The results are qualitatively very similar to those of reported $X\alpha - LCAO$, $SW - X\alpha$ (9), and $SCF - DV - X\alpha$ (10) calculations finding an open shell configuration with a $2t_1^4$ $HOMO$. However, our 'd-band width' is with 0.9 eV nearly the half of the results of those calculations, 1.6. eV. Referring to the results of an ab initio calculation (5), the d character orbitals are obtained as the lowest in contradiction to the nickel band structure (6). The Ni_4 cluster may serve as a model of a (111) surface of the nickel fcc crystal.

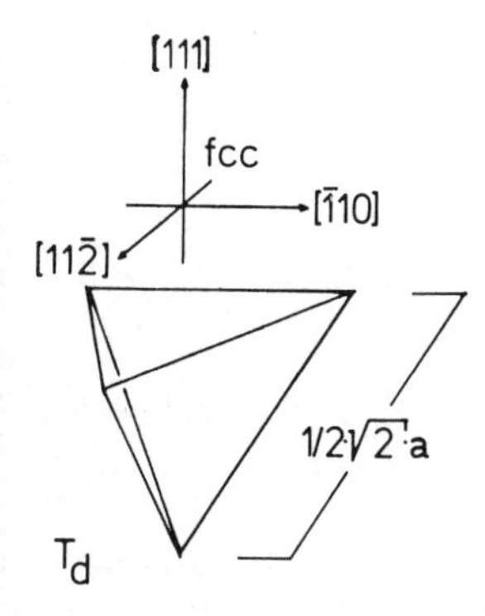

Fig. 1: Geometry of the tetrahedral Ni_4 cluster.

3.3 Nickel '(100) surface' cluster

As a model for a (100) surface, several clusters have been examined. Their geometries are shown in Figure 2: Ni_4 (atoms b to e), Ni_5 (atoms a to e), Ni_9 (atoms o, g, i, l, n, b to e), Ni_{13} (atoms f to o, b to e), and Ni_{14} (atoms a to o). Except for Ni_4 with a D_{4h} symmetry these clusters have C_{4v} symmetry. The key results of the calculations are also presented in Table 3.3.

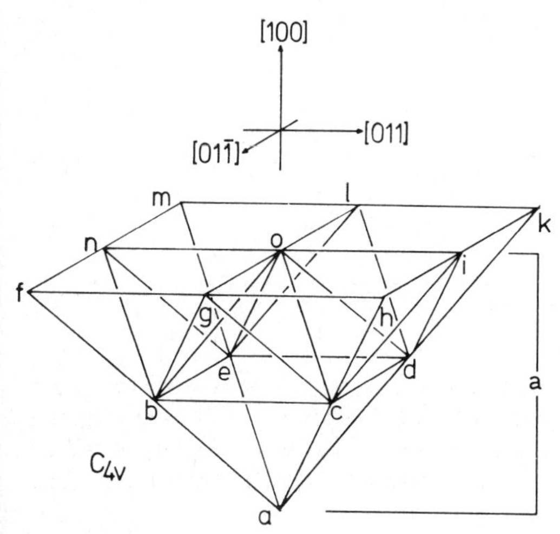

Fig. 2: Geometry of nickel '(100) surface' cluster(s).

Table 3.3 Ionization energies(eV), total energies(Ry), 'd-band width'(eV) of nickel clusters

Cluster (R=2.49Å)	Lowest orbital	Lowest d-orbital	HOMO	LUMO	Lowest unoccupied sp-orbital	Total energy (...per atom)	'd-band width'
$Ni_2, D_{\infty h}$	σ_g 9.125	π_u 8.682	π_g 8.458	σ_u 8.345	LUMO	-14.0005 (-7.0003)	0.224
Ni_4, T_d	A_1 12.221	A_1 9.063	T_1 8.168	T_2 4.925	LUMO	-29.3290 (-7.3323)	0.895
'(100) surface' clusters							
Ni_4, D_{4h}	A_{1g} 10.360	B_{1g} 9.062	B_{2g} 8.413	A_{2g} 8.125	E_u 6.542	-27.2665 (-6.8166)	0.937
Ni_5, C_{4v}	A_1 12.327	E 9.148	A_2 8.112	E 6.590	LUMO	-35.6474 (-7.1295)	1.036
Ni_9, C_{4v}	A_1 14.457	B_1 9.909	E 8.343	B_1 8.307	A_1 6.689	-65.3377 (-7.2597)	1.748
Ni_{13}, C_{4v}	A_1 20.494	B_1 10.740	A_1 8.567	E 8.509	A_1 7.534	-85.5969 (-6.5844)	2.548
Ni_{14}, C_{4v}	A_1 20.914	B_1 11.247	A_1 8.720	B_1 8.661	B_2 8.346	-91.0432 (-6.5031)	2.589
'(110) surface' clusters							
Ni_5, C_{2v}	A_1 11.591	B_2 10.081	B_1 8.430	B_2 8.374	B_1 5.969	-35.2316 (-7.0463)	1.747
Ni_8, C_{2v}	A_1 13.147	A_1 9.795	A_2 8.373	B_2 8.365	A_2 6.769	-55.8224 (-6.9778)	1.547
'bulk' clusters							
Ni_{13}, O_h	A_{1g} 16.226	T_{2g} 10.546	T_{2g} 8.378	A_{1u} 8.190	T_{2g} 6.167	-94.4746 (-7.2673)	2.356
Ni_{19}, O_h	A_{1g} 26.771	T_{2g} 12.196	E_g 8.626	A_{1g} 8.276	LUMO		3.570

The lowest orbital is of s character (ca. 93% s; 3% p; 4% d) for all cluster sizes. With increasing cluster size the 'd-band width' also increases due to a spread of the d levels. The *HOMO* is taken to be equal to the Fermi energy which is about $8.4 \pm 0.2eV$. The unoccupied d levels increase in number, too, yielding a 'conduction band'. The gap between 'd-band' and 'sp-band' amounts to ca. 0.3 eV for Ni_{14}.

Regarding the total energy per atom vs. cluster size, the absolute value increases from Ni_4 to Ni_9 whereas for Ni_{13} and Ni_{14} it decreases. This may be explained by the fact of a perhaps better geometrical arrangement of the nickel atoms. The total energy per atom favours a Ni_{13} cluster of O_h instead of C_4 symmetry. Table 3.4 lists the effective charges of representative atoms of the cluster and the net overlap charge between an atom and the other atoms of the cluster. The Ni_5 cluster firstly shows typical effects with respect to charges. The negative charge concentrates at the periphery while the positive charge accumulates in the central region. The charge fluctuations increase with cluster size. The overlap charge between an atom and the rest-cluster as a measure of bonding exhibits the reason for the decrease of the absolute value of the total energy per atom. In Ni_{14} and Ni_{13} net repulsive interactions occur.

3.4 NICKEL '(110) SURFACE' CLUSTER

Two clusters, Ni_5 and Ni_8, have been calculated in geometries simulating a (110) surface of a *fcc* crystal. Those are shown in Figure 3, Ni_5 (atoms *a* to *e*), Ni_8 (atoms *a* to *h*), and Table 3.3 lists the results of the present calculation.

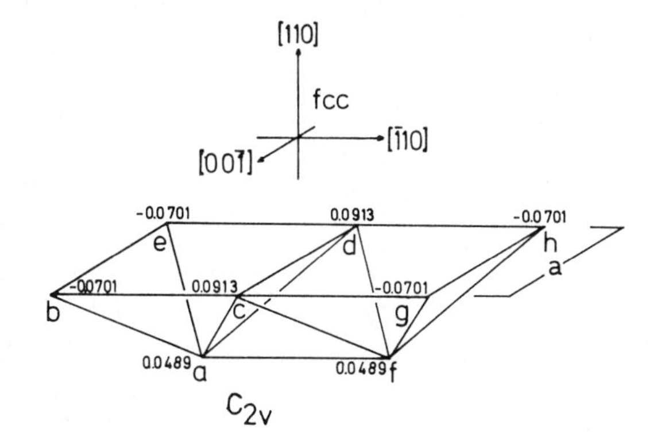

Fig. 3: Geometry of nickel "(110) surface" cluster(s)

At first sight there are two striking points. The absolute value of the total energy per atom and the d-band width decreases. The first result is comprehensible from the fact that the chosen geometry of Ni_8 is very unlikely for a free Ni_8 cluster. A more symmetrical grouping would yield a lower absolute value of the total energy like in the case of N_{13} in C_{4v} and O_h symmetry. The decrease in the 'd-band width' has its reason in a mixing of orbitals with mainly d and s character at the bottom of the d manifold. Hence a clear distinction between s and d-band is not possible. The advantage of the Ni_8 geometry consists in the low charge fluctuations because there are little differences between the atomic sites with regard to symmetry operations. Therefore the Ni_8 cluster may serve as a good model for a neutral nickel surface.

3.5 'BULK' NICKEL CLUSTERS

The Ni_{13} and Ni_{19} cluster shown in Figure 4, Ni_{13} (atoms *a* to *n*) and Ni_{19} (atoms *a* to *t*), represent the local symmetry of a nickel atom in a bulk nickel *fcc* crystal.

Table 3.4 Effective atomic charges and overlap charges of nickel '(100) surface' cluster

Cluster (C_{4v})	Effective atomic charge(s)					Net overlap charge of the ...th atom with the other cluster atoms				
	a[1)]	b[2)]	f[3)]	g[4)]	o[1)]	a[1)]	b[2)]	f[3)]	g[4)]	o[1)]
Ni_{14}	-0.2490	0.1444	-0.3278	0.3738	-0.5126	0.3724	0.4476	0.2144	-1.2060	0.0004
Ni_{13}	---	0.0275	-0.3233	0.4264	-0.5237	---	0.3036	0.3450	-1.2300	0.2384
Ni_9	---	-0.0046	---	-0.0513	0.2237	---	0.7836	---	0.5780	1.1592
Ni_5	0.0080	-0.0020	---	---	---	0.8616	0.2920	---	---	---
Ni_4, D_{4h}	---	0.0	---	---	---	---	0.3480	---	---	---

1) The notation is according to fig.3.2; 2) The atoms b,c,d,and e are equivalent; 3) The atoms f,h,k, and m are equivalent; 4) The atoms g,i,l,and n are equivalent.

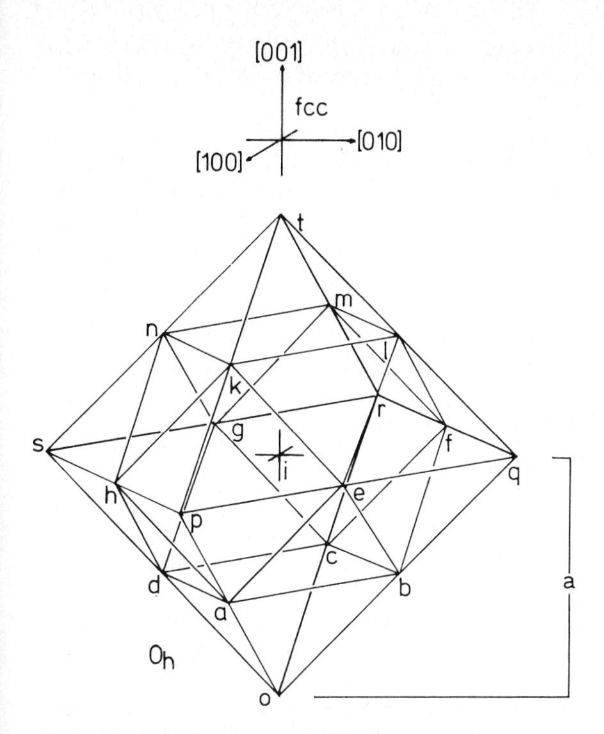

Fig. 4: Geometry of "bulk" nickel cluster(s)

Table 3.3 shows the computed results. The total energy of the Ni_{19} cluster has not been calculated since the Hartree-Fock ground state has not been reached so far according to convergence problems. The position and the character of the orbitals, however, is not significantly affected. The 'd-band width' is found to be 2.4 eV for Ni_{13} and 3.6 eV for Ni_{19}. The first ionization potential is about 8.5 eV. Comparing these results with a $SW - X\alpha$ calculation (11) on Ni_{13} yielding a 'd-band width' of about 2.5 eV and a $HOMO$ with an orbital energy of 5.6 eV the calculations are in good agreement taking into account that in $X\alpha$ approximation the absolute values of orbital energies are ca. 2 to 3 eV smaller than ionization energies. A reported HF calculation on large nickel clusters, Ni_{13} to Ni_{87}, (12) was performed under the assumption that the $HOMO$ is of 4s character as has been concluded from other small nickel cluster calculations (e.g. Ref. 5). This is, however, in contradiction to the results of $X\alpha$ calculations as well as the bulk band structure with the Fermi energy inside the d-band. The reason for the wrong result of the HF calculation is the largely different relaxation compared to Koopman's theorem between s and d electrons shifting the s-band above or at the top of the d-band. Considering the highest occupied MO of the d-band, the ionization potential is about 9.4 eV (Ni_6, Ref. 5) which is larger than the work function value. For that reasons this calculation is without any significance.

3.6 Dicussion and conclusions

The following general conclusions can be drawn from the various calculations of different types of clusters listed in Tab. 3.3. Clusters of different symmetry but the same size like Ni_4: D_{4h}, T_d; Ni_5: C_{2v}, C_{4v} and Ni_{13}: C_{4v}, O_h show pairwise similar trends. The lower symmetry species has a higher total energy and a larger 'd-band width'. The increase in 'd-band width' is due to the removal of degeneracy of orbitals. The 'd-band width' also increases with increasing cluster size approaching the range of bulk values. Experiments give a value of about 3.4 eV (13) whereas theoretical band structure calculations obtain a value of about 4.5 eV (6). The 'd-band width' of Ni_{19}, 3.6 eV, exceeds the experimental one which is assumed to be the lower limit of a true d-band width (13).

The first ionization energies are nearly constant in the range from 8.1 to 8.7 eV for all calculated clusters. Comparing this with first ionization energies of small nickel clusters, $Ni_2 - Ni_{23}$, estimated by photoionization experiments with values in the range of 5.6 to 6.4 eV (14) and the electronic work function of nickel ca. 5.2 eV, the calculated values are about 2.5 eV too large. On the other hand, other calculations (5,11,12) seemingly produce first ionization energies closer to the work function value. As mentioned, however, in chap. 3.5 the $SW - X\alpha$ calculations (11) gives orbital energies instead of ionization potentials and the HF calculations (5,12) compare 4s orbitals with the work function instead of the 3d which determine the Fermi energy. Taking this into account, the correct interpreted results of these calculations are in agreement with our results, and show the same deviation from the experimental values.

A density of states (DOS) curve was obtained from the discrete eigenvalue spectrum by using Gaussian functions. Figure 5 shows the DOS of Ni_{13} together with the d component. The main features of the bulk DOS (6,13) are represented. The shape of the DOS of the occupied states is, in most parts, governed by the d orbital contributions. The Fermi energy falls within this 'd-band'. Above the 'd-band', separated by a gap, the DOS is of sp character as in experimental and theoretical band structures. From this point of view, the clusters converge to bulk properties as illustrated by Figure 6.

The effective charges at different atomic sites to be a difficulty in modelling a neutral metal surface, since their values can considerably deviate from neutrality. This feature has to be considered as an artifact of the cluster model and has to be corrected in an appropriate way.

In conclusion, our calculations offer the possibility to produce by the $SCC - X\alpha$ method reliable clusters of sufficient size to model substrate surfaces for describing adsorption processes without too much effort in computation time.

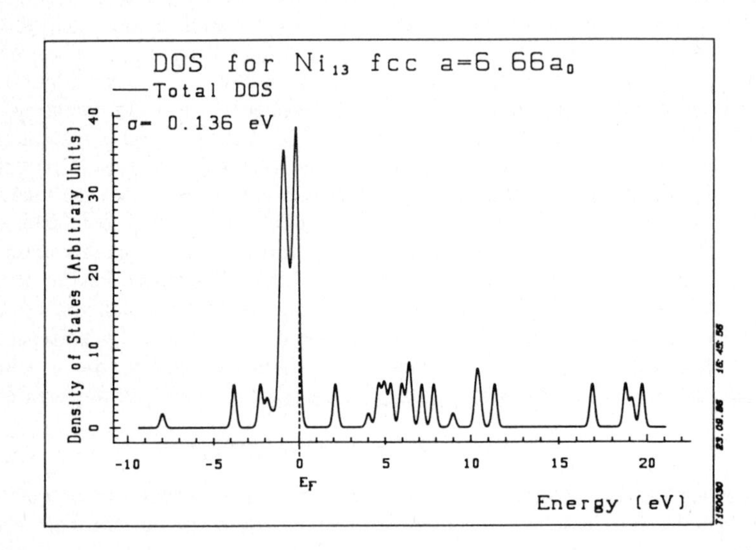

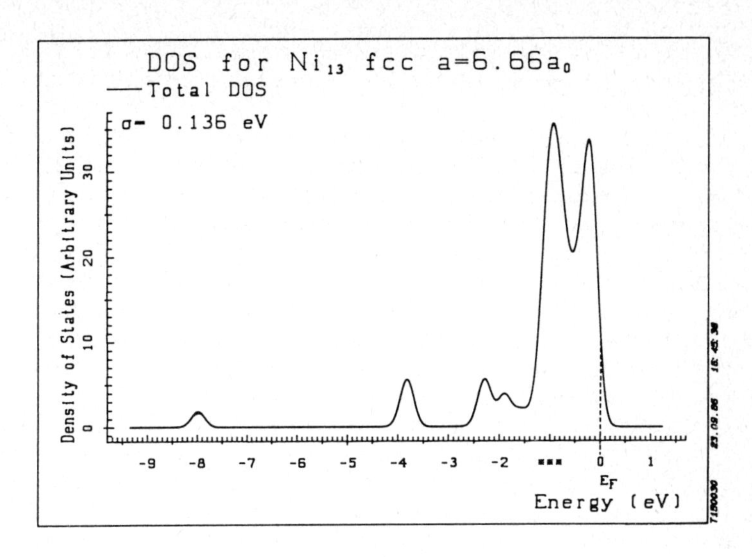

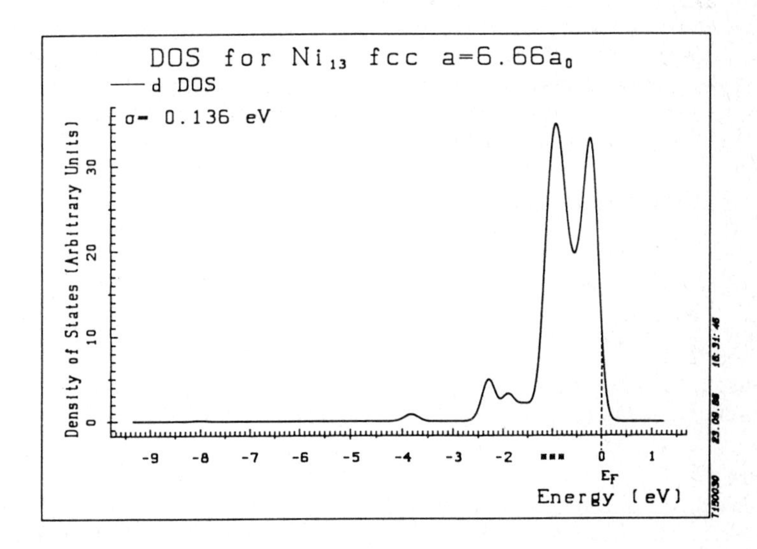

Fig. 5: Density of states of Ni_{13} .

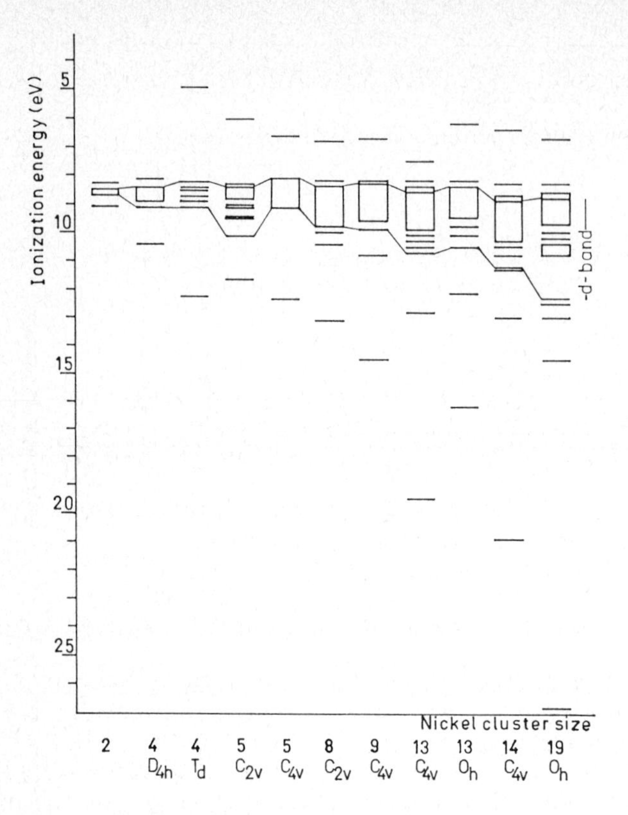

Fig. 6: Ionization energies vs. cluster size.

REFERENCES

0. Introduction

1. Hermann, K.: Phys. Bl. 36 (1980) 227
2. Johnson, K.H., Messmer, R.P.: J. Vac. Sci. Technol. 11 (1974) 236

1. Theory

1. Grodzicki, M.: J. Phys. B 13 (1980) 2683
2. Grodzicki, M.: Theorie und Anwendungen der Self-Consistent- Charge-$X\alpha$ Methode. Habilitationsschrift. Hamburg 1985
3. Johnson, K.H.: Adv. Quant. Chem. 7 (1973) 143
4. Baerends, E.J., Ellis, D.E., Ros, P.: Chem. Phys. 2 (1973) 41,51
5. Basch, H., Viste, A., Gray, H.B.: Theor. Chim. Acta 3 (1965) 458
6. Grodzicki, M.: Croatica Chem. Acta 57 (1984) 1125
7. Grodzicki, M., Walther, H., Elbel, S.: Z. Naturforsch. 39b (1984) 1319
8. Bläs, R., Grodzicki, M., Marathe, V.R., Trautwein, A.: J. Phys. B 13 (1980) 2693
9. Hütsch, M., Grodzicki, M.: Self-consistent charge $X\alpha$ calculations on small titanium compounds and TiO_2 clusters. (This publication)
10. Grodzicki, M., Hütsch, M., Kühnholz, O.: Comp. Phys. Comm. (submitted)

2. Computational Details

1. Briggs, D. (ed.): Handbook of X-Ray and Ultraviolett Photoelectron Spectroscopy. London: Heyden 1977
2. Kittel, C.: Einführung in die Festkörperphysik. 6. Aufl. München, Wien: R. Oldenbourg 1983

3. Clusters

1. Shim, I., Dahl, J.P., Johansen, H.: Int. J. Quant. Chem. 15 (1979) 311
2. Noell, J.Q., Newton, M.D., Hay, P.J., Martin, R.L., Bobrowicz, F.W.: J. Chem. Phys. 73 (1980) 2360
3. Harris, J., Jones, R.O.: J. Chem. Phys. 70 (1979) 830
4. Upton, T.H., Goddard III, W.A.: J. Am. Chem. Soc. 100 (1978) 5659
5. Basch, H., Newton, M.D., Moskowitz, J.W.: J. Chem. Phys. 73 (1980) 4492
6. Callaway, J., Wang, C.S.: Phys. Rev. B 7 (1973) 1096; Moruzzi, V.L., Janak, J.F., Williams, A.R.: Calculated Electronic Properties of Metals. New York: Pergamon Press 1978
7. Rösch, N., Rhodin, T.N.: Phys. Rev. Lett. 32 (1974) 1189
8. Moskovits, M., Hulse, J.E.: J. Chem. Phys. 66 (1977) 3988
9. Messmer, R.P., Lamson, S.H.: Chem. Phys. Lett. 90 (1982) 31
10. Holland, G.F., Ellis, D.E., Trogler, W.: J. Chem. Phys. 83 (1985) 3507
11. Messmer, R.P., Knudson, S.K., Johnson, K.H., Diamond, J.B., Yang, C.Y.: Phys. Rev. B 13 (1976) 1396
12. Upton, T.H., Goddard III, W.A., Melius, C.F.: J. Vac. Sci. Technol. 16 (1979) 531; Melius, C.F., Upton, T.H., Goddard III, W.A.: Sol. State. Com. 28 (1978) 501
13. Goldman, A., Donath, M., Altman, W., Dose, V.: Phys. Rev. B 32 (1985) 837; Himpsel, F.J., Knapp, J.A., Eastman, D.E.: Phys. Rev. B 19 (1979) 2919; Eastman, D.E., Himpsel, F.J., Knapp, J.A.: Phys. Rev. Lett. 40 (1978) 1514
14. Rohlfing, E.A., Cox, D.M., Kaldor, A.: J. Chem. Phys. 88 (1984) 1497

SELF-CONSISTENT CHARGE $X\alpha$ CALCULATIONS ON SMALL TITANIUM COMPOUNDS AND TiO_2 CLUSTERS

M.Hütsch and M.Grodzicki
I.Institut für Theoretische Physik der Universität Hamburg
Jungiusstr. 9, D-2000 Hamburg 36, West Germany

1. Introduction

The semiempirical self-consistent charge $X\alpha$-method (SCC-$X\alpha$)/1-3/ has turned out to be a promising tool for electronic structure calculations on large systems with up to 50 atoms and 200 basis orbitals or even more. It contains only a few atomic parameters with clear physical meaning/4/ so that they can be derived in principle from atomic data getting especially important in the case of transition metals where not enough experimental data exist for an extensive parameter fit.

In this paper at first we describe in some detail the determination of parameters for titanium from atomic data, secondly the fit to experimental data such as ionization potentials from photoelectron spectra, optical transition energies, 10 Dq values, and to other theoretical calculations as far as they are available, and finally calculations on other molecules having not been used for the optimization procedure to demonstrate the transferability of the obtained parameter set.

Afterwards we perform as the most interesting application some calculations on the electronic structure of larger TiO_2 bulk and surface clusters which can serve as substrate models for the study of the interaction of small molecules like H_2, O_2, CO and H_2O with a $TiO_2(110)$ surface.

2. Determination of parameters

The three types of atomic parameters entering the Hamiltonian are described in detail in ref.4. The atomic ionization potential ε_i^μ of atomic orbital i is assumed to be a quadratic function of the effective atomic charge Q_μ of atom μ:

$$\varepsilon_i^\mu = \varepsilon_{i0}^\mu + \varepsilon_{i1}^\mu * Q_\mu + \varepsilon_{i2}^\mu * Q_\mu^2$$

The leading charge independent ε_0 is determined from X-ray and UV-photoelectron spectra/6/ for the atomic 3p, 3d and 4s levels to be:

$$\varepsilon_0(3p)=2.83, \quad \varepsilon_0(3d)=0.59, \quad \varepsilon_0(4s)=0.50$$

(all energies are in Rydberg atomic units). The ε_1 and ε_2 parameters are obtained either from a quadratic fit to isoelectronic series obtained from optical data/7/ or from Hartree-Fock calculations on various low lying states of neutral atoms and ions /8/. ε_2 turned out to be nearly constant for one period and has been chosen for the first row transition elements as: $\varepsilon_2(3p,3d)=0.12$, $\varepsilon_2(4s,4p)=0.07$

Averaging the ε_1 values obtained from /7,8/ yields the result:

$$\varepsilon_1(3p)=0.60, \quad \varepsilon_1(3d)=0.55, \quad \varepsilon_1(4s)=0.43$$

In first row transition elements the atomic 4p level is unoccupied but can become important in molecular bonding. The only way of getting these 4p parameters is through the isoelectronic series of ref.7: $\varepsilon_0(4p)=0.25$, $\varepsilon_1(4p)=0.30$

The STO orbital exponents ζ_i^μ are assumed to be linear functions of Q_μ:

$$\zeta_i^\mu = \zeta_{i0}^\mu + \zeta_{i1}^\mu * Q_\mu$$

These parameters can be derived from Slater/9/, Burns/10/, from single-zeta Hartree-Fock calculations/8/ and from <r>-expectation values according to:

$$\zeta_{nl,0} = (n+1/2)/\langle r\rangle \qquad \text{or} \qquad \zeta_{nl,0} = ((n+1/2)*(n+1)/\langle r^2\rangle)^{1/2}$$

where $\langle r\rangle$, $\langle r^2\rangle$ are calculated either from nonrelativistic Hartree-Fock functions/8/ or relativistic Dirac-Fock functions/11/. For the 4s function the differences between all these methods are only minor and an average of $\zeta_0(4s)=1.20$ is obtained. Nearly the same is true for 3p leading to $\zeta_0(3p)=3.25$. But there are large differences for 3d ranging from $\zeta_0(3d)=2.71$ (from single-zeta HF-functions for the $3d^2 4s^2$ ground state) over $\zeta_0(3d)=2.16$ (from $\langle r^2\rangle$-expectation values for the $3d^2 4s^2$ ground state) to $\zeta_0(3d)=1.87$ (from $\langle r^2\rangle$-expectation values for the $3d^3 4s^1$ ground state) leading to an average value of $\zeta_0(3d)=2.34$ for the $3d^2 4s^2$ ground state. The charge-dependent ζ_1 is determined from the <r>-expectation values of the positive and negative ion, respectively(ref.2). As these values are nearly the same as <r> for the neutral atom, $\zeta_1(3p)$ is chosen to be zero. The values obtained for 3d and 4s are $\zeta_1(3d)=0.34$, $\zeta_1(4s)=0.30$. Again for the unoccupied atomic 4p level only empirical rules exist to determine the respective parameters. $\zeta_1(4p)$ is chosen to be the same as $\zeta_1(4s)$ and $\zeta_0(4p)$ to lie between $\zeta_0(3d)$ and $\zeta_0(4s)$ leading to $\zeta_0(4p)=1.75$.

The atomic potential parameter η_μ is also assumed to be a linear function of the effective atomic charge Q_μ: $\eta_\mu = \eta_{\mu,0} + \eta_{\mu,1} * Q_\mu$

and is the negative slope of $\ln \rho_{at}(r)$ which is approximately constant over the valence region where bonding takes place/5/. The other way is to obtain η_μ from <r>-expectation values/8,11/ in an analogous way as has been described for ζ, yielding

$$\eta_0 = 3.07, \qquad \eta_1 = 0.60 \quad .$$

Summarized, the differing results from the various recipes show that even determination of parameters from atomic data is neither straight-forward nor unique requiring them to be optimized in calculations on some small reference molecules for which data exist from experiment or other theoretical calculations. To this end in table 1 we give at first a summary of experimentally and theoretically investigated molecules and model compounds which illucidates the situation under which the parameter optimization has to be performed and which molecules can be used for it.

Since most of the listed molecules are d^0-compounds, the molecular valence orbitals show mainly ligand character and thus are almost insensitive to a variation of Ti-parameters. Consequently, from p.e.spectra little information can be extracted about the quality of the important 3d parameters, wherefore molecules and model compounds had to be chosen for parameter optimization with experimental and/or theoretical information on optical transition energies and 10 Dq values being available (see table 1). Some other molecules, where only few comparable data exist have been used as cross checks for the transferability and are treated in some detail in the next chapter together with those on which further calculations were performed.

The results of the parameter optimization compared to the available data on each molecule are given in tables 2-5. Though the optical transitions in $TiCl_4$ are too low in energy the overall agreement is reasonably good with respect to the level ordering being always correct (according to molecular orbital and ligand field theory/40/), to the ionization potentials, level splittings, valence band widths, optical transitions and 10 Dq values. As can be seen from table 2 this is not the case for most of the other calculations on $TiCl_4$ which calculate the t_2 level either to lie slightly below the a_1 level (SW-Xα/30,33/) or deeply below it (NDO-calculations/31/ especially/36/)

Table 1: Investgated titanium compounds

molecule	available data (exp. and theoret.)	parameter optim.	cross check	further calculations
Ti_2	different ground states from a Resonance Raman spectrum/14/, DV-Xα/12/, EHM/13/, LDF/15/			not calculated because of unknown distance and ground state
TiH	ground state, geom., electronic structure from CASSCF/CI /16/, ab initio(pseudopot.)/17/, ab initio/18/		X	dipole moment
TiO	ground state, geom., electronic structure, dipole moment from PES+ SCF/CI /19/, CASSCF/CI /20/, ab initio/21,22/		X	
TiF_2	geom. from IR-spectrum/23/			electr. structure, dipole moment
TiF_3	symm. from IR-spectrum/23/, electronic structure from ab initio /24/			not calculated because of inconsistent geom.
TiH_4	exist. from mass spectr./27/, electr. struct. from ab initio /25/, FSGO/26/		X	
TiH_3F	electronic structure from ab initio/28/		X	dipole moment
$TiCl_4$	electr. struct., optical trans. from PES/32-35/, opt. absorption + CNDO/36/, DV-Xα/29/, SW-Xα/30/, INDO/31/, EHM+SW-Xα/33/	X		
$TiBr_4$	electr. struct. from PES/32,33/, INDO/31/	X		
TiJ_4	electr. struct. from PES/32/, INDO/31/			not calculated because of strong S-O splitt.
TiF_6^{3-}	electr. struct., optical trans., 10 Dq from opt. absorption/39,40/ EHM/37,39/, ab initio/38/	X		
$TiCl_6^{2-}$	electr. struct. from Pariser-Pople-Parr/41/		X	
TiO_6^{8-}	electr. struct., optical trans., 10 Dq from X-ray emission and absorption/45/, XPS/46/, DV-Xα /42,47,48/, SW-Xα/43,44/	X		

Table 2: $TiCl_4$

method	ref.	VB-width[a)]	ionization potentials a_1	t_2	e	t_2	t_1	first L->M transition	10 Dq
exp.	32-36	2.2	13.96	b)		12.77	11.76	4.4	0.95
DV-Xα	29	1.9	14.1	13.8	13.0	12.7	12.2	4.9	0.7
SW-Xα	30	1.3	13.0	13.2	12.8	12.5	11.9	3.8	1.2
EHM	33	2.2	13.9	12.85	12.7	11.8	11.7	-	-
SW-Xα [c)]	33	0.7	8.23	8.37	8.16	7.53	7.52	-	-
INDO	31	5.2	13.77	17.74	13.31	12.83	12.54	-	-
SCC-Xα	this work	2.6	14.76	14.45	13.11	12.90	12.16	2.25	0.90

a) VB=valence band, b) broad band(>0.5 eV) centered at 13.25 eV
c) orbital energies (all energies are in eV)

Table 3: $TiBr_4$

method	ref.	VB-width[a)]	ionization potentials a_1	t_2	e	t_2	t_1	first L->M transition	10 Dq
exp.	32,33,35	2.5	13.04	12.43	12.00	11.68	10.70	-	-
INDO	31	5.25	12.89	16.01	11.80	11.31	10.76	-	-
SCC-Xα 4p only	this work	2.49	13.96	13.75	12.25	12.11	11.47	2.09	0.87
SCC-Xα 3p + 4p	this work	2.63	14.14	13.49	12.26	12.10	11.51	2.39	0.82

a) VB=valence band (all energies are in eV)

Table 4: TiO_6^{8-}

method	ref.	separation O2s - VB[a)]	VB-width	separation bond.-nonbond. VB-orbitals	first L->M transition	10 Dq
exp.	45 46	15.0	8.0 5.5	4.0 1.9	3.03	2.1 -
SW-Xα	43	11.5	5.5	1.7	2.5	3.5
SW-Xα	44	12.8	2.2	0.8	-	-
DV-Xα	42	12.2	2.9	1.3	-	-
DV-Xα	47,48	-	6.0	-	4.5	2.7
SCC-Xα	this work	11.7	7.0	4.2	2.72	2.6

a) VB=valence band (all energies are in eV)

Table 5: TiF_6^{3-}

method	ref.	separation F2s - VB[a)]	VB-width	separation bond.-nonbond. VB-orbitals	first L->M transition	10 Dq
exp.	39,40	-	-	-	> 6.25	2.12
EHM	39	-	4.4	2.5	6.12	2.17
semiemp.	37	18.0	3.0	0.3	b)	1.96
ab initio	38	-	-	-	-	2.13
SCC-Xα	this work	16.5	1.4	0.6	6.85	2.40

a) VB=valence band, b) wrong level ordering (all energies are in eV)

yielding a far too large valence band width of about 9 eV, predicting the first L->M transition to be the $t_1 \rightarrow a_1(4s)$ with 1.39 eV and giving the wrong sign and magnitude for 10 Dq (-6.5 eV). The same occurs in the case of TiO_6^{8-} where the SW-Xα and DV-Xα calculations/42-44/ also give the wrong ordering of valence band orbitals which is not seen from table 4.

Further improvement of our results is obtained when enlarging the basis set and not including the 3p atomic orbital into the core, treating it as a valence orbital instead. The resulting effects are displayed in table 3 for $TiBr_4$ as an example since they are the same in all other compounds. Nevertheless the 3p atomic orbital is re-included into the core since the results are sufficently accurate, too, and our main goal was to perform calculations on large clusters containing far more than one titanium atom.

The resulting standard parameter set (´STD´) for Ti is listed in table 6 together with the respective from atomic data (´atom.´). As these two sets differ only little it is clear that molecular orbital calculations can be done even with that ´atomic´ parameters to get a first idea of the bonding in unknown compounds.

Table 6: ´atomic´and ´STD´parameter sets

	ε_0		ε_1		ε_2		ζ_0		ζ_1		η_0		η_1	
	atom.	STD	atom.	STD	atom.	STD	atom.	STD	atom.	STD	atom.	STD	atom.	STD
3p	2.83	2.93	0.60	0.55	0.12	0.12	3.25	3.37	0.0	0.0	3.07	3.70	0.60	0.60
3d	0.59	0.59	0.55	0.44	0.12	0.12	2.34	2.40	0.34	0.10				
4s	0.50	0.46	0.43	0.43	0.07	0.07	1.20	1.05	0.30	0.40				
4p	0.25	0.20	0.30	0.40	0.07	0.07	1.75	1.90	0.30	0.10				

3. Calculations on other molecules

TiH

From the HF and CI calculations of references 16-18 a bond distance of 3.50 a.u. (1.85 A) for TiH is derived with a $^4\Phi$ molecular ground state corresponding to the electronic configuration $...6\sigma^2\ 1\delta^1 3\pi^1 7\sigma^1$. Starting with this information the results of table 7 are obtained. As compared to the HF and CI results from refs.16,17 there are some differences concerning the level ordering and the 3d occupancy. Both refs. are considering the 7σ level to be largely Ti 4s in character which corresponds to the (unoccupied) 8σ in our calculation as can be seen by comparing the eigenvectors of ref. 17 and the SCC-Xα AO-contributions given in table 7. On the other hand in both calculations the I.P.´s for the bonding 6σ and for the MO with predominant 4s character (8σ in this work,7σ in ref. 17) are in excellent agreement. We believe, however, that our level ordering should be more reasonable since a crystal-field splitting of the Ti 3d levels of more than 5 eV as following from ref. 17 seems to be unlikely.

Table 7: TiH

method	ref.	ionization potentials[a)]								
		$6\sigma^2$	$1\delta^1$	$3\pi^1$	$7\sigma^1$	8σ	4π	9σ		
ab initio	17	9.56	11.22	10.82	6.10	-	-	-		
SCC-Xα	this work	9.45	7.86	7.84	7.68	6.20	3.24	0.69		
		AO - contributions to MO (in %)							AO-occupations	
	Ti 4s	8	-	-	28	54	-	12	0.57	-
	4p	0	-	0	0	11	100	54	0.01	-
	3d	24	100	100	70	10	0	2	3.10	2.26[b)]
	H 1s	68	-	-	2	25	-	32	1.32	-

a) orbital energies in ref. 17
b) from CASSCF/CI of ref.16
(all energies are in eV)

effective charge :0.32
overlap population:o.52
dipole moment :4.28 D

TiO

The ground electronic state of TiO is known from experimental/19/ and theoretical studies/20-22/ to be a $^3\Delta$ state which can be described in terms of the electronic configuration $...8\sigma^2 3\pi^4 1\delta^1 9\sigma^1$ where the 8σ and 3π levels are composed mainly of O 2p atomic orbitals, the 1δ is a pure Ti 3d atomic orbital and the 9σ molecular orbital is mainly Ti 4s in character. The experimental equilibrium bond length is 1.6203 Å. The results obtained are listed in table 8 and have to be compared to the results from experiment and the HF and CI calculations.

From the CASSCF/CI calculation/20/ we obtain a detailed population analysis. They had 8 active electrons arising from the Ti $3d^3 4s^1$ and O $2p^4$ while treating the O 2s as inactive with a fixed occupation number of 2 while we obtain 1.77 . Apart from this both calculations are in close agreement as can be seen from table 9. As to the calculated ionization potentials our results are closer to experiment and for the other calculations the same may be true as has been stated already in the case of TiH.

Table 8: Ionization potentials,dipole moments for TiO

method	ref.	ionization potentials				dipole moment
		8σ	3π	1δ	9σ	
exp.	19	a)	10.57	8.20	6.82	-
Δ SCF	19	7.29	7.65	8.36	6.03	-
Δ SCF/CI	19	9.21	9.51	7.58	5.67	-
ab initio	21	10.47	10.50	9.67	5.13	5.93
SCC-Xα	this work	12.53	10.79	7.65	7.08	6.55

a) broad band (about 0.8 eV) centered at 11.5 eV
(energies are in eV, dipole moments in D)

Table 9: Populationanalysis for TiO

method	ref.	AO - occupations					effective charge on oxygen	overlap population
		Ti			O			
		4s	4p	3d	2s	2p		
CASSCF/CI	20	o.81	0.27	2.38	2.0	1.38(8) 3.10(3)	-0.55	1.20
SCC-Xα	this work	0.64	0.08	2.69	1.77	1.54(8) 3.20(3)	-0.59	1.10

TiF_2

For TiF_2 only the geometry is known from IR-spectroscopy/23/ to be of C_{2v} symmetry with a bond angle of 130 +/- 5° and an estimated bond distance of 1.90 Å. From this the results of tables 10,11 are obtained.

Table 10: TiF_2

	Ti			F	
	4s	4p	3d	2s	2p
AO-occupation	0.16	0.20	2.68	1.89	5.59
effective charge	+0.96			-0.48	
overlap population	0.60				
dipole moment	4.60 D				

Table 11: TiF_2

MO	a_1^2	b_2^2	a_1^2	a_2^2	b_1^2	b_2^2	a_2^2	a_1	b_1	a_1	b_2	a_1	b_1	a_1	b_2
I.P.	15.58	15.55	14.21	14.20	14.10	14.07	9.31	9.21	9.19	8.92	8.86	4.49	2.60	-1.03	-2.91
AO - contributions to MO (in %)															
Ti 4s	1	-	0	-	-	-	-	0	-	9	-	48	-	29	-
4p	0	3	0	-	0	0	-	0	0	0	4	37	95	36	63
3d	11	9	3	5	1	1	92	96	98	79	84	4	0	3	2
F 2s	4	5	0	-	-	0	-	0	-	0	0	6	-	18	15
2p	84	83	97	95	99	99	8	4	2	12	12	5	5	14	20

(all energies are in eV)

TiH_4

TiH_4 is known to be a d^0-compound with a $^1A_1$ molecular ground electronic state arising from the configuration $...4a_1^2\ 3t_2^6$ which are mainly levels of hydrogen character. The symmetry is tetrahedral and the calculated Ti-H bond length is 1.70 Å from a HF/CI calculation/25/. The results for this relatively unambiguous molecule are compared in

Table 12: Population analysis for TiH_4

method	ref.	Ti			H	Q(Ti)	Q(H)	overlap population
		4s	4p	3d	1s			
HF/CI	25	0.60	0.51	1.67	1.29	1.22	-0.29	-
ab initio (FSGO)	26	-	-	-	1.66	2.64	-0.66	-
SCC-Xα	this work	0.225	0.10	2.45	1.305	1.22	-0.305	0.43

tables 12,13. As in the previous molecules the differences in the atomic orbital occupations arise from starting with different Ti atomic configurations $3d^3 4s^1$ and $3d^2 4s^2$ respectively. The overall agreement with the HF/CI calculation is again reasonably good while the results of the FSGO calculation are completely off.

Table 13: Ionization potentials and AO-contributions for TiH_4

method	ref.	ionization potentials a_1^2	t_2^6	e	t_2	10 Dq
HF/CI[a)]	25	13.11	12.32	-	-	-
ab initio[a)] (FSGO)	26	7.32	9.82	-	-	-
SCC-Xα	this work	12.52	11.08	8.81	7.38	1.43
	AO - contributions to MO (in %)					
	Ti 4s	5.5	-	-	-	
	4p	0	1.5	-	7.5	
	3d	0	41.5	100	63	
	H 1s	94.5	57	-	29.5	

a) orbital energies (all energies are in eV)

TiH_3F

Since TiH_3F does not exist and only one ab initio calculation is reported/28/ it serves as a simple model for covalent bonding in transition metal compounds and the charge transfer between different ligands. We used a C_{3v} geometry with bond distances of Ti-H = 1.628 Å and Ti-F = 1.810 Å and all bond angles 109°28′ as in ref. 28. From this we have calculated the electronic structure, charge distributions and the dipole moment. The results are compared to those of ref. 28 in tables 14,15. As for the Ti occupations the same is true as for the previous molecules. But there is a remarkable difference in the effective charge on the F atom. For F is much more electronegative than H the effective charge on F could not be smaller than that on H which would lead to a dipole moment with the wrong sign. So we believe that our effective charges and hence the occupation numbers are much more reasonable.

Table 14: Ionization potentials, 10 Dq, dipole moment for TiH_3F

method	ref.	ionization potentials $7a^2$	$8a^2$	$3e^4$	$9a^2$	$4e^4$	$5e$	$10a$	$6e$	10 Dq	dipole moment
ab initio[a)]	28	36.26	12.19	11.79	11.84	11.11	-2.04	-2.50	-4.84	2.8	-
SCC-Xα	this work	34.48	16.42	14.76	12.89	11.61	9.13	8.00	7.32	1.8	1.44
character of MO		F 2s	F 2p		H 1s		Ti 3d				

a) orbital energies (energies are in eV, dipole moments in D)

Table 15: Population analysis for TiH_3F

method	ref.	Ti 4s	Ti 4p	Ti 3d	F 2s	F 2p	H 1s	effective charges Ti	effective charges F	effective charges H	overlap populations Ti-F	overlap populations Ti-H
ab initio	28	o.70	o.75	1.37	1.94	5.34	1.30	+1.18	-0.28	-0.30	0.76	0.35
SCC-Xα	this work	0.14	0.18	2.38	1.88	5.52	1.30	+1.30	-0.40	-0.30	0.71	0.35

$TiCl_6^{2-}$

So far there has been reported only one semiempirical calculation on $TiCl_6^{2-}$ using the Pariser-Pople-Parr (PPP) method which applies the ZDO-approximation/41/. Since this octahedral complex is a closed shell system leaving the central Ti atom with a formally empty d-shell (d^0) the electronic structure is expected to lie closer to that of TiO_6^{8-} than to TiF_6^{3-} with its singly occupied d-shell. The bond length in the $TiCl_6^{2-}$ ion is taken to be 2.35 Å as in ref. 41. Since the overall level ordering in ref. 41 and our results both agree with molecular orbital theory we compare only selected properties in table 16. As this table is compared with tables 4,5 for TiO_6^{8-} ,TiF_6^{3-} respectively it is seen that especially the E_g/10 Dq ratio lies close to that of the TiO_6^{8-} closed shell system and far away from that of the TiF_6^{3-} open shell system. Moreover, from crystal field theory there is a relation between the 10 Dq values of octahedral and tetrahedral fields to be $10\ Dq(T_d) = -4/9 * 10\ Dq(O_h)$ where the sign means the reversed ordering of levels. Since we calculated both $TiCl_4$ (see table 2) and $TiCl_6^{2-}$ this can be examined yielding $2.60 * 4/9 = 1.15$ which is very close to the 10 Dq for $TiCl_4$.

Table 16: $TiCl_6^{2-}$

method	ref.	VB-width[a)]	separation bond.-nonbond. VB-orbitals	first L->M transition	10 Dq
PPP	41	9.4	0.3	6.3	0.5
SCC-Xα	this work	3.46	0.74	2.20	2.60

a) VB = valence band (all energies are in eV)

4. Application to TiO_2 clusters

Since the TiO_6^{8-} cluster being a model for the local Ti coordination of bulk TiO_2 was already used besides other Ti compounds for parameter optimization, in this chapter we will focus mainly on the electronic structure of the thermodynamically most stable TiO_2(110) surface.

4.1 Bulk electronic structure

As to get more information about the bulk TiO_2 in the rutile structure where each

Ti atom is coordinated to six O atoms and each O atom is coordinated to three Ti atoms (see fig. 1) we have calculated the electronic structure of a $Ti_3O_{15}^{18-}$ cluster. This cluster can be considered as composed of three subclusters , TiO_6 joined by common O atoms . Since in this cluster the central O ion is coordinated to all three Ti atoms and all Ti atoms are coordinated to six O atoms it should give a better approximation to the bulk electronic structure than the fairly small TiO_6^{8-} cluster.

Figure 1: TiO_2(rutile) crystal structure - the unit cell

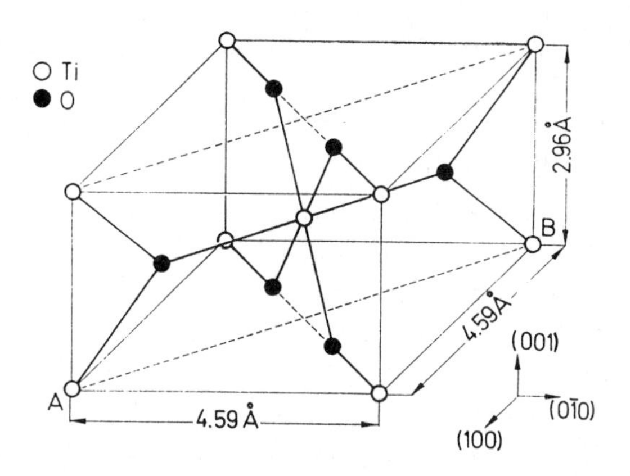

Plane A-B is the (110)-plane

Figure 2: TiO_2(110) surface cluster - Ti_7O_{24}

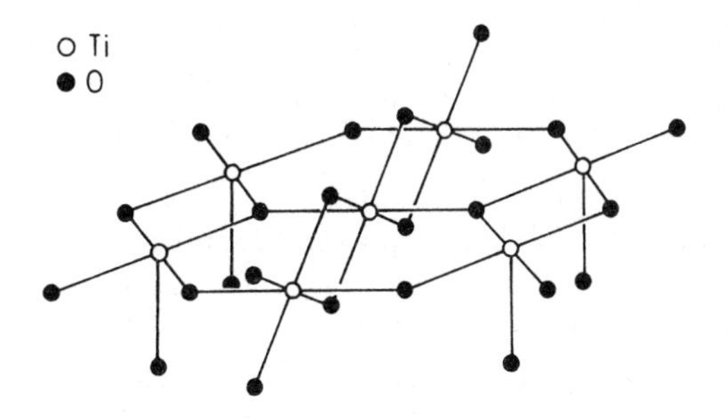

From the calculated level structure we obtained a density-of-states (DOS) by broadening each level by a Gaussian of half width $\sigma = 0.2$ eV which is shown in fig. 3. The overall level structure is similar to that of the smaller TiO_6^{8-} cluster. The Ti 3d levels are grouped into two narrow bands A and B which originate from the t_{2g} and e_g levels of O_h symmetry respectively. This means that the broadening of the d levels due to the interaction of different Ti atoms is relatively small compared to the crys-

Figure 3: Density-of-states (DOS) for TiO_2 bulk and (110)-surface clusters

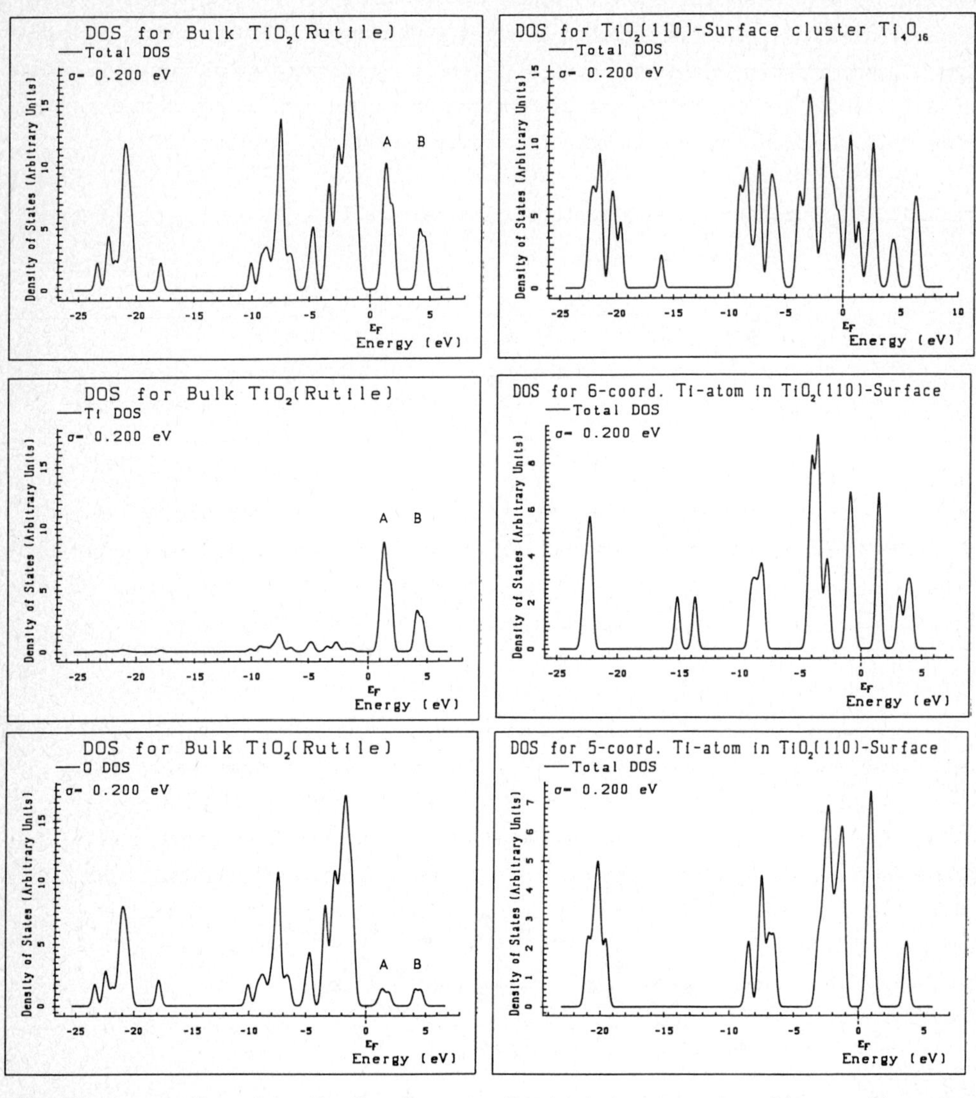

(Energies are referred to the Fermi energy E_f which is assumed to lie at midgap for bulk TiO_2)

a) Bulk TiO_2: total DOS from Ti_3O_{15}
b) " : Ti DOS "
c) " : O DOS "
d) Total DOS for: Ti_4O_{16} surface cluster
e) " :6-coord. surface atom
f) " :5-coord. surface atom

tal field splitting 10 Dq of the d levels. The energy difference between bands A and B ($\sim$10 Dq) is calculated to be 2.8 eV which corresponds to the 2.7 eV from the DV-Xα calculation/47/ and 2.1 eV from the X-ray absorption spectrum/45/. The energy gap between the Ti 3d and O 2p bands is about 0.5 eV smaller than that of the TiO_6^{8-} cluster due to the different effective charges occuring at the various oxygen ion sites in this cluster and the same is true in the DV-Xα calculation/47/. Nevertheless, the total state density curve in fig. 3a is quite similar to the X-ray emission and absorption spectra/45/ and the DOS curves for Ti (fig. 3b) and O (fig.3c) agree with that from the band structure calculation/49/.

4.2 Surface electronic structure

The structure of the $TiO_2(110)$ surface in the rutile crystal structure can be seen from fig. 2. There are two different Ti sites in this surface ; one is surrounded by five O atoms and the other by six O atoms where two are lying in the surface plane and two are placed 1.26 Å above and below this plane, respectively. For this reason in a first step we have calculated the electronic structure of two small clusters TiO_5 and TiO_6 corresponding to the two different Ti sites. Next we have calculated a $Ti_4O_{16}^{16-}$ surface cluster consisting of two TiO_5 and two TiO_6 subclusters joined by common O atoms in order to obtain a more accurate picture of the ideal surface. For these three clusters see also the DV-Xα calculations of /47,48/. In a last step a $Ti_7O_{24}^{20-}$ cluster had been calculated . Since this cluster contains all the possible adsorption sites on a $TiO_2(110)$ surface we are able to study the interaction of small molecules on different adsorption sites and inversion barriers between them whereas no direct comparison of the results could be made if they were obtained from calculations with different clusters. Since the calculations are still in progress we report the qualitative results only, obtained so far.

As to account for the changes of the Madelung potentials at different oxygen sites above and below the surface,the clusters are surrounded by hydrogen-like ´potential-atoms´´PA which interact with ´their´ oxygen atoms via a 1s wavefunction (saturating the dangling bonds), their effective charges (simulating the electrostatic potential of the ommitted ions outside the cluster), and each contributing one electron to the total cluster charge. In the case of our large $Ti_7O_{24}^{20-}$ cluster the latter leads to a neutral $Ti_7O_{24}PA_{20}$ from which we can obtain a first I.P. of about 7-10 eV which is in agreement with the experimentally obtained 9.54+/-0.10 eV for gaseous molecular TiO_2 from/52/.

For the small TiO_5 cluster there are only minor changes in the valence band region while in the TiO_6 cluster a group of 6 levels arising from the two outermost O atoms splits off from the top of the valence band and is driven into the bulk band gap (see figures 3e,f). The Ti 3d levels in TiO_6 into two groups, the lower three and upper two levels, i.e. the Ti ion feels the crystal field of nearly octahedral symmetry. On the other hand, a large splitting of the e_g levels can be found for the TiO_5 cluster. This is caused by the large differences in the electric field on the five-fold coordinated Ti atom compared to the octahedral field (see figures 3e,f).

Combining these clusters to the Ti_4O_{16} cluster these two effects (VB levels in the bulk band gap from outermost O atoms and large splitting of $e_g$levels for the five-fold coordinated Ti atoms) add to the intrinsic surface states of the $TiO_2(110)$ surface and causes the band gap at the surface to be about 1.8 eV smaller than the bulk band gap which is in agreement with the DV-Xα calculations /47,48/ and a band structure calculation/49/ (see fig. 3d). On the other hand, from UPS there is no experimental evidence for occupied surface states in the bulk band gap region/50,51/. For this reason the band structure calculation was repeated with the outermost oxygen atoms relaxed inward by about 0.20 Å which results in driving all of the occupied band gap states back below the bulk valence band edge/49/. Our calculation on the Ti_4O_{16} cluster with relaxed outermost O atoms yields the same result.

5. Conclusions

The calculations presented here mainly focus on three subjects. Firstly they describe the difficulties that arise in deriving parameters for semiempirical calculations on transition metal compounds and what is important in selecting compounds to which they can be fitted. Secondly in an extensive review on known titanium compounds nevertheless they show the possibility of yielding results comparable in quality to the results from the more sophisticated DV-Xα and CI calculations. Finally they show the capability of the SCC-Xα method of yielding reliable results for large size clusters with a justifiable amout of computing time, e.g. the $Ti_7O_{24}PA_{20}$ cluster with its 51 atoms and 179 basis orbitals converged after 26 iterations with 2 min 17 sec per iteration, i.e. 59 min 22 sec total computation time, on a Siemens 7.882/Fujitsu M-200 computer.

6. References

Parameter

/ 1/ M.Grodzicki;J.Phys. B13 (1980) 2683
/ 2/ R.Bläs,M.Grodzicki,V.R.Marathe,A.Trautwein;J.Phys. B13 (1980) 2693
/ 3/ M.Grodzicki,M.Hütsch,O.Kühnholz;Comp.Phys.Comm. (submitted)
/ 4/ O.Kühnholz,M.Grodzicki; 'Nickel Clusters as Surface Models for Adsorption', (in this volume)
/ 5/ M.Grodzicki;Croat.Chem.Acta 57 (1984) 1125
/ 6/ D.Briggs(Ed.):'Handbook of X-ray and UV Photoelectron Spectroscopy'; Heyden and Son,London(1978)
/ 7/ H.Basch,A.Viste,H.B.Gray;Theor.Chem.Acta(Berl.) 3 (1965) 458
/ 8/ E.Clementi,C.Roetti;Atom.Nucl.Data Tables 14 (1974) 177
/ 9/ J.C.Slater;Phys.Rev. 36 (1930) 57
/10/ G.Burns;J.Chem.Phys. 41 (1964) 1521
/11/ J.P.Desclaux;Atom.Nucl.Data Tables 12 (1973) 311

Ti_2

/12/ V.D.Fursova,A.P.Klyagina,A.A.Levin,G.L.Gutsev;Chem.Phys.Lett. 116 (1985) 317
/13/ A.B.Anderson;J.Chem.Phys. 64 (1976) 4046
/14/ C.Cosse,M.Fouassier,T.Mejean,M.Tranquille,D.P.DiLella,M.Moskovits; J.Chem.Phys. 73 (1980) 6076
/15/ J.Harris,R.O.Jones;J.Chem.Phys. 70 (1979) 830

TiH

/16/ S.P.Walsh,C.W.Bauschlicher,Jr.;J.Chem.Phys. 78 (1983) 4597
/17/ P.R.Scott,W.G.Richards;J.Phys. B7 (1974) 500
/18/ G.Das;J.Chem.Phys. 74 (1981) 5766

TiO

/19/ J.M.Dyke,B.W.J.Gravenor,G.D.Josland,R.A.Lewis,A.Morris;Mol.Phys. 53 (1984) 465
/20/ C.W.Bauschlicher,Jr.,P.S.Bagus,C.J.Nelin;Chem.Phys.Lett. 101 (1983) 229
/21/ K.D.Carlson,C.Moser;J.Chem.Phys. 46 (1967) 35
/22/ K.D.Carlson,R.K.Nesbet;J.Chem.Phys. 41 (1964) 1051

TiF_2,TiF_3

/23/ J.W.Hastie,R.H.Hauge,J.L.Margrave;J.Chem.Phys. 51 (1969) 2648
/24/ J.H.Yates,R.M.Pitzer;J.Chem.Phys. 70 (1979) 4049

TiH_4

/25/ D.M.Hood,R.M.Pitzer,H.F.SchaeferIII;J.Chem.Phys. 71 (1979) 705
/26/ E.R.Talaty,A.J.Fearey,G.Simons;Theor.Chem.Acta(Berl.) 41 (1976) 133
/27/ A.Breisacher,B.Siegel;J.Am.Chem.Soc. 85 (1963) 1705

TiH_3F

/28/ P.E.Stevenson,W.N.Lipscomb;J.Chem.Phys. 50 (1969) 3306

$TiCl_4$,$TiBr_4$,TiJ_4

/29/ T.Parameswaran,D.E.Ellis;J.Chem.Phys. 58 (1973) 2088
/30/ J.A.Tossell;Chem.Phys.Lett. 65 (1979) 371
/31/ D.R.Truax,J.A.Geer,T.Ziegler;J.Chem.Phys. 59 (1973) 6662
/32/ R.G.Egdell,A.F.Orchard;J.C.S. Faraday II 74 (1978) 485
/33/ P.Burroughs, S.Evans,A.Hamnett,A.F.Orchard,N.V.Richardson;
J.C.S. Faraday II 70 (1974) 1895
/34/ R.G.Egdell,A.F.Orchard,D.R.Lloyd,N.V.Richardson;
J.Electron Spectrosc. Relat. Phenom. 12 (1977) 415
/35/ P.A.Cox,S.Evans,A.Hamnett,A.F.Orchard;Chem.Phys.Lett. 7 (1970) 414
/36/ C.A.L.Becker,C.J.Ballhausen,I.Trabjerg;Theor.Chem.Acta(Berl.) 13 (1969) 355

TiF_6^{3-}

/37/ S.Yu.Shashkin,A.E.Nikiforov,V.I.Cherepanov;phys.stat.sol.(b) 97 (1980) 421
/38/ J.W.Richardson,D.M.Vaught,T.F.Soules,R.R.Powell;J.Chem.Phys. 50 (1969) 3633
/39/ H.D.Bedon,S.M.Horner,S.Y.Tyree,Jr;Inorg.Chem. 3 (1964) 647
/40/ C.J.Ballhausen,H.B.Gray;´Molecular Orbital Theory´,W.A.Benjamin Inc.,New York 1965

$TiCl_6^{2-}$

/41/ L.Oleari,E.Tondello,L.DiSipio,G.DeMichelis;Coord.Chem.Rev. 2 (1967) 421

TiO_6^{8-},TiO_2

/42/ T.Sasaki,H.Adachi;Intern.J.of Quantum Chem. 18 (1980) 227
/43/ J.A.Tossell,D.J.Vaughan,K.H.Johnson;Am.Mineral. 59 (1974) 227
/44/ F.M.Michel-Calendini,H.Chermette,P.Pertosa;Solid State Comm. 31 (1979) 55
/45/ D.W.Fisher;Phys.Rev. B5 (1972) 4219
/46/ S.P.Kowalczyk,F.R.McFeely,L.Ley,V.T.Gritsyna,D.A.Shirley;
Solid State Comm. 23 (1977) 161
/47/ M.Tsukada,C.Satoko,H.Adachi;J.Phys.Soc.Japan 47 (1979) 1610
/48/ M.Tsukada,C.Satoko,H.Adachi;J.Phys.Soc.Japan 44 (1978) 1043
/49/ R.V.Kasowski,R.H.Tait;Phys.Rev. B20 (1979) 5168
/50/ R.H.Tait,R.V.Kasowski;Phys.Rev. B20 (1979) 5178
/51/ V.E.Henrich,R.L.Kurtz;Phys.Rev. B23 (1981) 6280
/52/ D.L.Hildenbrand;Chem.Phys.Lett. 44 (1976) 281

HEAVY ION INDUCED DESORPTION FROM SOLID SURFACES

Y. Le Beyec
Institut de Physique Nucléaire, B.P. N°1, F-91406 Orsay

This lecture presents the main topics which have been recently investigated by a group at the Institut de Physique Nucléaire (IPN) in collaboration with other groups of research and visiting scientists. I will therefore cover various subjects ; complementary information will be obtained either from preprints available at IPN or from publications which should appear very soon. For the period 1985-86, the group was composed of S. Della-Negra, C. Deprun, Y. Le Beyec, B. Monart, K. Standing.

I have divided this talk into 5 parts :

1. Recent developments in instrumentation

2. Spontaneous desorption ionisation time of flight mass spectrometry

3. Erosion of ice by MeV particle bombardment

4. Influence of the charge state of incident ions on the secondary electron emission and secondary ion emission

5. Use of the H^+ secondary ion emission as a probe to measure the equilibrium charge state in a solid.

1. RECENT DEVELOPMENTS IN INSTRUMENTATION

Two ^{252}Cf plasma desorption mass spectrometers are in operation at the Laboratory. One [1] is used by external users to measure mass spectra of various compounds and also to study peculiar aspects of sample preparation like for example chemical effects enhancing the formation of protonated species in a mixture of small proteins [2]. The other one is used for technical improvements. An electrostatic mirror has been added to this T.O.F. mass spectrometer [3] and the mass resolution which is obtained now can reach M/Δ M 6000 as shown in fig. 1. The width of the T.O.F. peaks is around 3 or 4 nsec and this implies that many parameters are well adjusted (mechanical alignments, field uniformity inside the mirror, power supply stability,...). With a keV ion gun for desorption ionisation in a time of flight mass spectrometer, high mass resolution has also been recently achieved [4,5]. I would like however to insist on one important feature related to the use of a mirror. When a fragmentation of M^+ occurs in flight, the charged fragment keeps the velocity of the molecular ion but its energy is reduced $E = E_{total} \times m_1+/M$ and the time spent in the irror is smaller than M^+. The charged fragment species contribute to a peak in the reflex spectrum and the time calibration (and therefore the mass calibration) for this peak is different that for the intact molecular ion M^+. The only method to identify and to time calibrate these kinds of peaks is to use the coincidence method between neutral and ions [6]. The method allows also to identify the type of fragmentation occuring in flight.

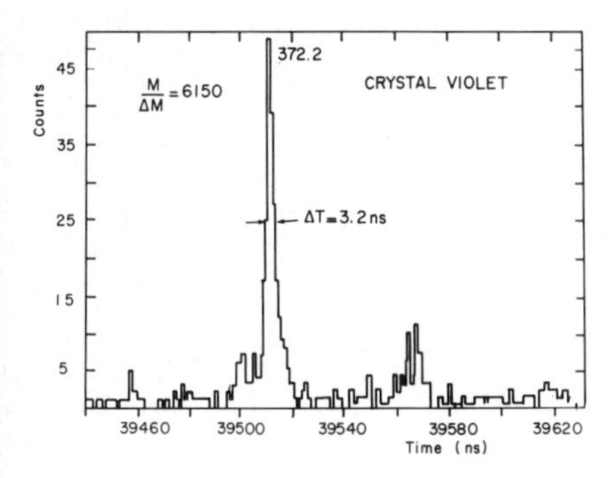

Fig. 1

In addition, I would like to mention the use of a simple detector based on ion-electron conversion [7,8,9]. Secondary electrons are emitted from a plane surface (coated by a CsI deposit) when this surface is hit by the secondary ions accelerated from the sample. These electrons are accelerated by a voltage (0.3 to 18 kV) and bent by a magnetic field onto a set of channel plates or a silicon detector. To detect the electrons directly emitted from the sample (see section 3) the Si detector can easily replaced the plane surface coated with CsI. Coincidence measurements can be made to study the response of various surfaces to the secondary ion with energy of 5 to 20 keV accelerated from the sample surface. It has been shown that the conversion ion-electrons acts as an amplifier [9].

2. SPONTANEOUS DESORPTION TIME OF FLIGHT MASS SPECTROMETRY

(in collaboration with G. Bolbach* and F. Rollgen**)

It was shown recently [10] that when a voltage of a few KV is established between a flat metallic surface covered with an organic or inorganic layer, and a thin grid parallel to the surface, electron and negative ions are simultaneously and "spontaneously" emitted from the surface. After traversing the flight tube, electrons and negative ions are observed by the same detector. Their time of flight differences are measured by a multistop time digitizer and the mass spectrum of the compounds on the surface is measured. The assumption was made that under field effects occuring at the surface, negative ions accompanied by electrons could be extracted directly from the sample surface.

A serie of experiments made in collaboration with F. Rollgen and G. Bolbach [11] has been performed to try to understand the mechanisms of emission. With a special source geometry and using two acceleration grids a very strong electron ion emission was observed so that mass spectra could be recorded in a few seconds. It has been concluded from these experiments that the sample surface is bombarded by primary particles coming from the grids. Therefore, we observe secondary ions due to keV particle induced desorption. The mechanism of primary emission is likely due to a field emission, but the type of charged species issued from the grid is still unknown. Based on this technique a very simple apparatus could be built for qualitative mass analysis.

---

* Institut Curie, Paris
** Institut fur Physik Chemie, Universität Bonn, R.F.A.

3. EROSION OF ICES BY MEV ION BOMBARDMENT

(in collaboration with J. Benit, J.P. Bibring, F. Rocard, Laboratoire R. Bernas, Orsay).

We have studied the erosion rates of ice films bombarded by various kinds of ions in the MeV energy range. Ions from the Orsay linear accelerator Ne, Ar, Kr at 1.1 MeV/u in their equilibrium charge states have been used. The disappearance of H_2O has been measured by on line Fourier transform infrared spectrometry [12]. Typical film thicknesses were in the range 1000-20000 Å.

It has been observed that 10^5 H_2O molecules/incident Kr ion are disappearing from the film. Molecules are either directly ejected or chemically transformed. Furthermore the linear dependence of the yields on film thickness is in agreement with a mechanism operating all along the ion tracks. The yields per unit thickness, for Ne, Ar and Kr are found to vary as $(dE/dx)^2$. In addition, the erosion of a double layer of D_2O ($\sim$2500 Å) covered by H_2O ($\sim$3000 Å) confirms the mechanism of erosion all along the track since the relative thicknesses of the two layers decrease at similar rates.

4. INFLUENCE OF THE CHARGE STATE OF INCIDENT IONS ON THE SECONDARY ION EMISSION AND SECONDARY ELECTRON EMISSION

(in collaboration with K. Wien*, G. Maynard**, C. Deutsch**)

After passing through a thin carbon foil, the beams from the linear accelerator exhibit a distribution of charge states (from $\sim 1^+$ to 10^+ for Ne, $\sim 6^+$ to 15^+ for Ar and $\sim 10^+$ to 26^+ for Kr). The charge state desired is selected by a magnet at a certain angle of elastic diffusion and the resulting low intensity beam is collimated to 1 mm diameter before stricking the target. The experimental set up has already been described elsewhere [13].

Targets consisted of thin layers of organic and inorganic compounds deposited (sublimation under vacuum or electrospay) on flat metallic surface (aluminized mylar or thin aluminium foils). The yield of secondary ions (per incident particle) has been measured as a

* Technische Hochschule, Darmstadt, RFA
** Laboratoire de Physique des Gaz et Plasma, Orsay

function of the charge state q_i of the incident ions. Fig. 2 shows the variation of the yield with q_i of Cs^+ ions emitted from a CsI target and H^+ ions from the same target. (H^+ comes from organic contamination at the target surface).

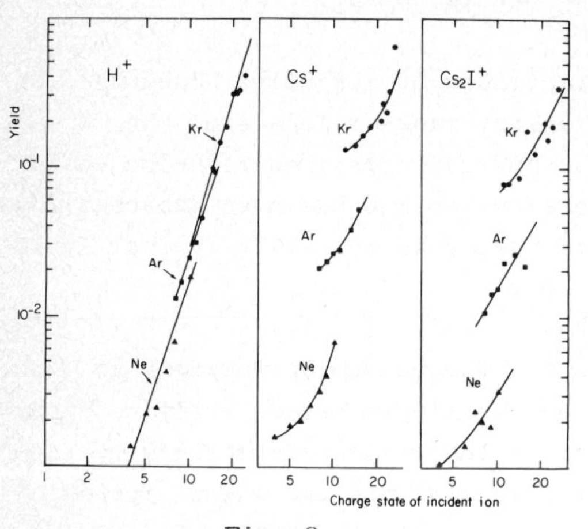

Fig. 2

The yield of H^+, $Y(H^+)$ varies as q_i^n with n 3 to 4 and does not depend on the type of primary ions. This behaviour indicates that H^+ ions originate very likely from the surface near the track. As seen later the H^+ yield can therefore be used as a sensitive probe of the projectile charge state at the surface. On the other hand the yield of Cs^+ (and also cesium iodide clusters) depends on the charge state and also on the bombarding ion. The shapes of three curves (Ne, Ar, Kr) are different with a steeper slope for the lightest ions. This is also observed in Fig. 3 showing the charge state dependence of the molecular ions M^+ of coronen ($C_{12}H_{12}$) emitted from a coronen target. When passing through a target the incoming projectile achieves an equilibrium charge state which depends on the nature of the projectile. For Ne, Ar and Kr ions at 1.16 MeV/u these equilibrium charge states which can be calculated [14] are respectively 6.8, 11 and 19. The desorption yield varies as $y \propto q_{eq}^n$ with n = 4 as shown in Fig. 2. However, the real dependence of y on the incident charge state q_i is hidden since cumulative effects due to charge variations inside the medium are influencing the emission of secondary ions from the surface.

In other words we may have the same incident charge state q_i for different bombarding ions crossing the surface but different values of equilibrium charge state q_{eq} inside. The energy lost dE/dx which is proportional to q^2 may be the same in the first one or two mono-layers but very different after a certain depht of penetration.

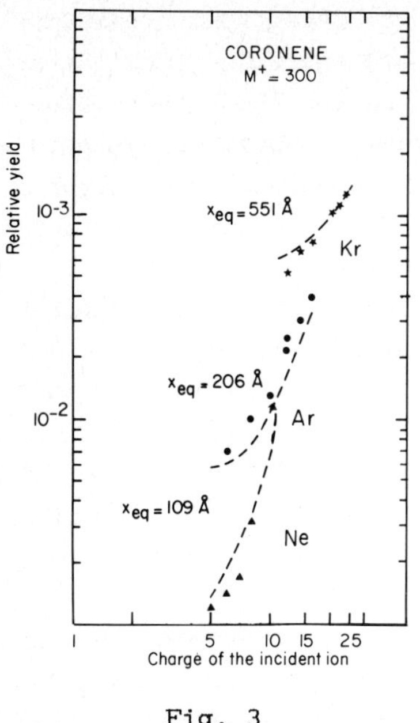

Fig. 3

In order to perform calculation on secondary ion yields it is therefore necessary :

i) to have a model for the transfer of energy to the surface,
ii) to know the variation of q as a function of the distance x of penetration in the medium [15].

We have used a simple approach and the recent calculation of G. Maynard and C. Deutsch [15]. We assume that the contribution to the yield dy from a distance x is dy $\propto$ cte (dE(x)/dx) dx, then

$$y \propto K_0 \int_0^{\infty} (K_1, q^2(x))\, q^2(x)dx$$

with a sharp cut off, at a distance x, the energy transferred to the surface is K_1 (dE(x)/dx) $\sim K_1 q^2(x)$. Until a certain depth x_{max}, the energy $K_1 q^2(x_{max})$ is useful for desorption. One has

$$y \propto K_0 \int_0^{x = x_{max} = K_1 q^2(x)} q^2(x)dx$$

and for $q = q_{eq}$, $x_{max} = K_1 q^2_{eq}$ and $y = K_0 K_1 q_{eq}^4(x)$.

This q_{eq} dependence is observed experimentally. The constance K_0K_1 can be deduced from the experimental value of yield, for Ne for example $y_{Ne}(q_{eq}) \propto$ cte q_{eq}^4 and K_1 is deduced from the fit of the Ne yield curve. This parameter is rather sensitive to the shape of the yield curve as shown by G. Maynard [15]. It is however remarquable that the parameters, fitted with the Ne results, apply to the Ar and Kr results. In Fig. 3, the dashed lines represent the calculation. Complete results will be published elsewhere. The charge state dependence of the secondary ion yield has been demonstrated by other groups [16-18]. Here we have extended to higher charge states and several incident ions with overlapping charge states. We conclude that a few hundreds of Angstroms below the surface are involved in the desorption processes.

In these experiments the number of electrons emitted simultaneously with the secondary ions have been measured. The data acquisition was made in the coincidence mode so that correlated spectra were stored event by event in the computer memory. It has been shown that the ion yield increases slowly with the number of electrons simultaneously emitted. Also the variation of the secondary electron yield with the incident charge state is very different from the S.I charge state dependence. This suggests that they arise from depths where the incident beam has achieved charge equilibrium.

5. USE OF THE H^+ SECONDARY ION EMISSION AS A PROBE TO MEASURE THE EQUILIBRIUM CHARGE STATE IN A SOLID

(in collaboration with K. Wien)

The yield of H^+ secondary ions is strongly dependent on the charge state of the primary ion when it crosses the surface. Targets of carbon, gold and nitrocellulose have been bombarded by Ar and Kr ions at 1.16 MeV/u with a range of charge state lying from 7^+ to 25^+ [19]. Fig. 4 shows the H^+ yields as a function of the incident charge state for a 1000 Å gold foil (black circle in the figure). The target is mounted on the special device which can be rotated through 180° to observe the ion ejected from the same 1 mm^2 target surface after the beam passes through the target material. The points denoted o in the figure correspond to H^+ yield measurements for primary ions leaving the target surface. It observed that for Ar and Kr ions as well, the yields are constant whatever the incident charge state. This implies that the equilibrium charge state has been achieved inside the target. The mean equilibrium charge state is 16.6^+ for Kr and 11.3^+ for Ar. The points Δ are yield measurements for H^+ ion emitted from the entrance side of the target but when the equilibrium charge state is achieved upstream in a similar gold foil. The primary ion reach the target foil about 70 nsec after the equilibrating foil. In this case the mean charge striking the gold target is about 3 charge (19.7) units larger. No such

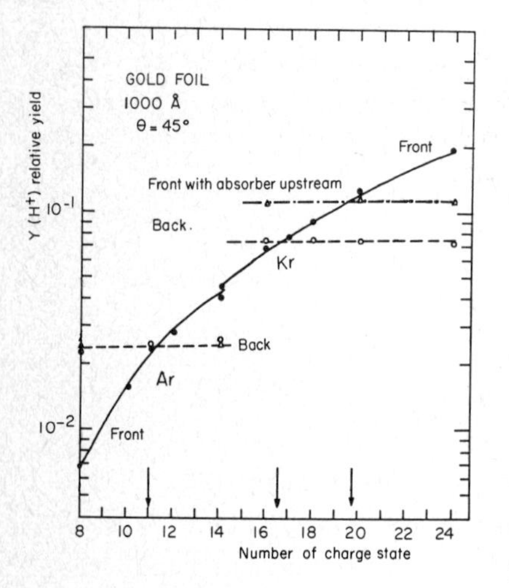

Fig. 4

difference was observed for ar ions of the same velocity. In the Betz and Grodzins model (20) it is assumed that Auger electron emission takes place after the projectile leaves the solid so the charge state on leaving the solid is lower than the charge state measured later. Our measurements show clear evidence for this process in the case of Kr.

The technique could be used also to obtain quantitative information on the projectile charge state in the interior of a solid.

References

1. S. Della-Negra, C. Deprun, Y. Le Beyec, Rev. Phys. Appl. 21 (1986) 401.
2. Y. Hoppilliard, this conference.
3. S. Della-Negra and Y. Le Beyec, Int. J. of Mass Spectr. and Ion Physics 61 (1984) 21.
4. X. Tang, R. Beavis, W. Ens, B. Schueler, K.G. Standing, Proc. of the ASMA Meeting (1986).
5. E. Niehuis, T. Heller, H. Feld and A. Benninghoven, Proc. Ion Formation from Organic Solids IFOS III, Springer proceedings in Physics, vol. 9 (1986) 198.
6. S. Della-Negra and Y. Le Beyec, Analytical Chemistry 57 (1985) 2035.
7. A.M. Zebelman, W.G. Meyer, K. Halbach, A.M. Poskanzer, R. G. Sextro, G. Gabor and D.A. Landis, Nucl. Instr. and Meth. 141 (1977) 429.
8. H. Danigel, H. Jungclas and L. Schmidt, Int. J. of Mass Spectr. and Ion Physics 52 (1983) 223.
9. S. Della-Negra, M. Dumail, Y. Le Beyec, Internal report IPNO-DRE-85-33.
10. S. Della-Negra, Y. Le Beyec, P. Häkansson, Nucl. Instr. and Meth. B9 (1985) 103.
11. S. Della-Negra, C. Deprun, Y. Le Beyec, B. Monart, K. Standing, F. Rollgen and G. Bolbach, to be published.
12. J. Benit, J.P. Bibring, S. Della-Negra, Y. Le Beyec, M. Mendenhall, F. Rocard and K. Standing, Nucl. Instr. and Meth. (in press).
13. S. Della-Negra, O. Becker, R. Cotter, Y. Le Beyec, B. Monart, K. Standing, K. Wien, Internal report IPNO-DRE-86-09, J. Phys. (in press).
14. K. Shima, T. Ishihara, T. Mikumo, Nucl. Instr. and Meth. 200 (1982) 605.
15. G. Maynard, this conference.
    G. Maynard, C. Deutsch, to be published.
16. E. Nieschler, B. Nees, N. Bishof, H. Fröhlich, W. Tiereth and H. Voit, Rad. Effect 83 (1984) 121.
17. P. Häkansson, J. Yayasinghe, A. Johansson, I. Kamensky and B. Sundqvist, Phys. Rev. Lett. 47 (1981) 1227.
18. C.K. Meins, J.E. Griffith, Y. Qier, M.H. Mendenhall, L.E. Seiberling and T.A. Tombrello, Rad. Eff. 71 (1983) 13.
19. S. Della-Negra, Y. Le Beyec, B. Monart, K. Standing and K. Wien, Internal report IPNO-DRE-86-20, to be published.
20. H.D. Betz and L. Grodzins, Phys. Rev. Lett. 25 (1970) 211.

Correlation Effects of Secondary Ions in ^{252}Cf-PDMS

Do clusters play an important role in the desorption process?

L. Schmidt and H. Jungclas
Kernchemie, Fachbereich Physikalische Chemie, Philipps-Universität,
D-3550 Marburg, F.R.G.

1. Introduction

Time-of-flight mass spectrometry can be applied to study the coincident desorption of secondary ions induced by high energy ions like ^{252}Cf fission fragments /1,2/. By means of a multistop TDC (time-to-digital converter) and an adequate software the detected ions per fission fragment are sorted and counted according to the multiplicity or to preset coincidence conditions (time windows). When a coincidence or anti-coincidence condition is applied, correlated desorption of different ions leads to different detection probabilities of these ions per impacting fission fragment. This variation of the detection probability can be quantified and expressed by a correlation coefficient. Correlation measurements for the desorbed ions of the organic compound Verapamil resulted significant correlation effects especially between the molecular ion and the main fragment ions /3/.

2. Fundamental aspects

The analysis of correlations between secondary ions by coincidence measurements is complicated by the fact that in an ionization event the coincidence probability of two ions species A and B is strongly dependent on the multiplicity of desorbed ions. The frequency distribution of the multiplicities, however, is in general not totally determined by the single ionization probabilities $p(I_n)$(I_n = one of N different ion species, $n \leq N$) but possibly also by variable features of the fission fragments like charge state, size, velocity and angle of incidence.

Otherwise, in case of independence (no correlation) the coincidence probability p(A,B) simply is the product of the single ionization probabilities:

$$p(A,B) = p(A)\ p(B) \quad (1)$$

This easily can be proved if equal ionization probabilities are assumed for all ion species ($p(I_n)=p$, $n \leq N$). In that case, the multiplicities automatically will result a binomial distribution (i.e. a Poisson distribution for large N, $p \ll 1$). But if the hypothetical case is considered that the multiplicity distribution is restricted e.g. to one certain multiplicity $M < N$ by some additional influence, equation (1) will no longer be valid. Provided that $p(A)=p(B)=M/N$ the coincidence probability then is calculated to be

$$p(A,B) = (M/N)\ (M-1)/(N-1) \tag{2}$$

and not $(M/N)^2$ as equation (1) would have resulted.
Of course equation (2) now has to be generalized in order to describe the true case of different ionization probabilities $p(I_n)$ for different ion species and to be independent from the true multiplicity distribution. Using conditional probabilities the coincidence probability for any multiplicity $M \geq 2$ can be written as

$$p_M(A,B) = p_M(A)\ p_M(B|A) \tag{3}$$

which generally is valid, even when correlation effects are present. This is also true for the anticoincidence probability ($\bar{A}$ means non A)

$$p_M(\bar{A},B) = p_M(\bar{A})\ p_M(B|\bar{A}) \tag{4}$$

But only in the case of independence the following equation is fulfilled:

$$p_M(B|A) = p_{M-1}(B|\bar{A}) \tag{5}$$

From equations (3) and (5) we finally get

$$p_M(A,B) = p_M(A)\ p_{M-1}(B|\bar{A}) \tag{6}$$

It easily can be realized that (2) is a special case of (6).

Now we are able to define and to quantify correlation effects: When equation (5) is found to be not fulfilled, there must be a correlation effect. The deviation between the left side ($c=p_M(B|A)$, coincidence measurement) and the right side ($a=p_{M-1}(B|\bar{A})$, anticoincidence measurement) can be used to define a correlation coefficient r /3/:

correlation ($c > a$): $r = 1-a/c$ ($0 \leq r \leq +1$)
anticorrelation ($c < a$): $r = c/a-1$ ($-1 \leq r \leq 0$)
independence ($c = a$): $r = 0$

The so defined r-values describe how much the desorption of an ion B is associated with the desorption of an ion A compared to the degree of association between the ion A and any other ion B.

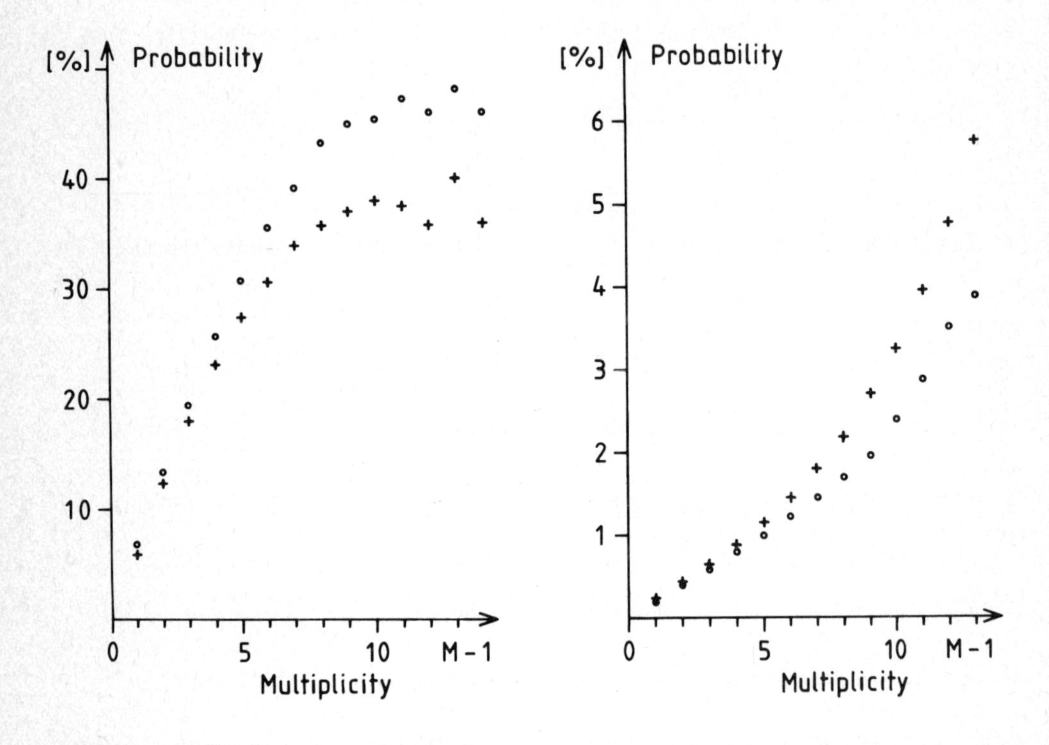

Fig. 1 Conditional probabilities for the fragment ion m/z 165 (Verapamil, positive ions) in coincidence (o $p_M(165|454)$) and in anticoincidence (+ $p_{M-1}(165|\overline{454})$) with the molecular ion.

Fig. 2 Conditional probabilities for background events at m/z ≈ 380 (Verapamil, positive ions) in coincidence (o $p_M(BG|454)$) and in anticoincidence (+ $p_{M-1}(BG|\overline{454})$) with the molecular ion.

3. Measurements

In Fig.1 an example of a positive correlation between fragment ion (m/z 165) of Verapamil and the quasimolecular ion (m/z 454) is demonstrated. The upper curve is showing the left side (coincidence measurement), the lower curve is showing the right side of equation (5) (anticoincidence measurement) as a function of the multiplicity. Both curves start at low multiplicities with linear slopes as expected. Thus, the relative difference (correlation effect) is constant first but becomes even greater at high multiplicities (saturation region).

The same measurements were done for an arbitrary background window and the quasi-molecular peak at m/z 454. The result (Fig.2) reveals a significant anticorrelation effect which is also more obvious at higher multiplicities. This underlines the competing situation between the large ions and the background, which is also suggested by the opposite curvatures of Fig.1 and Fig.2 (see also /3/).

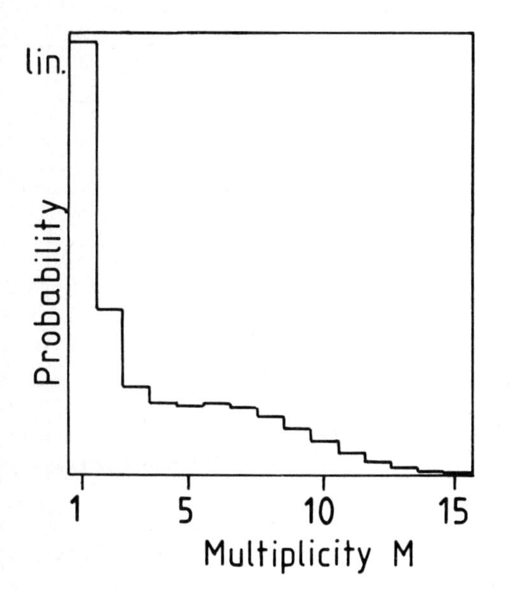

Fig. 3 Distribution of multiplicities (Verapamil, positive ions), fission fragments impacting on the front side of the sample (45°).

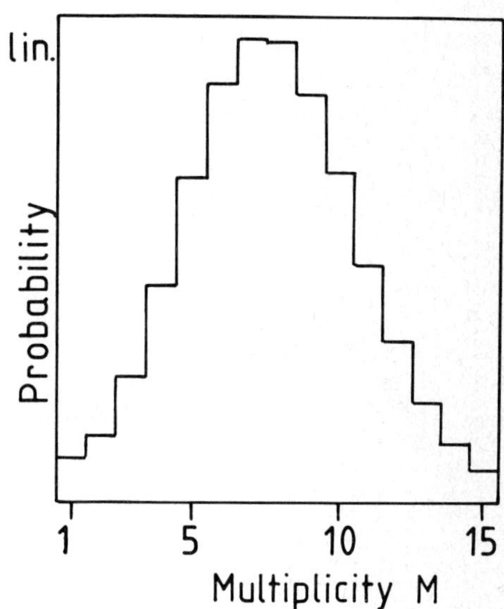

Fig. 4 Same as in figure 3 but with the condition that 1 of M detected ions is a molecular ion.

A completely different correlation effect arises when the Cf-source is installed in front of the sample at an angle of 45° instead of the more common application at the backside of the foil. The multiplicity distribution (Fig.3) then is obviously consisting of two components in contradiction to measurements of the same sample with the fission fragments coming from the backside. Setting a coincidence condition on the peak of quasimolecular ions only one component remains as shown in Fig.4. Additionally, the mass spectrum of events with multiplicity M=1 (Fig.5) reveals that the component of low multiplicities is contributing mainly to the background. These results suggest that there are two different types of ionization events: impacts desorbing ions which contribute to the peak information and impacts producing ions with a smeared energy-to-mass ratio. This conclusion is confirmed by a correlation analysis at a multiplicity which is expected to include contributions of both event types. The results listed in Table 1 for M=5 prove the strong correlation between all dominating ion species of the spectrum and their anticorrelation to background ions of comparable magnitude.

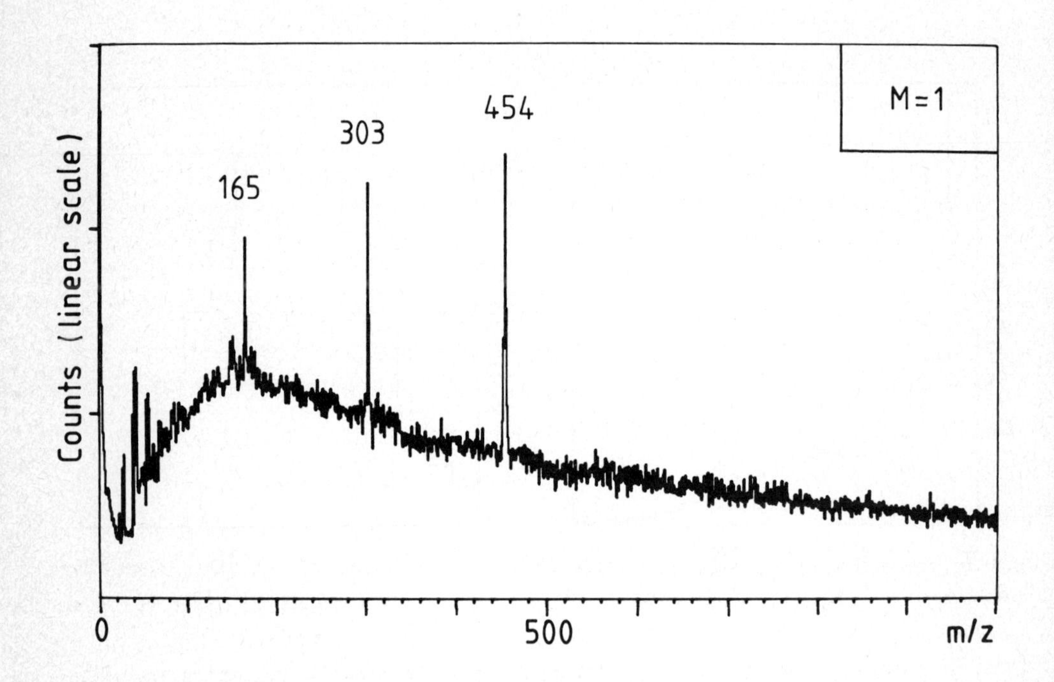

Fig. 5 Mass spectrum of Verapamil under the conditions of figure 3, however, just but events with multiplicity M=1 have been stored in the spectrum.

Table 1 Correlation coefficients of several ions and the background (BG) with the quasimolecular ion (m/z 454 ±1) of Verapamil at multiplicity M=5.

m/z	303	165	151	73	44	38	1	BG
r (%)	45 ±2	49 ±1	45 ±3	52 ±2	56 ±2	50 ±3	46 ±1	-54 ±4

The same effect was found also with other samples whenever the Cf-source was mounted in front of the sample and is therefore assumed to be fundamental. However, this has to be confirmed by further investigations.

4. Common results for two substances

Correlation measurements were not done with Verapamil only, but also with an electrosprayed sample of Thiamine, an organic compound which had been supposed to exhibit strong correlations and anticorrelations between the molecular ion and the two largest fragments in the positive ion spectrum /4/. Our results do not agree, however we found correlations quite similar to those found in the case of Verapamil. The common features of the correlations for both compounds are summarized in the following four points:

1. Significant positive correlation coefficients between quasimolecular ions and large fragment ions ($r \approx 10\% - 20\%$).
2. Small correlation coefficients between heavy ions and the light fragment or atom ions.
3. Hydrogen is desorbed almost completely independent from any other ion.
4. The background is partially anticorrelated especially to the large ions ($r \approx -10\%$).

5. Discussion

Looking for a simple explanation of the observed correlation effects (not regarding the possiblity of pseudo-effects) one has to ask first what conditions have to be fulfilled for the generation of correlations between secondary ions. The answer can be given in one essential statement: A necessary condition is that there are different origins for different ion species. These origins can be different locations on the sample surface: If the foil is not covered totally or if there are several not perfectly mixed compounds, different ions are desorbed from different sites leading to very strong correlations and anticorrelations between the different groups of ions. The certainly more interesting possibility is related to different ionization mechanisms for different ions. Then a positive correlation would mean either that some ions originate in the same mechanism or that at least two mechanisms for different ions are associated somehow. An anticorrelation can be induced either when at least two mechansims are competing with each other or when one ion competes with the generation of a another one within the same mechanism.

To give an example, imagine that as a consequence of the high energy and momentum transfer by an impact of a fast heavy ion rather intact ensembles of molecules (clusters), some of them charged or multiply charged, are desorbed. Of course these clusters will mostly be highly excited and metastable probably with a lifetime shorter than 1 ns. Imagine further that especially the multiply charged clusters will decompose by evaporation, fission or Coulomb explosion releasing some molecules and fragments at the same time, a few of them again in an ionized state. Now, if this process is preferably producing a certain group of ion species e.g. the molecular ions and large fragment ions the condition for a correlation effect is fulfilled as we can assume the following two mechanisms: the direct ionization desorbing mainly light fragments and atoms and the ionization by the decomposition of metastable intermediate clusters.

In addition some clusters will be likely to survive the first few nanoseconds after the impact and therefore contribute by subsequent decomposition to background events in the measured spectrum. These clusters are in competition with the short-living ones, if the number of metastable clusters per impact is relatively low. This will lead to an anticorrelation between large ions and background ions.

Though the observed correlation effects are consistent with the picture of highly excited metastable clusters as outlined above, it is not proved to be true, of course. On the other hand it could be worth while to consider the possibility that the decomposition of such short-living clusters might be the main ionization process for large organic ions in PDMS.

References

1 R.D. Macfarlane: Anal. Chem. 55, 1247 (1983)
2 S. Della Negra, D. Jacquet, J. Lorthiois, Y. Le Beyec, O. Becker and K. Wien: Int. J. Mass Spectrom. Ion Phys. 53, 215 (1983)
3 L. Schmidt and H. Jungclas: Springer Proc. Phys. 9, 22 (1986)
4 N. Fürstenau: Z. Naturforsch. 33a, 563 (1978)

Temperature Effects of Secondary Ion Emission

Observed with PDMS

R. Matthäus and R. Moshammer

Institut für Kernphysik der Technischen Hochschule
Schlossgartenstrasse 9, 6100 Darmstadt, West Germany

1. Introduction

So far, experimental results about the temperature dependence of fast heavy ion induced desorption have not been published. The aim of this report is to give a first survey of temperature effects on secondary ion emission in the high energy regime. In principle any change of surface structure or surface composition should be reflected in the secondary ions mass-spectra - also, if these changes are generated via heating or cooling the target.

An influence of the target temperature on the desorption process itself can be expected, if this process is a thermal volatilization from a hot zone surrounding the nuclear track. With respect to the thermal spike model of Sigmund and Szymonski [1] this temperature effect is rather weak, when target- and spike temperature are very different. As shown, the target temperatures of our experiments are much lower than any expected spike temperature.

The experiments are preformed with the two amino-acids Valine and Glycine, with Eu_2O_3 and Ice. All targets were investigated by ^{252}Cf - PDMS, i.e. by time-of-flight (=TOF) mass-spectrometry.

2. Experimental method

The main details of the experimental setup are shown in Fig. 1. The targets were prepared by evaporating the compound of interest onto a copper-plate. This plate was mounted on a metallic holder, which could be heated by an electric current or cooled by liquid nitrogen. The temperature ranged from -196°C to +350°C. The target was bombarded by fission fragments of ^{252}Cf under an angle of incidence of 30 degrees. Before hitting the target, the projectiles crossed a thin aluminium-foil to produce secondary electrons, which were accelerated by an electric potential in direction of the start detector. The positive secondary ions ejected from the target traversed the same electric potential and drifted to the stop detector of the TOF-spectrometer.

The TOF mass-spectra were analysed and stored by an on-line computer. The countrate of certain mass-lines was recorded separately as function of time together with the temperature, which was measured by a thermo-couple close behind the target-copper-plate.

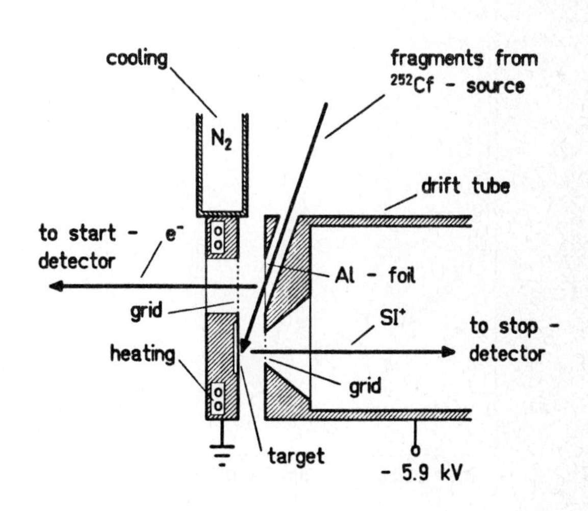

Fig. 1. Schematic drawing of the experimental setup (SI$^+$=Secondary ions)

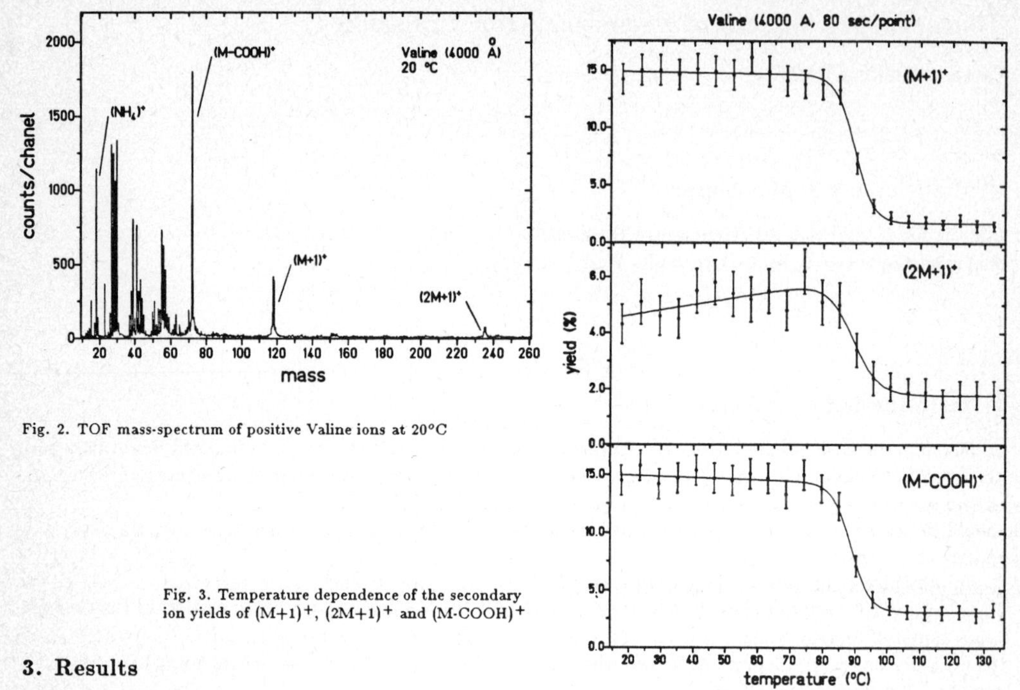

Fig. 2. TOF mass-spectrum of positive Valine ions at 20°C

Fig. 3. Temperature dependence of the secondary ion yields of $(M+1)^+$, $(2M+1)^+$ and $(M\text{-}COOH)^+$

3. Results

Valine and Glycine:
The mass-spectrum taken with Valine is presented in Fig. 2. The thickness of the sample was 4000 Å. The yields of three molecule specific ions are plotted as function of target temperature in Fig. 3. They keep nearly constant until 90°C, where the sample material is evaporated and the yields drop down to a certain back ground level. In this temperature range an increase of yields due to a temperature effect on the desorption process is certainly not detectable. The measurements with Glycine had a very similar result. The dropping of yields caused by sample evaporation occured at 105°C (sample thickness 5000 Å).

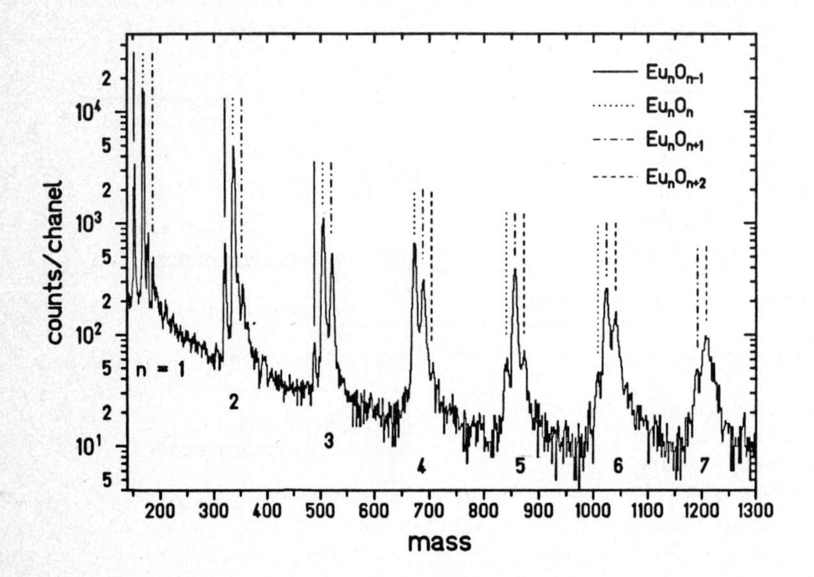

Fig. 4. Part of the positive secondary ion mass-spectrum of Eu_2O_3

Eu_2O_3:

Characteristical for positive mass-spectra of the Eu_2O_3 sample are series of cluster ions, partially shown in Fig. 4. Here each group consists of 3 or 4 peaks, which are build up by a pair of mostly unresolved mass-lines due to the Europium isotopes ^{151}Eu and ^{153}Eu. Only in the first group mass-lines could definitly assigned. At low cluster-number n also the protonated ion could be identified in the spectrum as, for instance, EuH^+ beside Eu^+ or $EuOH^+$ beside EuO^+. Eventually, also at higher n-values the observed lines correspond to the protonated ions. Bad mass resolution prevented a clear notation. We deduce from the cluster structure that Eu is inside the cluster ions bivalent, contrary to the sample ,which consists of Eu_2O_3. As seen in Fig. 4 and Tab. 1, mass-lines having a low number of extra O-atoms disappear with increasing n, the yields of ions with a higher number of O-atoms increase. For instance, the yield of $(EuO)O^+$ (or $(EuO)O^+$) first increases, reaches its maximum at n=5 and then decreases. The temperature dependence of $EuOH^+$ is presented in Fig.5 Up to 100°C the yield is constant, then it decreases from 12 to 5 percent till 330°C, where heating was stopped. While the target is cooling down, the yield of $EuOH^+$ remains unchanged. On the other hand, the yield of Eu^+ first went up and then stays on the high level during the temperature decrease. We assume, that the sample surface is dehydrated above 100°C.[3]

n=1 :	Eu^+	EuO^+	$(EuO)O^+$		
n=2 :	EU_2O^+	$(EuO)_2^+$	$(EuO)_2O^+$		
n=3 :	$EU_3O_2^+$	$(EuO)_3^+$	$(EuO)_3O^+$		
n=4 :		$(EuO)_4^+$	$(EuO)_4O^+$	$(EuO)_4O_2^+$	
n=5 :		$(EuO)_5^+$	$(EuO)_5O^+$	$(EuO)_5O_2^+$	
n=6 :		$(EuO)_6^+$	$(EuO)_6O^+$	$(EuO)_6O_2^+$	
n=7 :			$(EuO)_7O^+$	$(EuO)_7O_2^+$	$[(EuO)_7O_3^+]$

Tab. 1. Cluster-series of Eu_2O_3

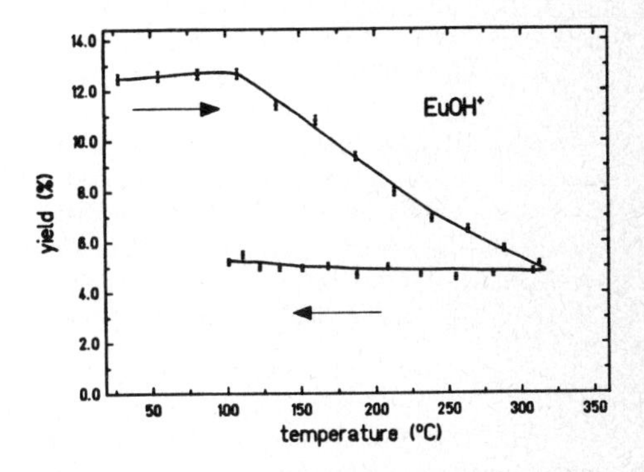

Fig. 5. Temperature dependence of $(EuOH)^+$

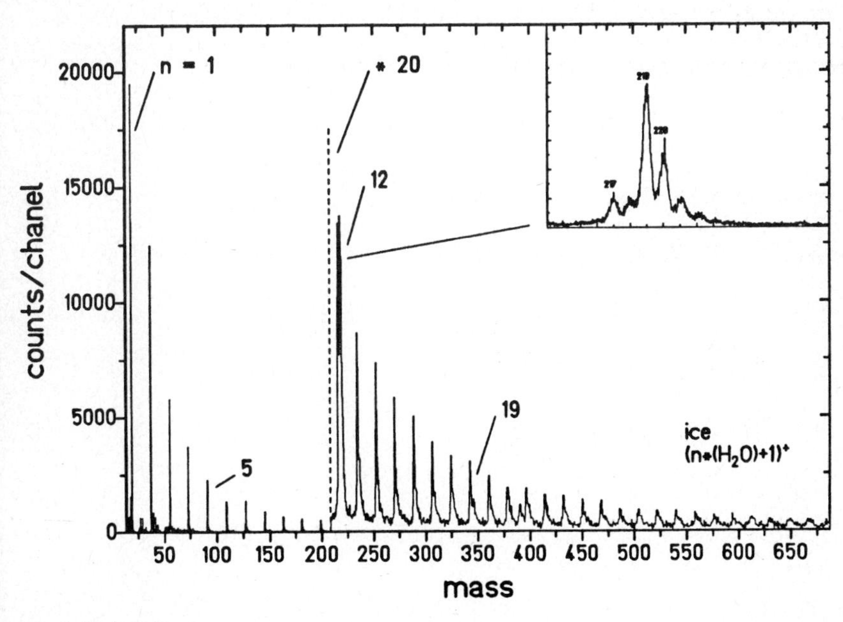

Fig. 6. Clusters of Ice

H_2O (Ice):
A film of Ice was produced on the copper-plate by blowing water vapor through a capillary onto the target substrate cooled down to the temperature of liquid nitrogen. A spectrum of positive ions ejected by ^{252}Cf fission fragments is given in Fig. 6. An impressive cluster series is observed up to a cluster number n = 38. The series has the structure $(H_2O)_nH^+$. With rising temperature the yields of H_2OH^+ and $(H_2O)_2H^+$ first decreases a little, but then at about -140°C increase considerable and drop down above -110°C (see Fig. 7). Our explanation is, that at - 190°C cluster emission occurs from an amorphous Ice surface, at -140°C the amorphous state is transformed into a cubic-crystalline structure (Literature on the preparation of Ice films and phase transition of Ice are given in Ref. [2]). Eventually the heat of recrystallisation released at -140°C supports the desorption-ionisation process. Above -100°C the sample is completely evaporated.

A rather intensive group of mass-lines is observed between m = 217 and m = 223 as illustrated in the enlarged po tion of Fig. 6. The origin of these ions could not be discovered so far. The peculiar temperature dependence of the most intensive line of the group is shown in Fig. 7. The yield vanishes at about +5°C.

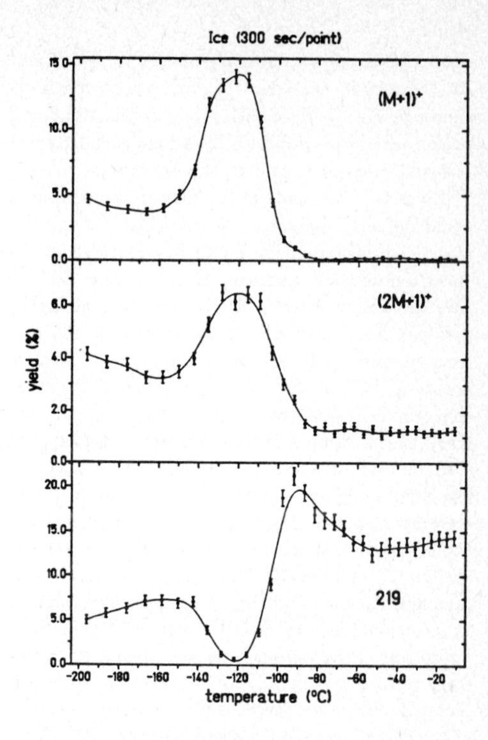

Fig. 7. Temeprature dependence of the secondary ion yields of $(M+1)^+$, $(2M+1)^+$ and mass 219

4. Conclusion

As demonstrated in the preceeding chapter secondary ions ejected from solids by fast heavy ion impact can be used as spectators of chemical and physical reactions at the surface, when these changes are induced by heating or cooling the sample. So far, a detailed explanation of the various effects reflected in the mass-spectra as function of target temperature can not be given. The cluster series observed with Eu_2O_3 and Ice seem to be an interesting subject for further studies.

5. References

1. P. Sigmund, M. Szymonski: Appl. Phys. A 33, 141 - 152 (1984)
2. P.V. Hobbs: Ice Physics, Oxford University Press (1974)
3. S.P. Sinha: Europium, Springer Verlag Berlin Heidelberg (1967)

THE USE OF NITROCELLULOSE BACKINGS IN 252-CALIFORNIUM PDMS

Per F. Nielsen
Institute of Molecular Biology
Odense University
5230 Odense M., Denmark

Biomolecules may be desorbed from a solid surface when bombardet with high energetic particles. The phenomenom has been utilized in Plasma Desorption Mass Spectrometry (PDMS), where in some cases fissionfragments from a decaying 252-Californium source interacts with a layer of biomolecules, and as a result intact molecular ions are desorbed from the sample surface. A mass spectrum may then be obtained using the time-of-flight method (1).

The results described in these notes have all been obtained using the commercially available PDMS instrument, BIN 10K, from BIO-ION, the details of which are described elsewhere (2).

In the original version the sample preparation consisted of dissolving the sample in question in a relatively volatile solvent (TFA, acetic acid or a mixture), and electrospraying it onto an aluminumfoil covered target.

Using this method of sample preparation impressive results have been obtained especially concerning molecular weigth determinations of larger peptides and proteins (5000 - 20000 a.m.u.)(4) . It has also been obvious,however, that only highly purified compounds could be analyzed succesfully. Low molecular weight impurities (especially Na^+ and K^+) quench the molecular ion yield, and from fig. 1 it can be seen that this yield decrease becomes more severe with increasing molecular weight of the sample. For compounds with molecular weights below 1000 -2000 a. m.u., the main effect of a high salt content is that the main peaks in the molecular ion region corresponds to cationized molecular ions ($(M+Na)^+$, $(M+K)^+$, $(M-H+Na+K)^+$ etc.). The absolute yield in this region, however, remains fairly constant. When the mass increases above 5000 - 10000 a.m.u., the effect is a more drastically reduced yield. It is evident, that in this region, the salt content must be kept low in order to obtain a usefull spectrum.

The introduction of a new sample application method, which utilizes the ability of nitrocellulose to bind larger biomolecules has made it possible to remove selectively unwanted low mass contaminants (3).

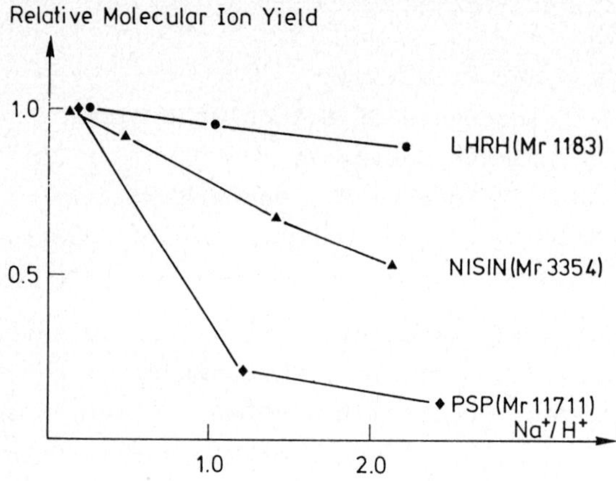

Fig. 1. Molecular ion yields of peptides of varying size as a function of the Na^+ content in the sample.

The two different sample application techniques used at present are shown in Fig. 2.

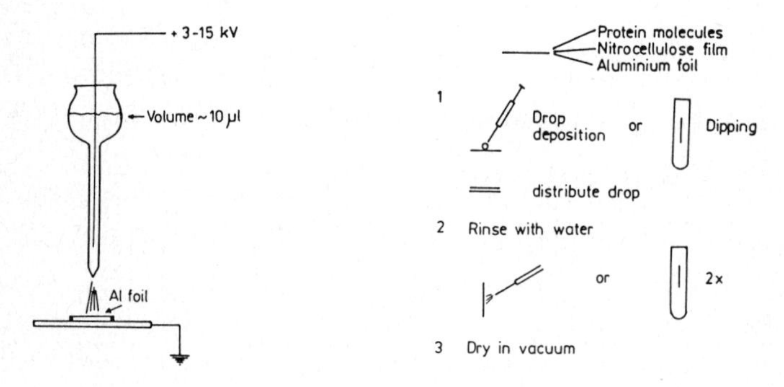

Fig. 2. Schematic presentation of the electrospray technique (a) and the adsorbtion technique (b).

The main advantages of using the adsorbtion technique may be summarized as follows:

I: Removal of low mass impurities "in situ" (on the target) is achieved

II: Sample application from relatively dilute watery solutions has become possible. Important when dealing with "real life samples".

III: The sensitivity is increased considerably

IV: Improvement of the obtained spectrum quality due a decrease in average internal energy of ions desorbed from nitrocellulose backings including:

a) increased molecular ion yield

b) increased abundance of multiply charged molecular ion species - important in the analysis of high molecular weight compounds as multiply charged ions are more easily detected due to higher energy aquired in the acceleration process

These points are illustrated in Fig. 3. Porcine Insulin (Mr. 5777.6 a.m.u.) was adsorbed to a nitrocellulose backing, but not rinsed. The resulting spectrum shows weak molecular ion peaks, and in the low mass region the peaks corresponding to Na^+ and K^+ are dominating. After rinsing the salt content is lowered drastically and intense molecular ion peaks appear.

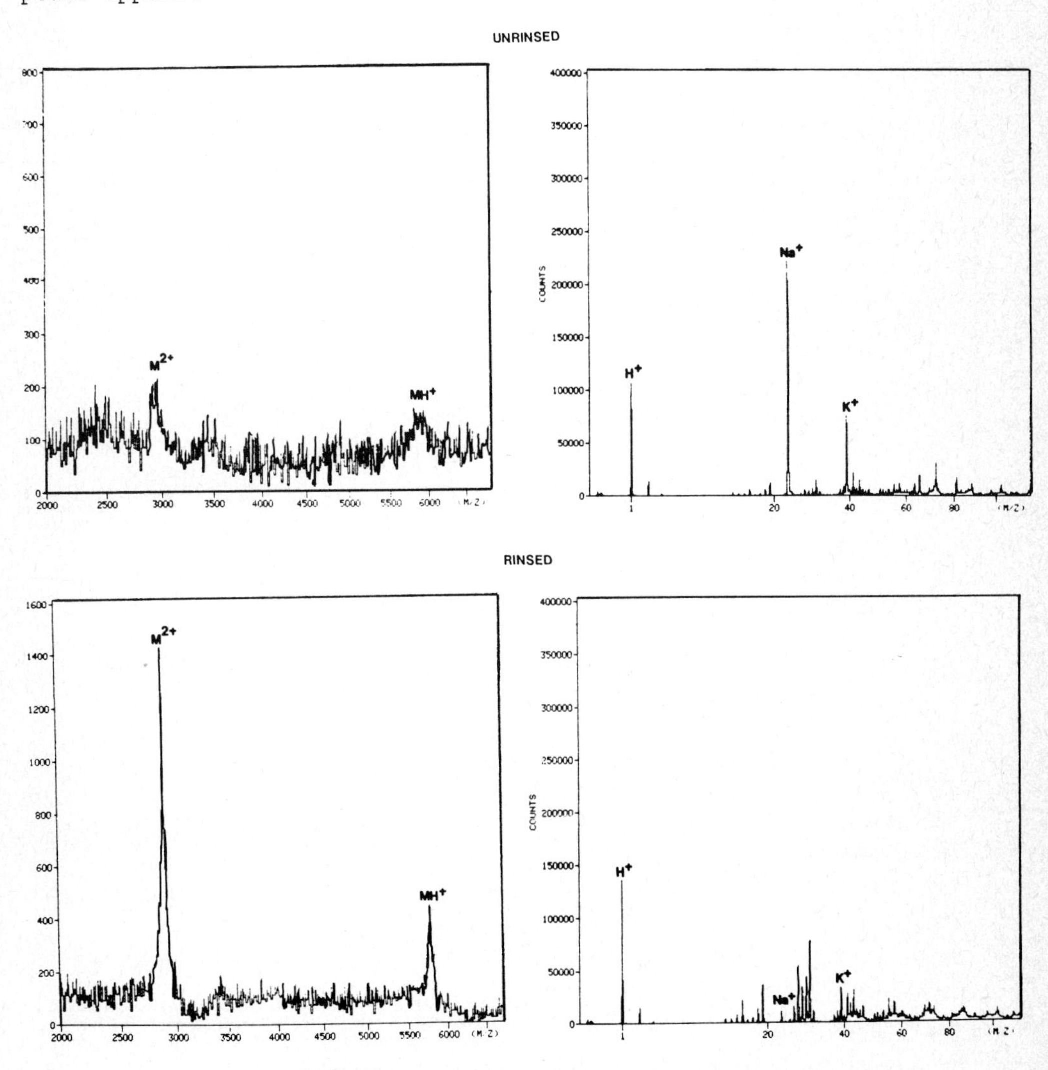

Fig. 3. Porcine insulin adsorbed to nitrocellulose. Spectre obtained before and after rinsing.

Thus the use of nitrocellulose backings has increased both the range of samples to which a 252-Cf PDMS analysis can be applied succesfully and the accuracy of the resulting molecular weight determinations.

One practical example of the usefullness of the adsorbtion technique is given in fig. 4.,where the spectrum (a: raw spectrum b: a smooth background has been subtracted)of a chlorophyll C binding protein is shown. The thus obtained molecular weight helped to clarify uncertanties concerning the amino acid sequence determination of this protein. A corresponding spectrum of an electrosprayed sample showed no molecular ion peaks.

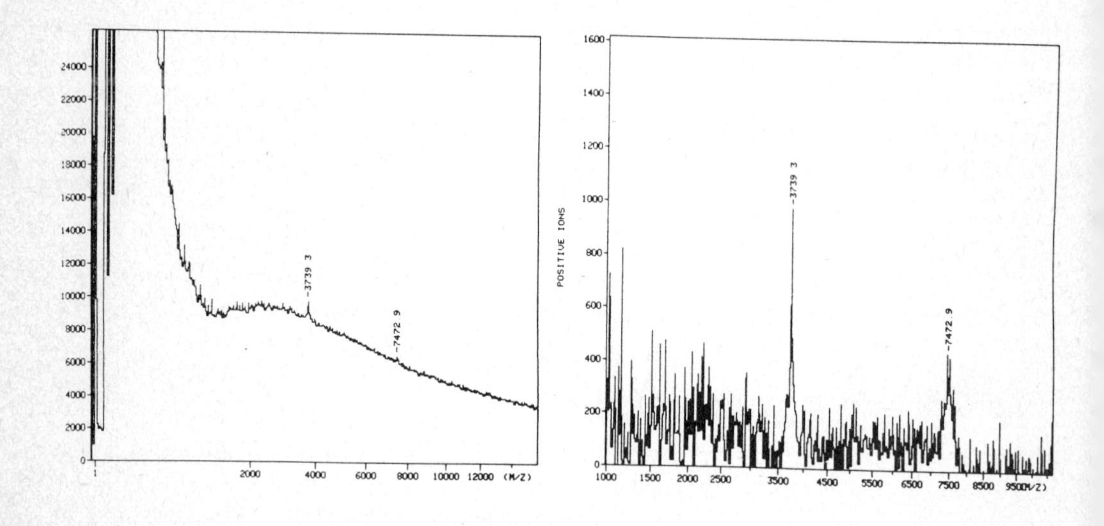

Fig. 4. PDMS spectrum of a chlorophyll C binding protein (Mr. 7473 a.m.u.)

References:

1. Torgerson,D.F., Skowronski,R.P. and MacFarlane,R.P.(1974): Biophys. Res. Commun. 60 616
2. Sundqvist, B., Håkansson, P., Kamensky, I.Kjellberg, J. and Roepstorff, P. (1984): Biomed. Mass Spectrom. 11 242
3. Jonsson, G.P., Hedin, A., Håkansson, P., Sundqvist, B., Säve, G., Nielsen, P.F., Roepstorff, P., Johansson, K. and Kamensky, I.(1986): Anal. Chem. 58 1084
4. Sundqvist, B., Hedin, A., Håkansson, P., Kamensky, I., Salehpour,M. and Säve, G. (1985): Int. Jour. Mass Spectrom. Ion Proc. 65 69

SAMPLE THICKNESS INFLUENCE ON THE DESORPTION YIELD OF SMALL BIO-MOLECULAR IONS

G Säve, P Håkansson and B U R Sundqvist
Tandem Accelerator Laboratory, Uppsala University,
Box 533, S-751 21 Uppsala, Sweden.

Fast ion induced desorption yields for positive secondary ions of biomolecules have been studied as a function of sample film thickness. The films were prepared with the spincoating technique and the film thicknesses were measured with ellipsometry. The secondary ions were mass analyzed with a time-of-flight technique. The sample thickness dependence was investigated for three different biomolecules; the amino-acid valine (MW=117) , the peptide LH-RH (MW=1182) and bovine insulin (MW=5733).

The influence of the sample thickness on fast-heavy-ion induced desorption yield of bio-molecules is of importance to know in applications of the plasma desorption mass spectroscopy technique,(PDMS)(1) and it may also give information on the mechanism for the desorption ionization process, i.e. if desorption is caused directly by the energy deposited by secondary electrons or if energy transport phenomena like thermal diffusion are playing a role.

In order to obtain samples with controlled thickness the spin coating technique was used. This method is superior to the electrospray method normally used in PDMS, in the production of smooth and homogeneous films of biomolecules. The spin coating technique is well-known and extensively used in the semiconductor industry for applying dopants or polymers on semiconductor surfaces and for electrode modifications in electrochemistry(2). Recently it was shown by us that biomolecules(3) such as amino-acids and proteins can also be spin coated to produce smooth and well defined films.

All three molecules were studied using two different series of spincasted samples with primary ions of 71.5 MeV $^{127}I^{12+}$ from the Uppsala EN-tandem accelerator. The spectra were subseqently analyzed with respect to the yield of the quasi molecular $(M+H)^+$ ion. The experimental results on yields of molecular ions as a function of film thickness for all three bio-molecules (valine, LH-RH and insulin) are very similar (see fig 1a,1b and 1c). The results have the same qualitative behaviour, namely an increase of the yields up to about 80-200 Å where the yield increases more slowly. For each molecule there is a specific "knee" thickness and then a slope of a weaker thickness dependence. The experimentally obtained "knee" thicknesses for valine, LH-RH and insulin are 120 ± 12 Å, 80 ± 8 Å and 200 ± 20 Å respectively.

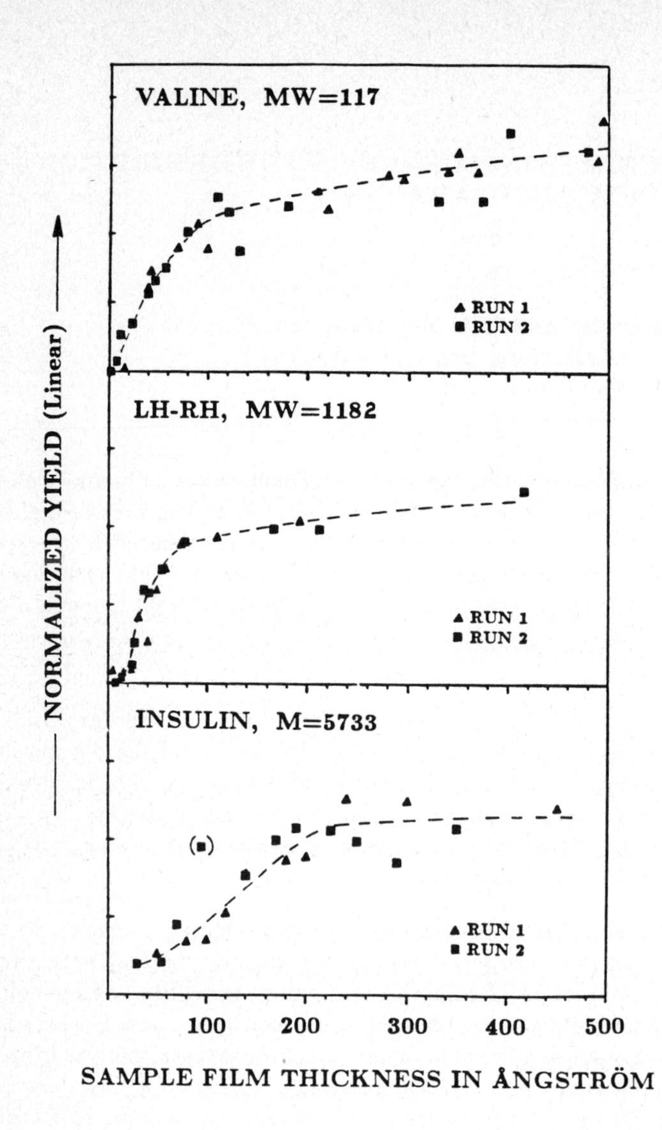

Fig.1 Desorption yield of the protonated full mass peak $(M+H)^+$ *for a) valine, b) LH-RH and c) bovine insulin.*

REFERENCES

1. D F Torgerson, R P Skowronski and R D MacFarlane
Biochem. Biophys. Res. Commun. **60** (1974) 616.

2. R W Murray, *in 'Electroanalytical Chemistry". Ed. A J Bard, Vol. 13, Marcel Dekker, New York, (1984) 192.*

3. U Jönsson, G Olofsson, M Malmquist, G Säve, J Fohlman, P Håknsson and Bo Sunqvist. *To be submitted to Anal Chem.*

CLUSTER PHYSICS — SOME REMARKS

E. R. HILF

Fachbereich Physik
Carl v. Ossietzky-Universität Oldenburg
P.O.B. 2503
D-2900 Oldenburg
FRG

I. INTRODUCTION

'Small is beautiful' is the title of a famous book of E.F. Schumacher [1] the subtitle 'a study of economics as if people mattered'. Likewise in physics the individualities of atoms or molecules and the dynamics of the interactions between them as well as with the outside world, come the more into focus, the more constrictions are released. In a solid these are the spatial periodicity of the lattice structure and its related binding, released by forming clusters of a small number of molecules or atoms. Likewise on this island we may have felt more excited being a small cluster of people with outdoor activities related to its finite area: its shape and its surface (beaches).

Thus whereas in scattering of molecules on one hand, and studying the intermolecular binding in a solid, each molecule can be treated as alike to the others, this is not so in a cluster with its rich variety of number (including zero) of concerting neighbours for a 'bond'.

Similarly with the small number of participants we had ample time to learn and discuss the different concepts and results of each other. In respecting and honouring them all, and their excellent professionality, I will restrict myself in this text to some few remarks, hommageing some related older but original and partly unpublished work of former collaborators of our group, H. Gräf and R. Böndgen.

II. CLUSTER FORMATION BY EXPANSION

The arsenal of production mechanisms of clusters has been enriched by PDMS and SIMS to access a broad range of the ratio of electronic to ionic temperature $\tau := T_e/T_i$ by varying appropriately the velocity and charge of the impinging ion. Far away from the ion track τ seems to be large, $\tau > 1$, whereas clusters collected close to the ion track, especially abundant for high Z-ions, seem to look like $\tau \sim 1$ and thus similar to clusters from the vastly explored adiabatic expansion technique [2], [3].

There, to study the formation of clusters one needs a detailed microscopic model of the process turning out the need for large computer calculations. For a first orientation, however, a simpler but analytical calculation may be in order.

For the formation of clusters from the expansion of a gas by a nozzle Soler, Garcia et al. [3] had given an analytically tractable model assuming that coagulation is the important process, skipping the monomer addition. With the assumption of adiabatic expansion, they got cluster-yield curves which could be fitted nicely by one parameter (per oven-conditions), which one may call the integral-collision cross-section along the flow path.

In 1985 R. Böndgen [4] set out with the opposite simplification, skipping the coagulation part but tracing the monomer addition and emission, and keeping the deviation from adiabaticity due to the released latent heat.

To begin with, the calculation of the hydrodynamic flow revealed that the condensation sets in at or even just after the end of the nozzle with a supersaturation which is then diminishing due to the released latent heat along the flow path.

Starting out with the wellknown Becker-Döring equation for the time-evolution of the abundance f of a cluster with g atoms,

$$\partial_t f = C_{g-1} f_{g-1} + E_{g+1} f_{g+1} - (C_g + E_g) \cdot f_g \qquad \text{for} \quad g \geq 2 \quad , \tag{1}$$

with the monomer addition (C) and emission (E) coefficients, one has to substitute the usual very simple boundary conditions by the correct constancy of total number of atoms,

$$d_t \sum_{g=1}^{\infty} g \cdot f_g(t) = 0 \quad , \tag{2}$$

yielding

$$d_t f_1 = -2C_1 f_1 + 2E_2 f_2 - \sum_{g=2}^{\infty} C_g f_g - \sum_{g=3}^{\infty} E_g f_g \quad . \tag{2a}$$

Since one aims finally at the formation of large clusters, g may be approximated to be a continuous variable. Then (1) turns into the Zeldovich-equation

$$\partial_t f = \partial_g \left(C \cdot n \cdot \partial_g \left(\frac{f}{n} \right) \right) \quad , \tag{1b}$$

after replacing the E_g by the use of the respective fictitious equilibrium abundance n. The boundary condition turns into an integro-differential equation

$$d_t f_1 = -C(1) \cdot f(1) \cdot \{1 - \frac{\partial_g \left(\frac{f}{n} \right)_{g=1}}{\frac{f}{n}}\} - \int_1^{\infty} dg \frac{f}{n} \partial_g (C \cdot n) \quad . \tag{2b}$$

To solve (1b) with (2b) approximately but analytically, several approximations have to be done.
1.) The boundary condition is approximated to

$$f_1(t=0) = \delta(g-1) \quad ; \qquad f(0,t) = 0 \quad . \tag{3}$$

2.) For the reasonable assumption of f being far from the respective equilibrium distribution n,

$$\frac{\partial_g f}{g} \gg \frac{\partial_g n}{n} \quad , \tag{4}$$

and no g-dependent drag of the flow of different g-species, $v(g,z) = v(z)$, the flow distance z can be substituted instead of the evolution time t,

$$\partial_z f(g,z) = \frac{1}{v(z)} \partial_g (C(g,z) \cdot \partial_g f) \quad . \tag{5}$$

The monomer addition coefficient may be assumed to separate as

$$C(g,z) = D(g) \cdot K(z) = \alpha(g,T) \cdot g^{2/3} \cdot v_{th} \cdot \phi_0 \quad , \tag{6}$$

where D is taken to scale as $g^{2/3}$ and v_{th} is the thermal velocity of cluster g at temperature T.

Substituting $\eta := \frac{8}{9} \cdot \int_{z_0}^{z} dz' K(z')/v(z')$ a suitable reformulation of the Zeldovich-equation and the boundary condition is gained,

$$\partial_t \psi(\eta, \tau) = \frac{1}{2} \left(\frac{1}{\eta} \cdot \partial_\eta \psi + \partial_{\eta\eta} \psi - \frac{1}{16\eta^2} \psi \right) \tag{1c}$$

with

$$d_\tau \psi(1,\tau) = -\frac{5}{6} \cdot \psi_1 + \frac{2}{3} \partial_\eta \psi \Big|_{\eta=1} - \int_1^\infty d\eta \psi \cdot \eta^{1/4} \quad . \tag{2c}$$

This set of equations can finally be separated by

$$\psi := F(\tau) \cdot L(\eta) \tag{8}$$

yielding

$$d_{\eta\eta} L + \frac{1}{\eta} d_\eta L + \left(k^2 - \frac{1}{(\psi\eta)^2} \right) \cdot L = 0 \quad , \tag{1d}$$

$$d_\tau F + \frac{k^2}{2} \cdot F = 0 \quad . \tag{2d}$$

Together these equations can be analytically evaluated [4], yielding

$$\psi(\eta,\tau) \simeq c_1(\tau) \cdot I_{1/4}(\eta/\tau) \cdot \exp(-\frac{1}{2}\eta^2/\tau) \quad , \tag{9}$$

basically a simple Besselfunction of order $\frac{1}{4}$.

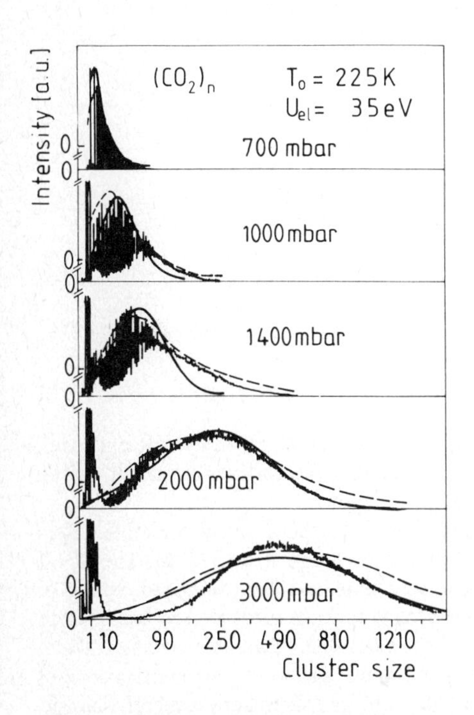

Fig. 1: Cluster abundance of $(CO_2)_g$ by nozzle expansion.
dashed line: coagulation only [3],
full line: monomer addition and emission [4].

Fig. 1 shows that both models (coagulation versus monomer addition and emissions) give a good fit to experiments.

Fig. 2 shows, however, the calculated integral cross-sections in both models as to the calculated one from the calculation of the thermodynamics along the hydrodynamic flow path. Details of the calculation are given in [4]. By inspection, the monomer-approximation seems to be closer to the curve calculated from the simple hydrogen model flow. From these models one may infer that a fullfledged, - but then necessarily numerical calculation has to take into account the coagulation of clusters as a necessary correction to the monomer model.

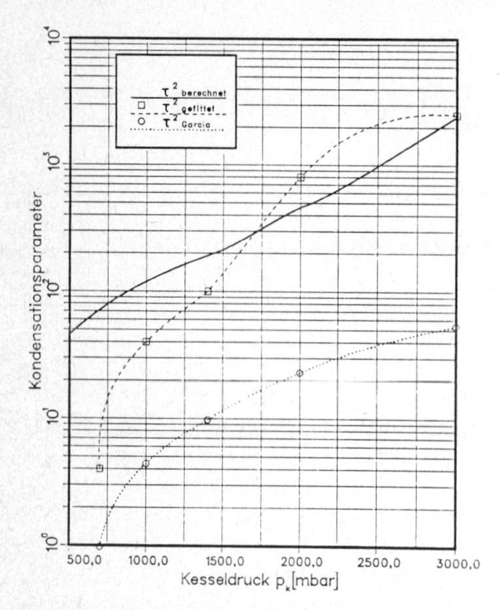

Fig. 2: Integral cross-sections along the flow path:
full line: from hydrodynamic calculations,
dotted line: from coagulation model [3],
dashed line: from monomer model [4].

III. ENERGY DENSITY FORMALISM FOR METALLIC CLUSTERS

As a second example of a maybe grossly oversimplified but therefore analytical approach to gain insight into cluster properties before starting by numerical calculations we dwell on the kinetic part of the energy density of the electrons of metallic clusters.

The properties of a **metallic cluster** of N atoms of atomic weight A are in principle determined by the coupled Schrödinger-equations for the N nuclei of charge $Z = A/2$ and the $N \cdot Z$ electrons. The enormous amount of the requested computing time and the calculated wealth of quantities is in remarkable contrast to the few global physical properties measured so far, — such as total binding energy, ionisation energy, Fermi energy of the electrons, mean atom configuration.

Thus a remarkable variety of approximations, simplifications and even parametrisations of the full quantum problem have been developed in the past. Let us follow here some of them to present a face-lift tuning on the simplest one, the energy-density functional.

To begin with, the kinetic energy of the nuclei is normally skipped, thus decoupling the system of nuclei as then fixed in configuration space (eventually studying the classical potential surface therein). This is the most treacherous step because of being taken from solid state physics with its tighter molecular bindings (no surface atoms) and its spatially hindered thus very restricted configuration space. The quantum interference of different configurations is a dominant feature of small clusters [5].

The second approximation deals with the system of $N * Z$ electrons. Instead of calculating the full $N \cdot Z$ wave function [6] calling this "ab initio", one may either expand the system of N fixed Coulomb potential wells $V(x_1 \ldots x_N)$ according to the range, ending up with a Woods-Saxon type $V(x_i)$, the range being the diameter of the cluster, plus potential "wobbles" $V(x_j; x_i)$ which are of the range of the intermolecular distance within the cluster. Skipping these latter terms the model is called Jellium-model and electron-wave functions have recently been studied in detail [7] with success, giving e.g. the Friedell oscillations of the electron density, near the surface, as well as magic numbers for ionisation energies if the Fermi energy hits a relatively well-bound eigenstate.

A further approximation in calculating the electron system's properties is known as energy density functional formalism since 51 years [8].

At each point in space the potential is assumed to be so flat, that the contributions of the electrons to the total energy and density can be approximated by the result taken by integrating the contributions from a free Fermi-gas of the same local Fermi energy in an infinite extended flat potential of the same depth taken per volume element. Then the total energy of the electron system and its total density $\rho(x)$ is given by the number of electrons, and the potential $V(x)$. Elegantly, the potential can be eliminated to give the total energy density as a functional of the density alone, $\epsilon[\rho]$, thus called local density approximation.

It applies for any Fermion quantum system if the derivative of the potential is small as compared to the Fermi wavelength, which may be reasonable for the least bound electron in a metallic solid. However, for a finite cluster, even with the Jellium approximation, the slopes of the potential at the surface are comparable to the Fermi wavelength and thus the semiclassical approximation to lowest order is apt not to work.

The traditional way is to keep the idea of a local approximation but expand the potential at each point in a Taylor-series. Knowing the solutions for a Fermi-gas for any Fermi energy in a flat potential, for a constant slope potential (giving Airy-functions), for a parabolic potential, etc. one again ends up in functionals $\epsilon[\epsilon_F, V(\underline{x}), \partial_{\underline{x}} V(\underline{x}), \partial_{\underline{xx}} V(\underline{x}), \ldots]$, $\rho[\epsilon_F, V(\underline{x}), \partial_{\underline{x}} V(\underline{x}), \partial_{\underline{xx}} V(\underline{x}), \ldots]$ for the local quantities ϵ and ρ, which can then be used in local quantities to be integrated over all $\underline{x}$ to give the desired observables. Depending on the number of derivatives included, the results are of 0th,1st,2nd,.. order as of a semiclassical approximation.

The obstacles are twofold: the semiclassical approximation expressions are increasingly complicated so that it gets quickly virtually impossible to analytically eliminate ϵ_F to get

$$\epsilon[\rho, V, \partial V, \partial\partial V, \ldots] \quad . \tag{10}$$

Secondly, as all semiclassical approximations of Quantum-mechanics the gradient series expansion is semidivergent, that is, including increasingly higher derivatives finally make the results less realistic [9].

A way to steer through between these two obstacles as a practical approach to gain observables somewhat realistically and still keep gradient corrections by introducing derivatives of the density into the energy functional is to replace derivatives of the potential, yielding

$$\epsilon[\rho, \partial_{\underline{x}}\rho, \partial_{\underline{xx}}\rho, \ldots; V] \quad . \tag{11}$$

To get an analytic expression for an energy-density functional, which according to the theorem of Kohn and Sham has in principle to exist, is a long standing goal.

Thorough studies for different proposals of gradient terms by a huge number of authors in the past 50 years have given confidence that in most cases the local $\epsilon[\rho]$ is a good approximation with regard to contributions from potential contributions [10] due to one or twobody forces forces, but that the main gradient contribution stems from the "long range" part of the Heisenberg uncertainty principle coming about here in that the curvature of the density is given by the Schrödinger equation by the gradient operator squared, i.e., the momentum-square of the Fermi energy. The kinetic energy contributions to the energy-density functional are known in two extremes, (for a proof see [11]),

1) high Fermi-momenta (semiclassical region) to be

$$\epsilon[\rho,(\nabla\rho)^2,\Delta\rho]-\epsilon[\rho]=+\frac{1}{36}\cdot\frac{(\nabla\rho)^2}{\rho}+\frac{1}{3}\Delta\rho+\ldots \tag{12}$$

and
2) large imaginary Fermi-momenta (tunneling region) to be

$$\epsilon[\rho,(\nabla\rho)^2]-\epsilon[\rho]=+\frac{1}{4}\cdot\frac{(\nabla\rho)^2}{\rho}+\ldots \tag{13}$$

originally proposed by v. Weizsäcker [8].

Discouragement in the past came about by attempting anyone of these expressions in both regions.

Instead of dealing with both (semidivergent) series (12, 13), a closed approximate expression has been proposed by H. Gräf [11], replacing the constants in front of each term once and for all by a function of the local Fermi kinetic energy. Such an approach has been called convergence acceleration. His result is

$$\epsilon[\rho,(\nabla\rho)^2\Delta\rho]-\epsilon[\rho]=\frac{1+c_1u}{36+4c_1u}\frac{(\nabla\rho)^2}{\rho}+\frac{1}{3+c_2u}\Delta\rho \tag{14}$$

with the adjusted parameters: $c_1\simeq 0.5$; $c_2\simeq 0.118$ and the definition of u to be approximated by

$$u:=|\nabla\rho|/\rho^{4/3}\quad . \tag{15}$$

This quantity measures how deep one is in the classical region ($u \ll 1$) or in the tunneling region (u large). As an example the minimization of the total kinetic energy $\int dV\epsilon[\rho]$ for a square box potential yields the density distribution to within 0.3% of the exact result as known here from adding the lowest N wave function contributions. Interesting enough, the Friedell-oscillations are gained (in contrast to (12) or (13)) by approximate intrusion of the Fermi wavelength via u. (Fig. 3).

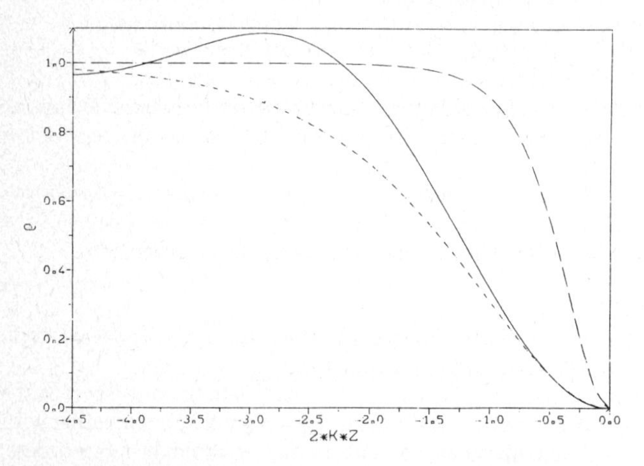

Fig. 3: Density distribution for a constant potential box.
full line: exact solution and (to within 3%) the result
of a variational procedure with (14),
dashed line: with v. Weizsäcker gradient term,
dotted line: with semiclassical approximation.

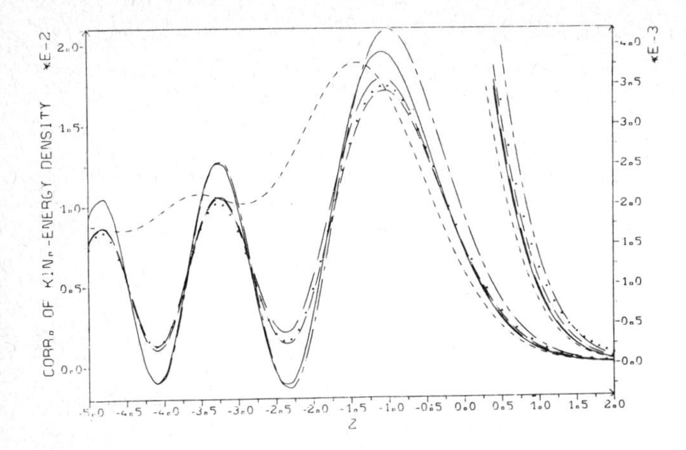

Fig. 4: Correction to $\epsilon[\rho]$ for a linear potential.
full line: exact solution,
long dashes: this work,
crosses and medium dashes: semiclassical expansion,
short dashes: v. Weizsäcker correction.

Fig. 4 shows for a linear potential the difference of $\epsilon[\rho,(\nabla\rho)^2,\Delta\rho]$ to the exact solution of a homogenous potential of the same size. Clearly, while (14) works well, neither the Weizsäcker-proposition (12) nor the semiclassical approach (13) to second order behaves reasonably. Interestingly enough, the addition of the 4th order term in (12) does not improve the result, — the second order seems to be the best semiclassical approximation before the divergence of (12).

The ansatz (14) is at present investigated for application in metal cluster physics.

ACKNOWLEDGEMENT

The fruitful collaboration with H. Gräf and R. Böndgen who did most of the calculations presented here, is acknowledged. The computations cited were done at the excellent Computing Center of the GSI (Gesellschaft für Schwerionenforschung, Darmstadt). The work had been supported in part by the German Bundesministerium für Forschung und Technologie.

REFERENCES

1. Schumacher, E.F., 'Small is beautiful', Publisher: Bond & Briggs Ltd., SBN 85634 012 X (1973)
2. Hagenah, O.F., Surf. Sci. 106 (1981) 101
3. Soler, J.M., Garcia, N, Echt, O., Sattler, K., Recknagel, E., Phys. Rev. Lett. 49 (1982) 1851
4. Böndgen, R., 'Bildung molekularer Cluster in adiabatischen Gasströmungen', Diplom-thesis, T.H. Darmstadt, April 1985
5. Franke, G., this volume
6. Koutecký, J., this volume
7. Ekardt, W., this volume
8. Weizsäcker von, C.F., Zeitschrift für Physik 96 (1935) 431
9. " 2nd Int. Workshop on Semiclassical Methods", Journ. de Physique C6 (1984) 45
10. Ludwig, S., Nucl. Phys. A219 (1974) 295
11. Gräf, H., Nucl. Phys. A343 (1980) 91

List of participants

E. Bominaar	Medizinische Universität Lübeck, Institut für Physik Ratzeburger Allee 160, D-2400 Lübeck, F.R.G.
C. Brechignac	Laboratoire Aime Cotton, Bat 505 Campus Orsay, 91405 Orsay Cedex, France
J.-P. Daudey	Université Paul Sabatier, Laboratoire de Physique Quantique, 118, route de Narbonne, 31062 Toulouse Cedex, France
W. Ekardt	Fritz-Haber-Institut der M.P.G., Department of Electron Microscopy, Faradayweg 4-6, D-1000 Berlin, F.R.G.
S. Elbel	Universität Hamburg, Institut für Anorg. u. Angew. Chemie, Martin-Luther-King-Platz 6, D-2000 Hamburg 13, F.R.G.
G. Franke	Universität Oldenburg, Fachbereich Physik, Carl-von-Ossietzky-Str. 9-11, D-2900 Oldenburg, F.R.G.
M. Grodzicki	Universität Hamburg, I. Institut für Theoretische Physik, Jungiusstr. 9, D-2000 Hamburg, F.R.G.
P. Håkansson	Uppsala University, Tandem Accelerator Laboratory, Box 533, S-75 121 Uppsala, Sweden
A. Hedin	Uppsala University, Tandem Accelerator Laboratory, Box 533, S-75 121 Uppsala, Sweden
F. Heijkenskjold	Uppsala Université, Tandem Accelerator Laboratory, Box 533, S-75 121 Uppsala, Sweden
E.R. Hilf	Universität Oldenburg, Fachbereich Physik, Carl-von-Ossietzky-Str. 9-11, D-2900 Oldenburg, F.R.G.
Y. Hoppilliard	Ecole Polytechnique, Laboratoire de Synthèse Organique, 91128 Palaiseaux Cedex, France
M. Hütsch	Universität Hamburg, I. Institut für Theoretische Physik, Jungiusstr. 9, D-2000 Hamburg, F.R.G.
R. Iffert	Universität Hannover, Theoretische Chemie, Callinstr. 3A, D-3000 Hannover 1, F.R.G.
L. Jansen	University of Amsterdam, Institute of Theoretical Chemistry, Nieuwe Achtergracht 166, 1018 WV Amsterdam, Netherlands
G. Jonsson	Uppsala University, Tandem Accelerator Laboratory, Box 533, S-75 121 Uppsala, Sweden
H. Jungclas	Philipps-Universität Marburg, Abteilung Kernchemie, Hans-Meerwein-Str., D-3550 Marburg, F.R.G.
E. Junker	Universität Bonn, Institut für Physikalische Chemie, Wegelerstr. 12, D-5300 Bonn 1, F.R.G.
F. Kammer	Universität Oldenburg, Fachbereich Physik, Carl-von-Ossietzky-Str. 9-11, D-2900 Oldenburg, F.R.G.

P.G. Kistemaker	FOM Inst. for Atomic and Molecular Physics, Department of Physics, Kruislaan 407, 1089 SJ Amsterdam, Netherlands
W.D. Knight	University of California, Department of Physics, Berkeley, CA 94720, USA
P. Koczon	Technische Hochschule Darmstadt, Institut für Kernphysik, Schloßgartenstr. 9, D-6100 Darmstadt, F.R.G.
J. Koutecký	Freie Universität Berlin, Institut für Physikalische Chemie, Takustr. 3, D-1000 Berlin 33, F.R.G.
O. Kühnholz	Universität Hamburg, I. Institut für Theoretische Physik, Jungiustr. 9, D-2900 Hamburg 36, F.R.G.
Y. Le Beyec	Institut de Physique Nucléaire IPN, B.P. N° 1, 91406 Orsay Cedex, France
R.R. Lucchese	Texas A & M University, Department of Chemistry, College Station, TX 77843, USA
G. Mahler	Universität Stuttgart, Institut für Theoretische Physik, Pfaffenwaldring 57, D-7000 Stuttgart 80, F.R.G.
V.R. Marathe	Medinzinische Universität Lübeck, Institut für Physik, Ratzeburger Allee 160, D-2400 Lübeck, F.R.G.
R. Matthäus	Technische Hochschule Darmstadt, Institut für Kernphysik, Schloßgartenstr. 9. D-6100 Darmstadt, F.R.G.
G. Maynard	Laboratoire de Physique des Gaz et des Plasmas, 91405 Orsay, France
R. Moshammer	Technische Hochschule Darmstadt, Institut für Kernphysik, Schloßgartenstr. 9. D-6100 Darmstadt, F.R.G.
B. Nees	Universität Erlangen, Physikalisches Institut, Erwin-Rommel-Str. 1, D-8520 Erlangen, F.R.G.
Per F. Nielsen	Odense University, Department of Molecular Biology, Campusvej 55, 5230 Odense M, Denmark
E. Nieschler	Universität Erlangen, Physikalisches Institut, Erwin-Rommel-Str. 1, D-8520 Erlangen, F.R.G.
F. Röllgen	Universität Bonn, Institut für Physikalische Chemie, Wegelerstr. 12, D-5300 Bonn, F.R.G.
G. Säve	Uppsala University, Tandem Accelerator Laboratory, Box 533, S-75121 Uppsala, Sweden
K. Sattler	University of California, Department of Physics, Berkeley, CA 94720, USA
H.J. Schellnhuber	Universität Oldenburg, Fachbereich Physik, Carl-von-Ossietzky-Str. 9-11, D-2900 Oldenburg, F.R.G.
J. Schulte	Universität Oldenburg, Fachbereich Physik, Carl-von-Ossietzky-Str. 9-11, D-2900 Oldenburg, F.R.G.

L. Schweikhard	Universität Mainz, Institut für Physik, Staudinger Weg 7, D-6500 Mainz, F.R.G.
L. Schmidt	Philipps-Universität Marburg, Atbeilung Kernchemie, Hans-Meerwein-Str., D-3550 Marburg, F.R.G.
O. Schrems	Universität Siegen, Institut für Physikalische Chemie, Adolf-Reichweinstr. 3, D-5900 Siegen 21, F.R.G.
L. Skála	Charles University, Faculty of Mathematics and Physics, Ke Karlovu 3, 12116 Prague, Czechoslovakia
B.U.R. Sundqvist	Uppsala University, Tandem Accelerator Laboratory, Box 533, S-75 121 Uppsala, Sweden
F. Träger	Universität Heidelberg, Physikalisches Institut, Philosophenweg 12, D-6900 Heidelberg, F.R.G.
M. Vollmer	Universität Heidelberg, Physikalisches Institut, Philosophenweg 12, D-6900 Heidelberg, F.R.G.
K. Wien	Technische Hochschule Darmstadt, Institut für Kernphysik, Schloßgartenstr. 9, D-6100 Darmstadt, F.R.G.
K.P. Wirth	Universität Bonn, Institut für Physikalische Chemie, Wegelerstr. 12, D-5300 Bonn, F.R.G.

Lecture Notes in Physics

Vol. 237: Nearby Molecular Clouds. Proceedings, 1984. Edited by G. Serra. IX, 242 pages. 1985.

Vol. 238: The Free-Lagrange Method. Proceedings, 1985. Edited by M.J. Fritts, W.P. Crowley and H. Trease. IX, 313 pages. 1985.

Vol. 239: Geometrics Aspects of the Einstein Equations and Integrable Systems. Proceedings, 1984. Edited by R. Martini. V, 344 pages. 1985.

Vol. 240: Monte-Carlo Methods and Applications in Neutronics, Photonics and Statistical Physics. Proceedings, 1985 Edited by R. Alcouffe, R. Dautray, A. Forster, G. Ledanois and B. Mercier. VIII, 483 pages. 1985.

Vol. 241: Numerical Simulation of Combustion Phenomena. Proceedings, 1985. Edited by R. Glowinski, B. Larrouturou and R. Temam. IX, 404 pages. 1985.

Vol. 242: Exactly Solvable Problems in Condensed Matter and Relativistic Field Theory. Proceedings, 1985. Edited by B.S. Shastry, S.S. Jha and V. Singh. V, 318 pages. 1985.

Vol. 243: Medium Energy Nucleon and Antinucleon Scattering. Proceedings, 1985. Edited by H.V. von Geramb. IX, 576 pages. 1985.

Vol. 244: W. Dittrich, M. Reuter, Selected Topics in Gauge Theories. V, 315 pages. 1986.

Vol. 245: R.Kh. Zeytounian, Les Modèles Asymptotiques de la Mécanique des Fluides I. IX, 260 pages. 1986.

Vol. 246: Field Theory, Quantum Gravity and Strings. Proceedings, 1984/85. Edited by H.J. de Vega and N. Sánchez. VI, 381 pages. 1986.

Vol. 247: Nonlinear Dynamics Aspects of Particle Accelerators. Proceedings, 1985. Edited by J.M. Jowett, M. Month and S. Turner. VIII, 583 pages. 1986.

Vol. 248: Quarks and Leptons. Proceedings, 1985. Edited by C.A. Engelbrecht. X, 417 pages. 1986.

Vol. 249: Trends in Applications of Pure Mathematics to Mechanics. Proceedings, 1985. Edited by E. Kröner and K. Kirchgässner. VIII, 523 pages. 1986.

Vol. 250: Lie Methods in Optics. Proceedings 1985. Edited by J. Sánchez Mondragón and K.B. Wolf. XIV, 249 pages. 1986.

Vol. 251: R. Liebmann, Statistical Mechanics of Periodic Frustrated Ising Systems. VII, 142 pages. 1986.

Vol. 252: Local and Global Methods of Nonlinear Dynamics. Proceedings, 1984. Edited by A.W. Sáenz, W.W. Zachary and R. Cawley. VII, 263 pages. 1986.

Vol. 253: Recent Developments in Nonequilibrium Thermodynamics Fluids and Related Topics. Proceedings, 1985. Edited by J. Casas-Vázquez, D. Jou and J.M. Rubí. X, 392 pages. 1986.

Vol. 254: Cool Stars, Stellar Systems, and the Sun. Proceedings, 1985. Edited by M. Zeilik and D.M. Gibson. XI, 501 pages. 1986.

Vol. 255: Radiation Hydrodynamics in Stars and Compact Objects. Proceedings, 1985. Edited by D. Mihalas and K.-H. A. Winkler. VI, 454 pages. 1986.

Vol. 256: Dynamics of Wave Packets in Molecular and Nuclear Physics. Proceedings, 1985. Edited by J. Broeckhove, L. Lathouwers and P. van Leuven. VIII, 187 pages. 1986.

Vol. 257: Statistical Mechanics and Field Theory: Mathematical Aspects. Proceedings, 1985. Edited by T.C. Dorlas, N.M. Hugenholtz and M. Winnink. VII, 328 pages. 1986.

Vol. 258: Wm. G. Hoover, Molecular Dynamics. VI, 138 pages. 1986.

Vol. 259: R.F. Alvarez-Estrada, F. Fernández, J.L. Sánchez-Gómez, V. Vento, Models of Hadron Structure Based on Quantum Chromodynamics. VI, 294 pages. 1986.

Vol. 260: The Three-Body Force in the Three-Nucleon System. Proceedings, 1986. Edited by B.L. Berman and B.F. Gibson. XI, 530 pages. 1986.

Vol. 261: Conformal Groups and Related Symmetries – Physical Results and Mathematical Background. Proceedings, 1985. Edited by A.O. Barut and H.-D. Doebner. VI, 443 pages. 1986.

Vol. 262: Stochastic Processes in Classical and Quantum Systems. Proceedings, 1985. Edited by S. Albeverio, G. Casati and D. Merlini. XI, 551 pages. 1986.

Vol. 263: Quantum Chaos and Statistical Nuclear Physics. Proceedings, 1986. Edited by T.H. Seligman and H. Nishioka. IX, 382 pages. 1986.

Vol. 264: Tenth International Conference on Numerical Methods in Fluid Dynamics. Proceedings, 1986. Edited by F.G. Zhuang and Y.L. Zhu. XII, 724 pages. 1986.

Vol. 265: N. Straumann, Thermodymamik. VI, 140 Seiten. 1986.

Vol. 266: The Physics of Accretion onto Compact Objects. Proceedings, 1986. Edited by K.O. Mason, M.G. Watson and N.E. White. VIII, 421 pages. 1986.

Vol. 267: The Use of Supercomputers in Stellar Dynamics. Proceedings, 1986. Edited by P. Hut and S. McMillan. VI, 240 pages. 1986.

Vol. 268: Fluctuations and Stochastic Phenomena in Condensed Matter. Proceedings, 1986. Edited by L. Garrido. VIII, 413 pages. 1987.

Vol. 269: PDMS and Clusters. Proceedings, 1986. Edited by E.R. Hilf, F. Kammer and K. Wien. VIII, 261 pages. 1987.